MILLIKAN
AND HER CRITICS

PHILOSOPHERS AND THEIR CRITICS
General Editor: Ernest Lepore

Philosophy is an interactive enterprise. Much of it is carried out in dialogue as theories and ideas are presented and subsequently refined in the crucible of close scrutiny. The purpose of this series is to reconstruct this vital interplay among thinkers. Each book consists of a temporary assessment of an important living philosopher's work. A collection of essays written by an interdisciplinary group of critics addressing the substantial theses of the philosopher's corpus opens each volume. In the last section, the philosopher responds to his or her critics, clarifies crucial points of the discussion, or updates his or her doctrines.

1. Dretske and His Critics
   Edited by Brian McLaughlin
2. John Searle and His Critics
   Edited by Ernest Lepore and Robert van Gulick
3. Meaning in Mind: Fodor and His Critics
   Edited by Barry Loewer and Georges Rey
4. Dennett and His Critics
   Edited by Bo Dahlbom
5. Danto and His Critics
   Edited by Mark Rollins
6. Perspectives on Quine
   Edited by Robert B. Barrett and Roger F. Gibson
7. The Churchlands and Their Critics
   Edited by Robert N. McCauley
8. Singer and His Critics
   Edited by Dale Jamieson
9. Rorty and His Critics
   Edited by Robert B. Brandom
10. Chomsky and His Critics
    Edited by Louise M. Antony and Norbert Hornstein
11. Dworkin and His Critics
    Edited by Justine Burley
12. McDowell and His Critics
    Edited by Cynthia Macdonald and Graham Macdonald
13. Stich and His Critics
    Edited by Dominic Murphy and Michael Bishop
14. Danto and His Critics, 2nd Edition
    Edited by Mark Rollins
15. Millikan and Her Critics
    Edited by Dan Ryder, Justine Kingsbury, and Kenneth Williford

MILLIKAN
AND HER CRITICS

EDITED BY
DAN RYDER, JUSTINE KINGSBURY,
AND KENNETH WILLIFORD

A John Wiley & Sons, Inc., Publication

This edition first published 2013
© 2013 John Wiley & Sons, Inc.

Wiley-Blackwell is an imprint of John Wiley & Sons, formed by the merger
of Wiley's global Scientific, Technical and Medical business with Blackwell Publishing.

Registered Office
John Wiley & Sons, Ltd., The Atrium, Southern Gate, Chichester, West Sussex, PO19 8SQ, UK

Editorial Offices
350 Main Street, Malden, MA 02148-5020, USA
9600 Garsington Road, Oxford, OX4 2DQ, UK
The Atrium, Southern Gate, Chichester, West Sussex, PO19 8SQ, UK

For details of our global editorial offices, for customer services, and for information about how to apply for permission to reuse the copyright material in this book please see our website at www.wiley.com/wiley-blackwell.

The right of Dan Ryder, Justine Kingsbury, and Kenneth Williford to be identified as the authors of the editorial material in this work has been asserted in accordance with the UK Copyright, Designs and Patents Act 1988.

All rights reserved. No part of this publication may be reproduced, stored in a retrieval system, or transmitted, in any form or by any means, electronic, mechanical, photocopying, recording or otherwise, except as permitted by the UK Copyright, Designs and Patents Act 1988, without the prior permission of the publisher.

Wiley also publishes its books in a variety of electronic formats. Some content that appears in print may not be available in electronic books.

Designations used by companies to distinguish their products are often claimed as trademarks. All brand names and product names used in this book are trade names, service marks, trademarks or registered trademarks of their respective owners. The publisher is not associated with any product or vendor mentioned in this book. This publication is designed to provide accurate and authoritative information in regard to the subject matter covered. It is sold on the understanding that the publisher is not engaged in rendering professional services. If professional advice or other expert assistance is required, the services of a competent professional should be sought.

Library of Congress Cataloging-in-Publication Data
Millikan and her critics / edited by Dan Ryder, Justine Kingsbury, and Kenneth Williford.
  p. cm.
 Includes bibliographical references and index.
 ISBN 978-0-470-65684-6 (cloth) – ISBN 978-0-470-65685-3 (pbk.)
 1. Millikan, Ruth Garrett. I. Ryder, Dan (Thomas Daniel) II. Kingsbury, Justine.
III. Williford, Kenneth.
 B945.M487M55 2013
 191–dc23
                                                             2012032812

A catalogue record for this book is available from the British Library.

Cover image: Photo of Ruth Millikan © Steve Pyke.
Cover design by Richard Boxall Design Associates.

Set in 10/12.5pt Ehrhardt by SPi Publisher Services, Pondicherry, India
Printed in Malaysia by Ho Printing (M) Sdn Bhd

1 2013

Contents

Notes on Contributors — vii

Foreword — ix
Daniel C. Dennett

A Millikan Bibliography — xiii

Introduction — 1
Dan Ryder, Justine Kingsbury, and Kenneth Williford

1 Toward an Informational Teleosemantics — 21
Karen Neander
Reply to Neander, by Ruth Millikan

2 Signals, Icons, and Beliefs — 41
Peter Godfrey-Smith
Reply to Godfrey-Smith, by Ruth Millikan

3 Millikan's Isomorphism Requirement — 63
Nicholas Shea
Reply to Shea, by Ruth Millikan

4 Millikan on Honeybee Navigation and Communication — 87
Michael Rescorla
Reply to Rescorla, by Ruth Millikan

5 Concepts: Useful for Thinking — 107
Louise Antony
Reply to Antony, by Ruth Millikan

6 Properties Over Substance — 123
Richard Fumerton
Reply to Fumerton, by Ruth Millikan

7	Millikan's Historical Kinds *Mohan Matthen* Reply to Matthen, by Ruth Millikan	135
8	Millikan, Realism, and Sameness *Crawford L. Elder* Reply to Elder, by Ruth Millikan	155
9	Craning the Ultimate Skyhook: Millikan on the Law of Noncontradiction *Charles Nussbaum* Reply to Nussbaum, by Ruth Millikan	176
10	Are Millikan's Concepts Inside-Out? *Jesse Prinz* Reply to Prinz, by Ruth Millikan	198
11	The Epistemology of Meaning *Cynthia Macdonald and Graham Macdonald* Reply to Macdonalds, by Ruth Millikan	221
12	Weasels and the *A Priori* *David Braddon-Mitchell* Reply to Braddon-Mitchell, by Ruth Millikan	241
13	All in the Family *Willem A. deVries* Reply to deVries, by Ruth Millikan	259
	Afterword *Ruth Millikan*	281
	References	282
	Index	292

Notes on Contributors

Louise Antony is Professor of Philosophy at the University of Massachusetts at Amherst. She has published many papers in philosophy of mind, philosophy of language, epistemology, and feminist theory, and has also edited *Philosophers without Gods* (2007), *Chomsky and His Critics* (2003), and (with Charlotte Witt) *A Mind of One's Own* (1993).

David Braddon-Mitchell is Professor of Philosophy at the University of Sydney. He is the author of numerous papers in philosophy of mind, philosophy of science, and metaphysics, and co-author (with Frank Jackson) of *The Philosophy of Mind and Cognition* (1996).

Daniel C. Dennett is Austin B. Fletcher Professor of Philosophy and co-director of the Center for Cognitive Studies at Tufts University. He is the author of *Breaking the Spell* (2006), *Freedom Evolves* (2003), *Darwin's Dangerous Idea* (1995), *Consciousness Explained* (1992), *The Intentional Stance* (1987), and many other books and papers in philosophy of mind, cognitive science, free will, philosophy of biology, and secularism.

Willem A. deVries is Professor of Philosophy at the University of New Hampshire. He is the author of *Wilfrid Sellars* (2005), *Knowledge, Mind, and the Given* (with Timm Triplett), and numerous papers in philosophy of mind, philosophy of language, metaphysics, epistemology, and the history of philosophy.

Crawford L. Elder is Professor of Philosophy at the University of Connecticut. He is the author of *Familiar Objects and their Shadows* (2011), *Real Natures and Familiar Objects* (2004), and a large number of papers in both metaphysics and philosophy of mind.

Richard Fumerton is F. Wendell Miller Professor of Philosophy at the University of Iowa. He is the author of *Epistemology* (2006), *Realism and the Correspondence Theory of Truth* (2002), *Metaepistemology and Skepticism* (1995), *Metaphysical and Epistemological Problems of Perception* (1985), and numerous papers in epistemology, philosophy of language, and metaphysics.

Peter Godfrey-Smith is Distinguished Professor of Philosophy at the Graduate Center, City University of New York. He is the author of *Darwinian Populations and Natural*

Selection (2009), *Theory and Reality* (2003), and *Complexity and the Function of Mind in Nature* (1996), as well as many papers in philosophy of biology and philosophy of science.

Cynthia Macdonald is Professor of Philosophy at the University of Manchester, and Professor Emeritus and Adjunct Professor of Philosophy at the University of Canterbury. She is the author of *Varieties of Things: Foundations of Contemporary Metaphysics* (2005) and *Mind-Body Identity Theories* (1989), the editor of a large number of collections, and has published many papers in philosophy of mind, philosophy of language, epistemology, and metaphysics.

Graham Macdonald is Professor Emeritus and Adjunct Professor of Philosophy at the University of Canterbury, and Research Fellow in the Department of Philosophy at the University of Manchester. He is the author (with Philip Pettit) of *Semantics and Social Science* (1981, reissued 2011), the editor of many collections, including *Emergence in Mind* (2010) (with Cynthia Macdonald) and *Teleosemantics* (2007) (with David Papineau), and has published numerous papers in philosophy of mind, metaphysics, philosophy of biology, and philosophy of social science.

Mohan Matthen is Canada Research Chair in Philosophy, Perception and Communication at the University of Toronto. He is the author of *Seeing, Doing and Knowing* (2007), and many papers in the philosophy of perception, philosophy of biology, and ancient philosophy.

Karen Neander is Professor of Philosophy at Duke University, and the author of a large number of papers in philosophy of mind and philosophy of biology. Her forthcoming book is entitled *Mental Representation: The Natural and the Normative in a Darwinian World*.

Charles Nussbaum is Associate Professor of Philosophy at the University of Texas, Arlington. He is the author of *The Musical Representation: Meaning, Ontology, and Emotion* (2007) and numerous papers in philosophy of mind, aesthetics, and Kant.

Jesse Prinz is Distinguished Professor of Philosophy at the Graduate Center, City University of New York. He is the author of *Beyond Human Nature* (2012), *The Emotional Construction of Morals* (2007), *Gut Reactions* (2004), *Furnishing the Mind* (2002), and many papers in philosophy of mind, cognitive science, moral psychology, and aesthetics.

Michael Rescorla is Associate Professor of Philosophy at the University of California, Santa Barbara. He is the author of numerous papers in philosophy of mind, philosophy of computation, philosophy of language, and epistemology.

Nicholas Shea is Reader in Philosophy at King's College, London. He is the author of *On Millikan* (2004) and a large number of papers in philosophy of mind, cognitive science, and philosophy of biology.

Foreword

Philosophy is a dirty job. Cleaning up the confusions and obstinate misunderstandings of less philosophically astute folks would be hard enough without having to contend with all the dust kicked up by the efforts of other philosophers, vying with each other to do the same job better. Then there is the relentless tug of received opinion, creating "fixed points" that are better ignored – if only you could persuade others of this. In this churning melee of would-be conceptual cleansers, Ruth Garrett Millikan stands out, quite literally, as a calm, resolute, resourceful defender and developer of a growing family of insights – the Millikan vision, you might call it – that puts a surprisingly large number of contentious and utterly central issues in a better light: How can our words have meanings? How can our brains represent the world? How can knowledge be acquired in perception and passed on in communication?

This volume reveals the range and power of her vision, while also highlighting just how difficult it is to keep the alluring misconceptions at bay. Millikan's "take" on the issues is typically so orthogonal to the prevailing presumptions that she has had to devise a special vocabulary to keep her readers from falling back into the bad habits of thought she is exposing: *Normal* (with its capital N) and *proper functions* and the *concept/conception* distinction, to name some key examples. Some of these innovations have worked better than others, and she has revised Millikanese over the years to deal with the most persistent miscommunications. (We can expect more improvements in the future. Her much-misunderstood use of the term *empirical concept* may soon be replaced by a neologism: *unicept*. Stay tuned.)

There is nothing new about philosophers insisting on creating proprietary idiolects – think of Kant, Hegel, and Heidegger – but unlike some of the others, she takes on the burden of *explaining* why her innovations are good moves, instead of simply *brandishing* them, as philosophers often do, leaving the task of comprehension as an exercise for the uninitiated. This constructive spirit is well exhibited in the essays here and especially in Millikan's responses, but one can hear a few echoes of the ferocious reactions her work triggered in the early days. I recall all too well the colossally rude dismissiveness she encountered in some quarters when she first presented her revolutionary arguments, so I particularly relish my role here in setting the stage for her vindication. On the strength of this volume I would say that she has finally succeeded in domesticating her critics, always setting a good example of how to conduct oneself in discussion.

I have often told the tale of how I came to learn of Ruth and her work, and I gather the tale has taken on a life of its own, as retold tales do, with mutations and embellishments – including some of my own, I now see – so this is a good place to try to set down the unvarnished truth, as I reconstruct it thanks to the fact that my correspondence over the years now resides, alphabetized in yearly cartons, in Tufts' library. It was short work, recently, for an archivist to extract all my early correspondence with Ruth, from her very first letter of February 12, 1979:

> Dear Professor Dennett:
> Enclosed is a manuscript completed just before reading your ... *Brainstorms*. Both my colleagues here at Connecticut and I had considered this paper of mine to be hopelessly maverick.... But the orientation is strikingly like that of *Brainstorms*, certainly if you contrast it with current approaches in the philosophy of language.... My view of representations has a different slant from yours. I am not sure *how* different in the end, but would be immensely pleased to have your view of the matter....
>
> Sincerely,
> Ruth Garrett Millikan

Hopelessly maverick, but strikingly like my own orientation? I was not sure I wanted to give this unsolicited manuscript from an unknown author so much as a skim. It could well be an unintended *reductio ad absurdum* or parody – the line is sometimes hard to draw – of my work. The letter and manuscript had been forwarded to me in Oxford, where I was spending a sabbatical, so months had passed in those pre-email days, but skim it I eventually did, and wrote back in May, apologizing for not doing it justice because of deadlines I faced: "I could easily see the convergence in our work that prompted you to write, and my initial reaction is that your fundamental idea about the evolution of linguistic features is perfectly plausible and well worth pursuing." Not much encouragement, I now see, but enough, thank goodness, to keep me on her mailing list. Her swift reply revealed that the "warm response" of Philip Pettit, who was visiting at UConn then, "has encouraged me that it may be of some value." Good eye, Philip! Soon (October 25, 1979) she began to "inundate" me with drafts. Later, in March, 1980, she reciprocated my cautious blessing of her work by providing a much needed boost to me, after I had sent her a draft of "Beyond Belief," which was tormenting me that year.

> Dear Dan Dennett,
> Thank you for the packet of ideas. The argument that *de re* and *de dicto* aren't two kinds of belief seems to me to be right on and to cut through to what needs to be explained *much* better and more fully than has been done before....

Thank you, Ruth! I needed that. The letter went on for pages of insightful questions and criticisms that preview some of her more recent discussions of the relations between representations and the world they mesh with. ("Those possible worlds. I sense that you have your doubts too? One trouble with them is that they offer no friction, no resistance, hence no foothold or 'constraints'. Are worlds with phlogiston and witches in them possible worlds?") She ended the letter by urging me *not* to read the manuscript she had recently sent me, of which she was "ashamed," and especially to ignore her definition of "proper function" therein. She was replacing it with a better version ("medium bulky" – uh-oh) that would soon be sent to me.

The next mailing, in November of 1980, included drafts of the first chapters of a book, "… now (provocatively) titled *Language, Thought and Other Biological Categories*." (The title was, of course, perfect. That's exactly what the Millikan vision is about.) She added:

> I find that I have to rewrite and rewrite and, especially, *expand* before anything that I have to say becomes understandable. Exhausting work! And painful, since I have so much already thought out and drafted that I want to communicate about but that will undoubtedly have to go through a long long series of revisions first.

The bulky chapters began arriving. I read them in the summer of 1981 and thought they were terrific, though in need of further editing. With Ruth's permission, I sent them to my friend and publisher, Harry Stanton at Bradford Books, MIT Press, urging him to triangulate my enthusiasm with at least two other referees, since maybe I was hopelessly maverick myself, and shouldn't be trusted. Fred Dretske and Hector-Neri Castañeda were the other philosophers Harry consulted – a wide tripod indeed – and they were equally enthusiastic, so Bradford Books became the home of her landmark book. I did some energetic editorial work on it with her and wrote the foreword. Ruth and I had still never met, or even, I think, talked on the phone. When she learned that we were finally going to meet at an APA division meeting cocktail party, she sent me a note warning me – in case I had been encouraging her over the years under the misapprehension that I was taking a sweet young thing under my arm! – that she was older than I was, a mother with grown children. I had already figured that out; nobody could have written LTOBC without mastering – *overcoming* might be a better term – the literature, and spending years of hard thinking putting the pieces in the right places.

And ever since she built her wonderful Millikan machine, it has proved to be a prodigious generator of further philosophical enlightenment, requiring some maintenance and improvement to be sure, but as robust and extensible as any explanatory system in philosophy. The key to its power lies in its unwavering – and demanding – biological naturalism; she never lets you forget that minds aren't magic, that they have to have evolved just the way hearts and livers had to evolve. Many would-be naturalists among philosophers of mind and language have underestimated the necessity of seeing these issues from the *reverse engineering* perspective of evolution by natural selection: What do these things have to do to earn their keep, and how could they possibly come to do it?

Just as important, what *needn't* they do? Much philosophical theorizing attributes stupendous powers to mental events and dispositions that are strictly gratuitous, which is a good thing, since they are almost certainly impossible. I have come to recognize in recent years that perhaps the central revolutionary idea Darwin gave us, his "strange inversion of reasoning" (Dennett 2009), is the idea of *competence without comprehension*. We tend to think – especially if we are philosophers – that competence must *flow from* comprehension, that first we need to comprehend *in order* to be competent. ("Meaning rationalism" was Millikan's term for this ubiquitous conviction in LTOBC.) No, it's the other way around, actually: our comprehension is the product of cascades of semi-comprehending, pseudo-comprehending, uncomprehending competences that we are endowed with, first by natural selection and then by learning and cultural – especially linguistic – redesign.

Perhaps Millikan's key insight is that our ability to identify and reidentify things and properties in the world, without which we could not acquire the knowledge and comprehension we have, starts with an innate perceptual proclivity to (try to) identify distal things, a competence that we exercise without knowing – or having to know – how or why it works. Upon this evolved foundation the structure of the distal world shapes our empirical concepts, letting us learn what we ought to learn first about the world we live in. As she says, "... the ability to reidentify things that are objectively the same when we encounter them in perception is the most central cognitive ability that we possess" (OCCI, 109), and this ability – having, in her sense, a *concept* of a thing – does not depend on any particular *conception* (definition, intension, mode of presentation, etc.) we may have of it.

As she has insisted, "Failure to account for our capacity to represent individuals in language and thought has been, perhaps, the most serious failing common to contemporary naturalist theories of content" (VOM, 43), and it has been her achievement to repair that failing. The result is not a minor difference in outlook among philosophers of language and mind: either you are with Millikan – you get it, and see that one way or another a brand of "teleofunctionalism" is the only way to make sense of meaning – or you are doomed to recycle the pre-Darwinian fantasies that have continued to beguile so many deep thinkers for more than a century after the *Origin of Species*. Millikan may be the maverick, but the hopelessness lies on the other side of the fence.

Daniel C. Dennett
Blue Hill, ME

A Millikan Bibliography

1969 *Empirical Identity*. Dissertation in philosophy, Yale University (Sterling Library).
1979 "An Evolutionist Approach to Language," *Philosophy Research Archives*, 5(4).
1983 "Dennett's Rational Animals: How Behaviorism Overlooked Them," *Behavioral and Brain Sciences*, 6: 372–373.
1984 *Language, Thought, and Other Biological Categories: New Foundations for Realism*. Cambridge, MA: MIT Press. ("LTOBC")
1984 "Naturalist Reflections on Knowledge," *Pacific Philosophical Quarterly*, 65(4): 315–334.
1986 "Thoughts Without Laws: Cognitive Science With Content," *Philosophical Review*, 95: 47–80.
1986 "The Price of Correspondence Truth," *Noûs*, 20(4): 453–468.
1986 "Metaphysical Antirealism?" *Mind*, 95(380): 417–431.
1986 "Of What Use Categories?" *Behavioral and Brain Sciences*, 9(4): 163–164.
1987 "Review of Christopher Hookway's *Minds, Machines, and Evolution*," *Noûs*, 21(2): 95–98.
1987 "What Peter Thinks When He Hears Mary Speak," *Behavioral and Brain Sciences*, 10(4): 725–726.
1989 "Biosemantics," *Journal of Philosophy*, 86: 281–297.
1989 "In Defense of Proper Functions," *Philosophy of Science*, 56(2): 288–302.
1989 "An Ambiguity in the Notion "Function'," *Biology and Philosophy*, 4(2): 172–176.
1990 "Truth Rules, Hoverflies, and the Kripke–Wittgenstein Paradox," *Philosophical Review*, 99(3): 323–353.
1990 "Compare and Contrast Dretske, Fodor and Millikan on Teleosemantics," *Philosophical Topics*, 18(2): 151–161.
1990 "The Myth of the Essential Indexical," *Noûs*, 24(5): 723–734.
1990 "Seismograph Readings for Explaining Behavior" (review of Fred Dretske's *Explaining Behavior*), *Philosophy and Phenomenological Research*, 50(4): 807–812.
1990 "Clarifications on *Language, Thought and Other Biological Categories*," *Annals of Scholarship*, 7: 147–149.

1991	"Speaking Up for Darwin," in B. Loewer and G. Rey (eds.) *Meaning in Mind: Fodor and his Critics*. Oxford: Blackwell, 151–164.
1991	"Perceptual Content and Fregean Myth," *Mind*, 100(4): 439–459.
1992	"Review of Robert Cummins' *Meaning and Mental Representation*," *Philosophical Review*, 101(2): 422–425.
1992	"Review of Jerry Fodor's *A Theory of Content*," *Philosophical Review*, 101(4): 898–901.
1993	*White Queen Psychology and Other Essays for Alice*. Cambridge, MA: MIT Press. ("WQ")
1993	"Propensities, Exaptations, and the Brain," in WQ: 31–50.
1993	"A Philosophical Essay on Ethology and Individualism in Psychology: What is Behavior (part 1) and The Green Grass Growing All Around (part 2)," in WQ: 135–170.
1993	"White Queen Psychology; or, The Last Myth of the Given," in WQ: 279–363.
1993	"Knowing What I'm Thinking Of," *Proceedings of the Aristotelian Society*, suppl. vol. 67: 109–124.
1993	"Explanation in Biopsychology," in J. Heil and A. Mele (eds.) *Mental Causation*. Oxford: Oxford University Press, 211–232.
1993	"On Mentalese Orthography," in B. Dahlbom (ed.) *Dennett and his Critics*. Oxford: Blackwell, 97–123.
1993	"Content and Vehicle," in N. Eilan, R. McCarthy, B. Brewer (eds.) *Spatial Representation*. Oxford: Blackwell, 256–268.
1994	"On Unclear and Indistinct Ideas," in J. Tomberlin (ed.) *Philosophical Perspectives* vol. 8. Atascadero, CA: Ridgeview Publishing, 75–100.
1995	"Reply: A Bet with Peacocke," in C. Macdonald and G. Macdonald (eds.) *Philosophy of Psychology: Debates on Psychological Explanation*. Oxford: Blackwell, 285–292.
1995	"Pojecia Syntetyczne: Filozoficzne Rozwazania o Kategoryzacji," *Roczniki Filozoficzne*, 43(1): 165–180.
1996	"Pushmi-pullyu Representations," in J. Tomberlin (ed.) *Philosophical Perspectives* vol. 9. Atascadero CA: Ridgeview Publishing, 185–200.
1996	"On Swampkinds," *Mind and Language*, 11: 103–117.
1996	"Varieties of Purposive Behavior," in R. Mitchell (ed.) *Anthropomorphism, Anecdotes, and Animals*. Albany: SUNY Press, 189–197.
1997	"Images of Identity: In Search of Modes of Presentation," *Mind*, 106: 499–519.
1997	"Troubles with Wagner's Reading of Millikan," *Philosophical Studies*, 86(1): 93–96.
1998	"Language Conventions Made Simple," *Journal of Philosophy*, 95(4): 161–180.
1998	"Cognitive Luck: Externalism in an Evolutionary Frame," in P. Machamer and M. Carrier (eds.) *Philosophy and the Sciences of Mind*. Pittsburgh: Pittsburgh University Press and Konstanz: Universitätsverlag Konstanz, 207–219.
1998	"A Common Structure for Concepts of Individuals, Stuffs and Basic Kinds: More Mama, More Milk and More Mouse," *Behavioural and Brain Sciences*, 22(1): 55–65.
1998	"With Enemies Like These I don't Need Friends: Author's Response," *Behavioural and Brain Sciences*, 22(1): 89–100.

1998 "Proper Function and Convention in Speech Acts," in L. E. Hahn (ed.) *The Philosophy of Peter F. Strawson: The Library of Living Philosophers*. LaSalle, IL: Open Court, 25–43.
1998 "How We Make Our Ideas Clear," The Tenth Annual Patrick Romanell Lecture, *Proceedings and Addresses of the American Philosophical Association*, 72(2): 65–79.
1998 "A More Plausible Kind of 'Recognitional Concept'," in E. Villanueva (ed.) *Concepts, Philosophical Issues* vol. 9. Atascadero, CA: Ridgeview Publishing, 35–41.
1998 "Review of Peter Godfrey-Smith's *Complexity and the Function of Mind in Nature*," *Philosophy of Science*, 65(2): 375–377.
1999 "Wings, Spoons, Pills and Quills: A Pluralist Theory of Functions," *Journal of Philosophy*, 96(4): 191–206.
1999 "Historical Kinds and the Special Sciences," *Philosophical Studies*, 95: 45–65.
1999 "Reply to Boyd," *Philosophical Studies*, 95: 99–102.
1999 "On Sympathies with J. J. Gibson, and on Focusing Reference," replies to Treffner and Saidel, *Behavioral and Brain Sciences*, 22(4): 732–733.
2000 *On Clear and Confused Ideas: An Essay about Substance Concepts*. Cambridge: Cambridge University Press. ("OCCI")
2000 "Naturalizing Intentionality," in B. Elevitch (ed.) *Philosophy of Mind, Proceedings of the Twentieth World Congress of Philosophy* vol. 9. Charlottesville: Philosophy Documentation Center, 83–90.
2000 "Representations, Targets and Attitudes (discussion with Robert Cummins)," *Philosophy and Phenomenological Research*, 60(1): 103–111.
2000 "Reading Mother Nature's Mind," in D. Ross, A. Brook, and D. Thompson (eds.) *Dennett's Philosophy: A Comprehensive Assessment*. Cambridge, MA: MIT Press, 55–75.
2001 "What has Natural Information to do with Intentional Representation?" in D. Walsh (ed.) *Evolution, Naturalism and Mind*. Cambridge: Cambridge University Press, 105–126.
2001 "Purposes and Cross-Purposes: On the Evolution of Language and Languages," *Monist*, 84(3): 392–416.
2001 "The Language–Thought Partnership: A Bird's-Eye View," in H. J. Glock (ed.) *Language and Communication*, 21: 157–166.
2001 "Cutting Philosophy of Language Down to Size," in A. O'Hear (ed.) *Philosophy at the New Millennium*, Royal Institute of Philosophy Supplementary Series. Cambridge: Cambridge University Press, 125–140.
2001 "The Myth of Mental Indexicals" (revised version of "The Myth of the Essential Indexical"), in A. Brook and R. DeVidi (eds.) *Self-Reference and Self-Awareness: Advances in Consciousness Research* vol. 11. Philadelphia: John Benjamins, 163–177.
2001 "A Theory of Representation to Complement TEC," *Behavioral and Brain Sciences*, 24(5): 894–895.
2002 "Biofunctions: Two Paradigms," in R. Cummins, A. Ariew, and M. Perlman (eds.) *Functions: New Readings in the Philosophy of Psychology and Biology*. Oxford: Oxford University Press, 113–143.

2003 "In Defense of Public Language," in L. M. Antony and N. Hornstein (eds.) *Chomsky and his Critics*. Oxford: Blackwell, 215–237.

2003 "Teleological Theories of Mental Content," in L. Nadel (ed.) *The Encyclopedia of Cognitive Science*. New York: Macmillan.

2003 "Vom angeblichen Siegeszug der Gene und der Meme" ["On the Rumored Takeover by the Genes and the Memes"] in Becker et al. (eds.) *Gene, Meme und Gehirne. Geist und Gesellschaft als Natur*. Frankfurt am Main: Suhrkamp, 90–111.

2004 *Varieties of Meaning*. Cambridge, MA: MIT Press. ("VOM")

2004 "On Reading Signs: Some Differences between Us and the Others," in D. K. Oller and U. Griebel (eds.) *Evolution of Communication Systems: A Comparative Approach*. Cambridge MA: MIT Press.

2004 "Existence Proof for a Viable Externalism," in R. Schantz (ed.) *Current Issues in Theoretical Philosophy II: The Externalist Challenge. New Studies on Cognition and Intentionality*. New York: de Gruyter, 227–238.

2005 "The Son and the Daughter: On Sellars, Brandom, and Millikan," *Pragmatics and Cognition*, 13: 59–72.

2005 *Language: A Biological Model*. Oxford: Clarendon Press. ("LBM")

2005 "On Meaning, Meaning, and Meaning," in LBM: 53–76.

2005 "Some Reflections on the Theory Theory–Simulation Theory Debate," in S. Hurley and N. Chater (eds.) *Perspectives on Imitation: From Mirror Neurons to Memes* vol 2. Cambridge, MA: MIT Press, 182–188.

2005 "Why (Most) Concepts are not Categories," in H. Cohen and C. Lefebvre (eds.) *Handbook of Categorization in Cognitive Science*. Amsterdam: Elsevier, 305–315.

2006 "Styles of Rationality," in M. Nudds and S. Hurley (eds.) *Rationality in Animals*. Oxford: Oxford University Press, 117–126.

2006 "Useless Content," in G. Macdonald and D. Papineau (eds.) *Teleosemantics*. Oxford: Oxford University Press, 100–114.

2006 "Précis of Language: A Biological Model" and replies to reviewers, *SWIF Philosophy of Mind Review* 5(2), http://www.swif.uniba.it/lei/mind/swifpmr.htm.

2007 "An Input Condition for Teleosemantics? Reply to Shea (and Godfrey-Smith)," *Philosophy and Phenomenological Research*, 75(2): 436–455.

2007 "Précis of Varieties of Meaning: The Jean Nicod Lectures 2002" and responses to reviews by Bermúdez, Recanati, Rosenberg, and Taylor, *Philosophy and Phenomenological Research*, 75(3).

2008 "A Difference of Some Consequence between Conventions and Rules," *Topoi*, 27: 87–100.

2008 "Biosemantics," in B. McLaughlin (ed.) *The Oxford Handbook in the Philosophy of Mind*. Oxford: Oxford University Press, 394–406.

2009 "Embedded Rationality," in M. Aydede and P. Robbins (eds.) *The Cambridge Handbook of Situated Cognition*. Cambridge: Cambridge University Press, 171–181.

2010 "On Knowing the Meaning; with a Coda on Swampman," *Mind*, 119(473): 43–81.

2010 "It's Likely Misbelief Never has a Function," *Behavioral and Brain Sciences*, 32(6): 529–530.
2010 "Gedacht wird in der Welt, nicht im Kopf" (interview with Markus Wild), *Deutsche Zeitschrift für Philosophie*, 58(6): 981–1000.
2011 "Loosing the Word–Concept Tie," *Proceedings of the Aristotelian Society*, suppl. vol. 85: 125–143.
2011 "Die eingebettete Vernunft," *Deutsche Zeitschrift für Philosophie*, 59(4): 483–496.
2011 "Commentary on Pautz," *Consciousness Online 3*, http://consciousnessonline.files.wordpress.com/2011/02/commentary-on-pautz-ruth-millikan2.pdf.
2012 "Natural Signs," in S. B. Cooper, A. Dawar, and B. Löwe (eds.) *How the World Computes: Seventh Conference on Computability in Europe* (CiE 2012), *Lecture Notes in Computer Science*. Heidelberg: Springer.
2012 "What's Inside a Thinking Animal," *XXII Deutscher Kongress für Philosophy, Proceedings Welt der Gründe, Kolloquium 19, Action and decision in non-human animals: Do animals live in the space of reason?*: 889–893.
2012 "Spracherwerb ohne eine Theorie des Geistes" ["Language Acquisition Without a Theory of Mind"], in A. Burri (trans.) *Biosemantik. Sprachphilosophische Aufsätze*. Frankfurt: Surkamp Verlag: 85–115.
2012 "Are There Mental Indexicals and Demonstratives?" *Philosophical Perspectives*, 26(1).
Forthcoming "Troubles with Plantinga's reading of Millikan," *Philosophy and Phenomenological Research*.
Forthcoming "Accidents," 2012 John Dewey Lecture (Central APA), *Proceedings and Addresses of the American Philosophical Association*.
Forthcoming "Teleosemantics," in B. Kaldis (ed.) *Encyclopedia of Philosophy and the Social Sciences*. Thousand Oaks, CA: Sage.
Forthcoming "Natural Information, Intentional Signs and Animal Communication," in U. Stegmann (ed.) *Animal Communication Theory: Information and Influence*. Cambridge: Cambridge University Press.
Forthcoming "An Epistemology for Phenomenology?," in R. Brown (ed.) *Consciousness Inside and Out: Phenomenology, Neuroscience, and the Nature of Experience*. Berlin: Springer.
Forthcoming "The Tangle of Biological Purposes that is Us," in B. Bashour and H. Muller (eds.) *Contemporary Philosophical Naturalism and Its Implications*. London: Routledge.
Forthcoming "Confessions of a Renegade Daughter," in J. Shea (ed.) *Sellars and His Legacy*. Oxford: Oxford University Press.
Forthcoming "What do Thoughts do to the World? Deflating Socially Constituted Objects," in M. Gallotti and J. Michael (eds.) *Studies in the Philosophy of Sociality, Volume I*. Heidelberg: Springer.

INTRODUCTION

DAN RYDER, JUSTINE KINGSBURY, AND KENNETH WILLIFORD

Ruth Millikan's work is notable for its originality, scope, and coherence, and for its unwavering naturalism, with her focus on the proper functions of our cognitive and linguistic mechanisms as the unifying thread. Besides her meticulously worked-out and comprehensive theory of mind and language, she has made important contributions to philosophy of biology (especially to the discussion of biological functions), epistemology (especially on the nature of empirical knowledge), and metaphysics (especially the metaphysics of natural kinds).

Millikan's first book, *Language, Thought and Other Biological Categories* ("LTOBC"), appeared in 1984. It was followed by a steady stream of articles (some of them collected in *White Queen Psychology and Other Essays for Alice* ("WQ," 1993) and by three further books: *On Clear and Confused Ideas: An Essay About Substance Concepts* ("OCCI," 2000), *Varieties of Meaning* ("VOM," 2004), and *Language: A Biological Model* ("LBM," 2005). (A bibliography of Millikan's works can be found on p. xiii.) LTOBC was striking both for the novelty of its ideas and for its density. Many of the views defended in Millikan's copious later philosophical output, along with the arguments for those views, were already there in the 333 pages of LTOBC. It is unsurprising that many found the book hard to get to grips with. It earned her a reputation for being a *difficult* philosopher; a reputation that those who have read her later, more expansive work know to be undeserved, or at least greatly exaggerated.

This volume contains thirteen new essays on the work of Ruth Millikan, along with Millikan's replies to each. The philosophical range of the contributors is a testament to the breadth of Millikan's contribution to philosophy. Karen Neander in "Toward an Informational Teleosemantics," Peter Godfrey-Smith in "Signals, Icons, and Beliefs," Nicholas Shea in "Millikan's Isomorphism Requirement," and Michael Rescorla in

"Millikan on Honeybee Navigation and Communication" take up issues concerning Millikan's teleological account of mental content. In "Concepts: Useful for Thinking," Louise Antony takes exception to various features of Millikan's account of concepts. Richard Fumerton's "Properties over Substance" is a critical examination of Millikan's claim (in OCCI) that substance concepts are ontologically and epistemologically prior to other concepts. Mohan Matthen in "Millikan's Historical Kinds" and Crawford Elder in "Millikan, Realism, and Sameness" discuss Millikan's account of natural kinds; in Elder's case this is on the way to considering whether or not Millikan is a realist about objects. Charles Nussbaum writes about a Millikanian approach to naturalizing the law of noncontradiction ("Craning the Ultimate Skyhook: Millikan on the Law of Noncontradiction").

Millikan has a radically externalist view about meaning, and a number of the essays are about matters arising from this. Jesse Prinz, in "Are Millikan's Concepts Inside-Out?," argues for a breaking down of the externalism/internalism divide. Cynthia and Graham Macdonald in "The Epistemology of Meaning" suggest a way in which teleosemantics can be made compatible with the existence of something like Fregean senses, while David Braddon-Mitchell in "Weasels and the *A Priori*" would like to bring Millikan and two-dimensionalism closer together. Finally, Willem deVries, in "All in the Family," discusses the way in which both Millikan's and Robert Brandom's theories are, despite their striking differences, recognizably descendants of the views of their teacher, Wilfrid Sellars, and urges a rapprochement that would bring both closer to their roots.

Proper Functions

The function of a thing, in one sense, is what that thing is *for*, what it is supposed to do. One of the homes for such teleological concepts is in biology. For example, the function of the heart is to pump blood, and the function of the liver is to remove toxins from the bloodstream; arguably what makes something a heart or a liver is its function. Central to Millikan's philosophy of mind and language is the idea that representations and the mechanisms that produce them have this kind of function as well; hence the provocative title *Language, Thought and Other Biological Categories*.

In the decade or so before the publication of LTOBC, Larry Wright (1976) defended a naturalistic account of teleological functions, and Robert Cummins argued in response that only artifacts have teleological functions. What sense, Cummins asked, can be made of the claim that an object is *supposed to do* something unless that object was made or designed by someone for that purpose? In opposition, Cummins defended a non-teleological, dispositional account of functions in biology, whereby a function is anything a mechanism does that contributes to the capacities of the larger system to which it belongs. A particular capacity is privileged only by our interests as investigators.

An important difference between Cummins-functions and teleological functions is that if something has a Cummins-function, it must actually perform it – the Cummins-function of a thing is always something that it actually succeeds in doing. The teleological function of something, in contrast, is something that it *should* do but not necessarily something that it actually does. While biologists and physiologists are certainly interested in Cummins-functions, when they talk about *malfunction* (a sub-class of the cases in which something has a function which it fails to perform) and when they investigate evolutionary etiology in order to explain why some characteristic is present (Why do I have a heart? To circulate my blood), they seem to be talking about teleological functions. However, Cummins' challenge needs a response: What is it for something to have a teleological function, absent any creator or designer who made that thing for a purpose?

Millikan's account of teleology is not intended as an analysis of either an everyday or a biologist's notion of function. Rather, she introduces a theoretical definition of what she calls *proper functions*, a definition whose usefulness is to be judged by the explanatory success of the account of mind and language in which it plays a central role. Nevertheless, it meets Cummins' challenge. Millikan's (like Wright's) is an etiological account of function, according to which the proper function of my heart, for example, is to do whatever it is that the hearts of my ancestors did that contributed to the survival and reproduction of my ancestors and consequently the survival and proliferation of hearts of that type:

> To put things very roughly, for an item A to have a function F as a "proper function," it is necessary (and close to sufficient) that one of these two conditions should hold. (1) A originated as a "reproduction" (to give one example, as a copy, or a copy of a copy) of some prior item or items that, *due* in part to possession of the properties reproduced, have actually performed F in the past, and A exists because (causally historically because) of this or these performances. (2) A originated as the product of some prior device that, given its circumstances, had performance of F as a proper function and that, under those circumstances, normally causes F to be performed by *means* of producing an item like A. (Millikan 1989b, 288–289)

Functions that fall under (1) above are *direct* proper functions; those that fall under (2) are *derived* proper functions. Proper functions can be *relational*; for example, a chameleon's pigment-changing mechanism has the (direct) proper function of bringing about a certain relationship between the skin of the chameleon and the surface the chameleon is on (in order to perform the further function of concealing it from predators). Continuing with the same example, a particular skin pattern has a proper function *derived* from the proper function of the pigment-changing mechanism that produces it: the proper function of enabling the chameleon to escape detection by predators. The pattern can have a derived proper function even if no chameleon has ever displayed precisely that pattern before.

A consequence of all this is that representations themselves can have proper functions, not just devices that produce them; *learned* representations can have proper functions, not just innate ones; and *unique* never-before-produced representations can have proper functions, not just representation types that are reproduced from earlier types. The caricature of Millikanian teleosemantics as applicable only to frogs reflexively snapping at flies is very far from the truth, something that all our contributors fully appreciate.

Representations: The Basic Teleosemantic Framework

On Millikan's view, different types of representations are characterized by their proper functions, derived from the proper functions of the mechanisms that produce them: the representation producers. Representations are also *used* by other mechanisms: their consumers. (According to Millikan, the proper functions of its consumers are the most important in determining a representation's content.) In his contribution, Peter Godfrey-Smith raises questions about the producer–consumer pairing ubiquitous in Millikan's treatment of representations, from states of bacterial magnetosomes to human linguistic utterances.

Representations fall into two broad types (not mutually exclusive): imperative and indicative. *Imperative* representations have the proper function of enabling a representation consumer to bring about a particular state of the world. For example, a representation in motor cortex may have the proper function of getting the rest of the motor system to bring about a certain arrangement of the limbs. More accurately, it is supposed to bring about a certain arrangement of the limbs according to a particular "mapping rule," whereby variations in the state of the cortex are supposed to produce corresponding variations in the state of the limbs. It is helping to effect a mapping according to this rule that is selected for, and therefore characterizes a proper function of motor cortex. This means that the representational system is *productive*: it can represent (and have the function of producing) unique states of the limbs never before produced in its ancestors. In his contribution, Nicholas Shea addresses this kind of productivity.

In contrast with imperative representation, "… what makes an item an *indicative* [representation] has nothing to do with what it does; it has to do only with the *conditions under which* it Normally does whatever it does." Roughly speaking, "Normal conditions" are those that must be mentioned in a minimal, general explanation of how the representation[1] has managed to perform its job in the past. That job is to enable its consumers to perform *their* proper functions. Among the normal conditions for proper performance of the consumer's function will be what the indicative representation represents. The thought is that indicative representations may be used for all sorts of things, but they will in general not be effective unless they stand in the right relationship to the world.

For example, beavers slap their tails in order to signal danger, and send the colony to safety. The representation producer is the slapping beaver (or its slap-producing mechanism), and the representation consumers are the hearing beavers (or their response mechanisms). The proper function of the slap is to enable the hearers to elude danger – this is the effect that resulted in the survival and proliferation of the slapping–fleeing combination, and it is thus the proper function of the consumer mechanism. A Normal condition for performing this function is that the slap coincide with the presence of danger, and it is this particular Normal condition that Millikan identifies as what the slap represents. It forms part of a Normal explanation for how the hearers have eluded danger in the past. (In her contribution, Karen Neander disputes this Normal conditions based analysis, arguing that sensory representations have the function of carrying information.)

---

[1] Or rather its ancestral incarnations.

A Normal explanation for how the hearers have eluded danger includes other Normal conditions, however. In order to flee, the hearing beavers need adequate oxygen, for example. Why does the slap only represent danger, among all of these Normal conditions? It is danger that the slap "maps" onto, not the presence of oxygen. The time and place of the slap vary, not with anything to do with oxygen, but with the time and place of danger. As in the motor cortex example, all representation involves a "mapping rule," whereby transformations of the representation map onto variations in the environment represented. Bee dances are perhaps a more obvious example. Different bee dances are systematically related to each other, and the states of affairs onto which they map are similarly systematically related to each other. Transformations of the dance "correspond one-to-one to transformations of the location of nectar relative to hive and sun" (Millikan 1986, 74). Again,

> ... [w]hich mapping rule (which transformation correlation) is the relevant one to mention – which rule determines what the dance represents – is quite obvious. This rule is determined by the evolutionary history of the bee. It is that in accordance with which the dance must map onto the world in order to function properly in accordance with a Normal explanation, or, what is the same, in order that the mechanisms within watching bees that translate (physicist's sense) the dance pattern into a direction of flight should perform all of their proper functions (including getting the bees to nectar) in accordance with a Normal explanation. (Millikan 1986, 78–79)

It is a Normal condition for proper performance of a watching bee's nectar-finder that the dancing bee produce a dance that maps onto the current location of nectar in a certain way. That way is defined by the mapping rule, which is a correspondence between a system of possible representations and related group of possible states of affairs in the world. According to Millikan, this mapping rule plays a part in causally explaining the selection of the producer–consumer system. While the presence of oxygen may also be a Normal condition, it does not feature in any such mapping rule, therefore neither the beaver slap nor the bee dance represent anything about oxygen.

Many primitive representations are simultaneously indicative and imperative: Millikan calls these "pushmi-pullyu representations." Both beaver tail slaps and bee dances are examples. The slap both says that there is danger, and tells other beavers to take cover; the dance both tells of the location of nectar, and commands the bees to go there. The story for more complex representational systems (such as human mental and linguistic representational systems) is more of the same, except that the indicative and imperative functions are more often separated. Desires and commands are imperative representations, so their function is to cause their consumers to bring about a certain state of affairs. That state of affairs is the content of the desire, or the meaning of the command. Beliefs and indicative utterances are indicative representations, so their proper functioning depends on their mapping onto the world in the right way – the way that historically has enabled their consumers to perform their proper functions. The basic teleosemantic framework reappears in many forms throughout Millikan's work.

Informational teleosemanticists think that representation-producing systems have the function of producing states that carry natural information, that representations have the function of carrying information, and that the content of a particular representation is the information it has the function of carrying. Millikan thinks otherwise: although true representations usually do carry natural information, that is not their function. The proper function of a representation is what it is designed to effect, to cause; whether it carries information is a matter of its origins and its current relations.

In "Toward an Informational Teleosemantics," Karen Neander argues that at least for sensory indicative representations, informational teleosemantics can be made to work. Sensory systems, she suggests, have response functions: functions to respond to something by doing something else. Their function, according to Neander, is to produce representations that carry natural information.

Millikan, in her response, resists this move. The case she focuses on is that of a true representation that has not been produced in a Normal way. For example, someone may infer on the basis of mistaken evidence that Dan Dennett is in the next room. This belief is not a natural sign of that state of affairs, it does not carry natural information about it; nevertheless, it may happen to be true that Dan Dennett is in the next room. In cases like this, where an indicative representation is true by accident, the representation consumer doesn't care. It performs its function equally well however the representation is produced. Therefore, argues Millikan, the representation-producing mechanism can't have been selected for producing true representations *in a particular way*, but only for producing true representations, i.e., representations that map onto the world in the particular way required for interpretation by the consumer:

> ... if producing *true representations* is what *representation-producing systems* are selected for, informational teleosemantics can't be quite right. The problem to be solved by teleosemantics is not just how false representations are possible, but also how representations can sometimes be representations and be true despite having been arrived at through accident or malfunction.

Peter Godfrey-Smith's essay "Signals, Icons, and Beliefs" starts with the observation that Millikan's account of content is a version of what he calls the sender–receiver approach: any entity that has semantic content has it as a consequence of its relations to a producer on one side and an interpreter or consumer on the other. Godfrey-Smith discusses sender–receiver systems in the light of recent work by Brian Skyrms and others on how such systems come into existence and are maintained. He then discusses the content of "messages" or "signals" in such systems, comparing Skyrms' information-theoretic account with Millikan's account. Skyrms' account falls victim to the standard objection to simple informational accounts of content: it fails to allow for the possibility of misrepresentation. Millikan's account does not suffer from this problem, because for Millikan, the content of a "signal" is (roughly) the state of affairs that must obtain for the receiver of that signal to use it successfully. But Godfrey-Smith points out a different problem for Millikan: that this focus on the receiver/consumer of a representation in some cases yields content-attributions that are "at odds with reasonable-looking intuitions about content." The cases in question are ones in which the content of a representation on Millikan's "consumer-determined" view is not the same as the state of affairs which triggers the production of the representation: R is a response to state of affairs X, but R is useful only because X is reliably correlated with state of affairs Y because X and Y have a common cause. On Millikan's view, the content of R is Y, as Godfrey-Smith puts it, "no matter how remote Y is from the producer's ability to directly track the world." (Paul Pietroski [1992] gives a striking example in which a creature responds to dawn light with a representation meaning, according to Millikan's theory, "no predators this way.") Godfrey-Smith remarks that sometimes this fits with how we usually think of content, and sometimes it does not.

Godfrey-Smith explores how Millikan's account plays out in the case in which producers, representations, and interpreters are all internal to an organism. Raising the possibility that our mental mechanisms may not in fact have a straightforward sender–receiver structure, he considers some psychological research into internal maps (in rats): the evidence (along with a computational model of the rat's mapping system due to Reid and Staddon 1997) suggests that there is no internal map-reader distinct from the map. Godfrey-Smith suggests that this might be the situation with mental representations more generally: the distinction between producer, sign, and consumer might not be clear-cut.

Millikan accepts this last conclusion: there need be no "reader" of our mental representations, and producers and consumers are mechanisms or systems that "can overlap in their components and may operate in several capacities even at the same time." Her response to the query about Pietroski-type cases is that there is no general problem here: there are many intuitive cases where a sign and the state of affairs it is about have a common cause, for example beliefs about the future. The reason the Pietroski case jars with intuition is not this, but because as described, it involves an illicit attribution of subpersonal content to a whole creature (among other reasons).

Thought is productive: we can have an indefinitely large variety of thoughts, including novel ones with no history. Millikan accounts for the productivity of thought by appealing to isomorphisms (one to one mappings) between systems of representations and the systematically related states of affairs that they represent. In "Millikan's Isomorphism Requirement," Nicholas Shea argues that the relation of isomorphism is insufficiently demanding to play the role that Millikan assigns to it. He intends this as a friendly amendment, however, since there's something else that does the job.

Millikan has presented her theoretical picture as though a mere mathematical correspondence – an "isomorphism" is her term – between representations and affairs represented plays a causal/explanatory role. But, Shea points out, mere mathematical correspondence is cheap, and explains nothing. Rather, it is the *natural relations* among representations and the corresponding natural relations existing among states of affairs that do the explaining. An isomorphism does obtain, but comes along for free.

In her reply, Millikan generally accepts this point, saying that what she had in mind by "isomorphism" was indeed a correspondence between natural relations (call this a "natural isomorphism"). She is less concessive toward a related point that Shea makes, concerning the role of isomorphism in securing representational productivity. Shea points out that, depending on how the consumer system is structured, there is no necessity that relations among possible representations or relations among possible represented states of affairs be *natural* relations. Arbitrary mappings can work just as well, as long as the representation consumer is built to take advantage of them. However, it is only when the relations are natural that the representational system will exhibit productivity, allowing for never-before-represented states of affairs to be represented for the first time in the history of the species.

Millikan agrees with the link between natural relations and productivity, but she argues that Shea's "arbitrary" mappings *do* in fact exhibit productivity. Her key move here is to show that Shea's examples display a natural isomorphism after all. It is an isomorphism that falls under what she calls "substitutional correspondence" (which Shea would consider arbitrary) rather than a more robust "projectable correspondence" (which Shea would call non-arbitrary). Shea, she argues, mistakenly sees productivity only when the correspondence is "projectable." There is a *natural* isomorphism even when the correspondence is merely substitutional.

A final dispute between Shea and Millikan resides in whether a producer–consumer system typically makes use of natural isomorphism as opposed to making use of simple correlational information. In the bee dance case, for instance, Shea argues that the system does not make use of the fact that similar bee dances indicate similar distances away; an arbitrary mapping, with no similarity structure, would do just as well for the job bees give it, which depends only on correlational information. Millikan disagrees (as with Neander); Shea, she says, is focusing on the producer's job, and not the consumer's. Correlational information does not explain how the consumer manages to do its job, because accidentally true representations (e.g., representations that are not causally related to their represented states of affairs, and so do not carry information about them) serve the consumer just as well as information-carrying true representations. "Truth" here picks out the mapping according to the natural isomorphism (which may be merely substitutional) that causally explains selection of the producer–consumer system.

Millikan's theory assigns truth conditions to (indicative) representations by singling out a particular mapping rule or natural isomorphism that helps explain selection of that particular producer–consumer system. It is important to Millikan's view that there is continuity between what she sees as simple representational systems, such as bacteria representing the direction of low oxygen, or bees representing the relative location of nectar, and more complex representational systems like our own.

Michael Rescorla challenges this picture in "Millikan on Honeybee Navigation and Communication." His central challenge comes from a contrast between the biological and psychological sciences. Rescorla points out that Millikan takes representations to be theoretical posits: she is not analyzing the use of the term "representation," whether by ordinary folk or by scientists. If representations are theoretical posits they must earn their explanatory keep. Rescorla argues that they do not do so for simple representational systems. In particular, there is no explanatory need to refer to truth conditions for bees and bacteria. By contrast, there is such a need in scientific psychology, and this fundamental difference is present for all to see within scientific practice. In effect, Rescorla argues that Millikan is likely to be mistaking a superficial similarity between us and the bees and bacteria (functional isomorphisms) for a deep similarity showing our internal states to be of the same kind, a similarity that she mistakenly identifies with the having of truth conditions. Instead, for a state to have truth conditions is probably something completely different.

Millikan replies that a substantial part of her corpus may be thought of as a detailed attempt to show that there really is no gulf here. (Indeed, she originally set out to account for truth conditions in *language*, and the application of the view she developed to simple systems was an unintended byproduct.) In addition, she criticizes Rescorla (and Tyler Burge [2010], who has raised similar questions) for setting invisible goalposts: they want to be shown how truth conditions play an explanatory role in bees and bacteria, *as they do in psychology*, but then they fail to say, non-circularly, what this supposed explanatory role in psychology consists in. As far as she can see, she has satisfied this requirement: on her account, the having of truth conditions (understood as Normal functional isomorphisms) explains how an individual's behavior coordinates with its environment, while accounting for a representation's intentional nature, that is the possibility that it can fail to map.

Concepts

Over time, you have learned lots of things about cats, sometimes in a way that allows you to give voice to this knowledge, and sometimes more implicitly. However, each time you have learned something new, whether by observing a cat or hearing someone talk about one, in order to incorporate this new knowledge correctly into your cognitive system, you must have made use of your ability to identify the proper subject matter: cats. Similarly, when you make use of your knowledge in your practical and theoretical dealings with cats, you must first recognize that it is *cats* that you are dealing with. This achievement is far from trivial, since cats can manifest themselves in myriad ways: as ginger or tabby, as fluffy or sleek, yowling or especially silent, and even through the medium of someone else's speech. In your empirical dealings with the world, this ability to recognize something – an object or a real kind or a property or a relation – as the *same* again is absolutely fundamental, and it is the focus of the attention Millikan pays to concepts.

On Millikan's view, "basic empirical concepts" just *are* abilities to reidentify. Your concept of cats is an ability to reidentify the biological cat kind, implemented by some mechanism that has that particular reidentification as its proper function. Your concept of *square* is an ability to reidentify square things from many distances, under many lighting conditions, by eye or by touch. In the context of judgment, the essence of the act of reidentifying is the performance of inferences where the concept acts like a middle term in a syllogism.

In turn, a person's ability to do A consists in the disposition to succeed in doing A if they try "… under the conditions that accounted for their … past successes in doing A" (OCCI, 62). These conditions are "Normal conditions" for exercise of that ability, that is, conditions needed in order to explain how exercises of the mechanisms responsible for that disposition had succeeded in those past cases that are responsible for their having been maintained or selected for. Thus an empirical concept is a disposition to successfully reidentify something in Normal conditions. In her contribution, Louise Antony criticizes both Millikan's historical analysis of abilities, and her identification of concepts with abilities to reidentify rather than with mental words in the language of thought. She also addresses Millikan's generalizations concerning the mechanism by which reidentification is achieved, defending the identity judgment against Millikan's attacks.

A reidentification function depends on there being something objective to reidentify. The most fundamental entities that both humans and other animals capable of reidentifying, Millikan calls *substances*. (In his contribution, Richard Fumerton questions the epistemological priority Millikan assigns to substance concepts.) A substance is anything that retains its properties over space and/or time for some underlying reason; thus, things "about which it is possible to learn from one encounter something about what to expect on other encounters" (OCCI, 2). On this definition, a wide range of things can be substances, including individuals, kinds, stuffs, and event types, though the underlying reasons why these retain their properties across encounters with them can vary widely. Different samples of water have similar characteristics because of the stability of water's chemical structure, commonalities across cats are explained more by common ancestry, whereas an individual object, like a child's teddy bear, retains its properties because of basic physical conservation laws. In his contribution, Mohan Matthen offers an alternative to Millikan's account of historical kinds like cats, whereas Crawford Elder takes Millikan to task for failing to maintain a proper realism about individuals.

There are various ways in which the ability to reidentify can be foiled; one of these is when a cognitive system confuses two substances or properties as being the same. The result is a confused or equivocal concept (hence the title *On Clear and Confused Ideas*). For example, someone's reidentificatory mechanism might collect up information from both jadeite and nephrite, forming the equivocal concept of jade. There is no guarantee that one's concepts are not confused in this way, but there is a cognitive mechanism whose function is to prevent it. When a concept is confused, like the jade concept, contradictions will tend to arise when the concept is applied in perception and inference. Therefore the cognitive mechanism that irons out contradictions serves to eliminate equivocal concepts. In this way, coherence acts as a test for unequivocal correspondence. In his contribution, Charles Nussbaum examines this key but underappreciated aspect of Millikan's theory, with an eye on a Millikanian explanation for how the law of noncontradiction gets its grip on the world.

In "Concepts: Useful for Thinking," Louise Antony takes issue with several features of Millikan's picture of substance concepts. Antony thinks of concepts as vocabulary items in the language of thought (LOT), and she takes Millikan's view of substance concepts as abilities to reidentify to be inconsistent with the LOT view. Antony argues that concepts have a wider range of functions than Millikan allows for. In particular, Millikan is mistaken when she says the fundamental role of concepts is to allow for reidentification and thus the collection of information; instead, concepts allow us to *think* about things. In Millikan's failure to see this, and in her preference for an ability analysis of concept possession, Antony detects an underlying behavioristic bias. In reply, Millikan argues that the *good* parts of the LOT are fully consistent with an understanding of concepts as abilities, and Millikan (she says "Of course!") wants to emphasize concepts' important role in thinking; cf. their above-mentioned role in enabling inferences.

On the LOT view, syntactic forms serve as modes of presentation in a roughly Fregean explanation of opacity. In place of equivocal concepts, the LOT account posits the existence of false identity judgments, e.g., where the mental word JADEITE and the mental word NEPHRITE feature in the judgment JADEITE is NEPHRITE. There can also be true identity judgments, of course. Millikan disparages this view in OCCI, putting in its place an account she also finds in Strawson (1974), whereby acts of identification are not to be assimilated to judgments. Instead, they involve a fundamental process of "sameness marking," whereby there is some sort of marker whose function is to make the cognitive system treat a spread of information as pertaining to the same thing. A helpful image here is of two folders being merged: instead of an identity judgment, there is a change in the underlying representational language of the cognitive system. Antony denies that the LOT theory ultimately collapses into the Strawsonian view, as Millikan claims. On the contrary, the LOT view is perfectly coherent, and (she notes in passing) it offers the best explanation of the opacity phenomena as well. In response to Antony, Millikan argues that the identity of LOT syntactic vehicles (which is what accomplishes sameness marking) must ultimately be defined functionally. If that is correct, her original remarks concerning LOT still hold, since the supposed identity judgment would functionally merge what was previously two LOT terms. (She addresses Fregean treatments of opacity in her replies to Prinz and the Macdonalds, below.)

Antony and Millikan also participate in an exchange concerning Millikan's historical analysis of abilities (Antony favors a dispositional account), and equivocal concepts (Antony thinks Millikan's theory of confused concepts is itself confused). Millikan replies that Antony's critiques here are based on a misunderstanding of her views, and suggests that Antony's dispositional analysis is itself equivocal: it incorrectly sees true abilities where there is only accidental success.

Richard Fumerton's essay "Properties over Substance" is a critical examination of Millikan's claim (in OCCI) that substance concepts are ontologically and epistemologically prior to other concepts. He distinguishes four different ways in which the conceptual priority claim can be understood:

1   We typically acquire substance concepts before we acquire other concepts.
2   Other concepts can be analyzed, at least in part, into substance concepts (and not vice versa).
3   We can recognize substances much more easily than we can recognize things that belong to other categories.
4   We typically develop a vocabulary for labeling substances before we develop a vocabulary for labeling other things.

Fumerton notes that Millikan provides empirical evidence for (4), and points out that our use of common nouns might be a disguised way of talking about things having properties, and that in any case, it isn't clear that (4) provides reason to believe any of the other three priority claims. Fumerton suggests that what we grasp first and recognize most easily might be complexes of properties, and concludes by wondering whether there might in fact be only a terminological difference between this and Millikan's view.

Millikan responds by defending the view that "... one might recognize a substance owing to the fact that certain of its properties are manifesting themselves to one's senses, but do so without employing concepts of those properties." The defense is that the content of a concept is determined by the use to which it is put. What makes a concept a

MAMA concept is that the systems that use it won't function properly unless its tokening corresponds to the presence of Mama. Even though a baby may recognize Mama by her smell, her voice, etc., responding to those things serves no purpose *except* to make her respond appropriately to Mama. In this sense, our MAMA concept is prior to our concepts of the properties that enable us to recognize her.

On Millikan's view, biological (and other functional) kinds are historical kinds, and similarities between members of a biological kind (a species, for example) are to be explained by the historical relations they bear to one another. In "Millikan's Historical Kinds," Mohan Matthen defends the view that biological kinds are historical kinds against Michael Devitt's (2008) arguments for the revival of biological essentialism, arguing that although, as Devitt suggests, the similarities between members of a species may be able to be ahistorically explained in structural terms, polymorphism (the division of a species into stable dissimilar sub-kinds such as the sexes) cannot. Polymorphism can only be explained by certain historical relations, Matthen argues, and since these same historical relations are relevant also to similarities, it is clear that it is these historical relations (not the structural similarities) that define species membership. Millikan, on Matthen's view, overlooks the importance of polymorphism.

Matthen then argues that there is one respect in which biological kinds may be less historical than Millikan thinks. In cases in which there are two speciated descendant populations that turn out to be able to interbreed, he suggests, it is a mistake to rule out by definition the possibility that they belong to the same species – scientists do not treat the question as though it is settled by definition.

Millikan responds that the issue of how species should be delineated is not one that she wants or needs to take a stand on. However, as she says, she does use species as examples of historical kinds, and she does think that all of the actual groups which have been proposed in recent times as species have been historical kinds. The members of a species are alike because of certain historical relations that hold between them, and it is those relations, rather than the resulting similarities, which make them members of the same species. She suggests here that in cases of polymorphism, "each of the sexes [for example] within a sexual species forms a historical kind nested within the historical kind that is the species itself." In response to the case of the two speciated descendant populations which can interbreed, she suggests that the two populations are part of the same historical kind (as well as each separately constituting a different historical kind), since they do share a history which explains their similarities, both having split off from the same parent population: she declines to take a stand on the question of which of these historical kinds constitutes the species.

Crawford Elder, in "Millikan, Realism, and Sameness," takes up the subtitle of LTOBC, *New Foundations for Realism*, and argues that Millikan provides only half the foundation. He endorses Millikan's realist account of kinds as the only viable one, but contends that the apparatus she presents in chapters 16 and 17 meant to ground realism about individual objects and their persistence through time ultimately fails. Realism about objects requires that they have mind-independent origins and extinctions. According to Millikan's view, however, there can be large numbers of overlapping objects, including (for instance) "this adolescent," "this human being," and "this mass of human tissue," each of which has *different* temporal beginnings and endings due to their differing persistence conditions. As a consequence, Millikan says "a great deal of room is left for decision on our part as to how to divide the world into ... temporally extended wholes" (LTOBC, 293). This, Elder maintains, is a pretty stark statement of *anti*-realism about object persistence. He recommends that Millikan adopt a view whereby the membership conditions for natural kinds double as persistence conditions for their members, along with independently motivated tighter conditions on natural kindhood that will eliminate the problematic overlaps entailed by her current view.

The main thread of Millikan's reply says that the overlaps are not, in the end, problematic for realism. On her view, the persistence of objects is to be treated exactly parallel to her treatment of the identity of a historical kind across its members: "The individual remains the same individual because it hangs on to its properties over time in accordance with natural conservation laws or other causal mechanisms." Yes, there are many overlapping objects, but this does not endanger their mind-independence. Nor does it mean that we typically make some arbitrary intentional decision concerning what we're talking or thinking about, as the quotation in the previous paragraph seems to suggest. Linguistic and conceptual functions are naturally selected so as to pick out certain individuals rather than others. Typically, this "picking out" will be rather vague; while there are natural divisions where many properties change at once (e.g., death), our selected linguistic and conceptual purposes are served even though the natural demarcation is not fully determinate. There is a healthy vagueness in our thought and talk about the many overlapping, but nevertheless real, individuals in the world.

Charles Nussbaum's "Craning the Ultimate Skyhook: Millikan on the Law of Noncontradiction" explores some less charted territory for Millikan, in particular concerning what she ought to say about a naturalistic ground for the law of noncontradiction, as well as how to treat modal notions. He begins by asking where noncontradiction's ontologically legislative authority comes from. Why does the world invariably obey this law? After showing that a number of theories (including modal realism and ersatizism) fail to answer the question, he explains what he takes to be Millikan's answer. According to Millikan, properties come in ranges of contraries, i.e., there is a determinable with a range of determinates each of which is contrary to all the others. This contrariety is a matter of natural necessity: determinates under a determinable are incompatible by natural law. Negation is, fundamentally, what asserts this contrariety, whereas the principle of noncontradiction represents it more generally, as a hypothesis "about a certain kind of skeletal structure to be sought" ("Reply to Nussbaum," this volume). This necessity is what explains the "grip" of logic on the world. It isn't that logic constrains the world's structure; instead it reflects it. Millikan identifies exactly what logic reflects: natural property incompatibility within a categorical structure (i.e., some ontology or other involving specific substances and the property ranges [determinables] that characterize them).

Up to this point, Millikan largely accepts Nussbaum's description and extension of her account of noncontradiction. This is to get rid of the Myth of the Logical Given. But she is less inclined to accept Nussbaum's proposal for a gradualist account of how human beings acquired their capacity to reason according to noncontradiction, now seen as representing a very general worldly structure. She doubts Nussbaum's suggestion that this capacity is somehow derived from similar sensitivities to contrary spaces in perceptual processing, because perception does not have subject–predicate structure. Rather, non-contradiction has the status of a high-level theoretical postulate, albeit one built in by evolution. (Nussbaum also invites Millikan to say a few words about the modalities, and she obliges, driving an interesting wedge between necessity, which she of course takes to be grounded in real-world structure, and possibility, which she does not. "P is possible" has a merely epistemic function: it serves only to move the mind in a certain direction, and has no truth condition.)

Externalism, Language, and Meaning Rationalism

On Millikan's view, there are usually an indefinitely large number of ways of reidentifying something, depending on the circumstances, and different people (including different experts) may diverge drastically in the particular ways in which they reidentify that thing or property. In Millikan's terminology, they may differ drastically in their "conceptions" of it. Consider how a child might identify a weasel compared to how a paleontologist might, and then consider how the ways of identifying a weasel that are available to Helen Keller will be different again. Nevertheless, the conceptions found in the child, the paleontologist, and Helen Keller (not to mention the changing conceptions in a single person through time) all share a function, a particular reidentification function. What they all share – a concept, or "unicept" (see Dennett's foreword) – is externalistically individuated or relational. In his contribution, Jesse Prinz argues that Millikan's externalism is overblown, and that internal psychological factors make a greater contribution to concept identity than Millikan acknowledges.

The diversity of reidentification methods is of fundamental importance to Millikan's corpus, from her discussions of perception (perceptual constancy mechanisms) to language, and we now turn to the proper functions of the latter. The function of a language form is what it does that explains its proliferation and survival, of course. This is the language form's "stabilizing function." Like internal representations, some language forms perform their proper function involving cooperating producers (speakers) and consumers (hearers) by mapping onto the world according to a particular mapping rule – these are "semantic mapping rules" or simply "semantic mappings." Semantic mappings determine truth conditions (for indicative forms) or satisfaction conditions (for imperative forms), but two linguistic forms might share a truth condition while differing in semantic mapping (e.g., "It's raining" vs. "Rain is falling here now.")

Stabilizing function, semantic mapping, and truth conditions are thus three different (though related) aspects of linguistic meaning.[2] Much of Millikan's philosophy of language may be understood as an application of these aspects of meaning to various language forms. For example, the form "A is B" has a stabilizing function to get hearers to merge

---

[2] Unfortunately, in LTOBC Millikan used the term "sense" for semantic mapping, and "intension" for conception, a potential source of confusion. (Still more confusing, having introduced "unicepts," she has recently been calling conceptions "input methods.") In this volume, we avoid the older LTOBC terminology.

their concept files for A and B, as described earlier, but its truth condition is that "A" and "B" refer to the same thing. It is not part of the stabilizing function of "A is B" that the hearer form a belief about words, however. By contrast, that *is* part of the stabilizing function of "'A' has the same referent as 'B'." In addition, these two language forms differ in their semantic mapping rules. For example, "'A' has the same referent as 'B'" has a place occupied by the term 'referent', and it mentions (rather than uses) the terms 'A' and 'B'. "A is B" has neither of these features.

Millikan wields this three-pronged account of meaning to arrive at a comprehensive and nuanced treatment of a wide variety of language forms, including some hotly debated in the philosophy of language like definite descriptions, indexicals and demonstratives, propositional attitude reports, "exists," and "means." Each of these language forms and all of its subtypes, down to univocal terms, forms a lineage propagated by copying through years, centuries, or millennia of language use, forming a complex tree (or web) of descent similar to that found in the biological world. If a proper function of a language form is inherited via one of these lineages, that function is an aspect of the expression's linguistic meaning. If not – if, for example, the proper function in question is derived from the speaker's purpose alone – then that aspect is pragmatic. Contextual factors may play a role in either.

The proper function of an ordinary referential term will be to cause a representation of that term's referent in the hearer's mind (as part of a particular true belief, if the term is used in an indicative sentence). Details of the hearer's conception of that referent will normally be irrelevant to successful coordination between speaker and hearer; rare exceptions are when a definition must be carefully passed down from user to user (perhaps "royal flush"). As a result, conceptions may exhibit little or no overlap among competent members of the same language community, nor do they need to in order for the relevant language forms to perform their proper functions. The only rules that a speaker must grasp are mapping rules, and this grasp may be minimal: the fix that a particular language user has on a term's referent may amount to as little as simply knowing the term. This knowledge is the entering wedge that allows the language user to begin gathering information about the referent – filling the relevant file folder – upon hearing someone use the term in an indicative sentence.

It follows that in Millikan's philosophy of language, there is nothing like Fregean sense. "The meanings that characterize the public part of a language," she says, "are fully extensional" (LBM, 54). Both Jesse Prinz and Cynthia and Graham Macdonald take issue with Millikan's eschewal of Fregean senses. As he does in the conceptual case, Prinz argues that psychological factors play a more important role than Millikan acknowledges, while the Macdonalds are more concerned with the role senses play in rationalizing behavior. They fear that Millikan's view makes our knowledge of our own meanings so impoverished that it is incompatible with our status as rational agents.

David Braddon-Mitchell comes at the issue from two-dimensional semantics. He takes much of Millikan's picture on board, but tries to fit it with conceptual analysis in the form of internally available analytic knowledge of meaning. Finally, Willem deVries argues that a speaker's grasp of the rules of language is of fundamental importance in securing genuine intentionality, semantics, and reasoning, notions that have their home only within

Sellars' "space of reasons." Millikan, he says, neglects this aspect of Sellars' thought in her exclusive concentration on the causal order.

Jesse Prinz, in "Are Millikan's Concepts Inside-Out?," introduces the label "Outerism" to refer to a bundle of views which are often held together, as they are in Millikan's case: "Millikan is a realist, an externalist, a naturalist, an opponent of Fregeanism, and a crusader against Meaning Rationalism." Why "Outerism"? Because "… it shifts attention away from inner psychological resources and explains mental content by appeal to features of the mind-independent world." Prinz argues that the inner/outer distinction has been overblown, that the truth lies somewhere in between, and that Millikan's view is less Outerist than her rhetoric would lead one to believe. Instead, she is (or should be) an "inside-outerist": one who accepts that mental content is determined partially by what is inside the head, and partially by what is outside. In particular, it must be admitted that sometimes (especially when we run into unfamiliar cases) we are forced to make conscious decisions about what is going to fall under a particular concept we are using. For example, consider someone whose head is shaved except for a single hair standing out in bas-relief: Is he bald? Is someone with a split-brain one person, or two? These are ultimately matters for more-or-less arbitrary decision. As a result, while Prinz acknowledges that conceptions are variable among people (*contra* Frege), he argues that they do in fact play a role in fixing reference, even if only within an individual's idiolect.

Millikan staunchly defends a full-blown externalism in response to Prinz's proposed compromise, an externalism to which she was committed before Putnam's and Kripke's publications defending the orientation in the 1970s. (In fact, she complains that Putnam and Kripke's methodology of considering our intuitions about how the terms "meaning" and "reference" apply in possible cases is fundamentally internalist [LBM, ch. 7].) Millikan concentrates on two mistakes she takes Prinz to be making: first, he mixes up words and concepts. Yes, features of communicative contexts influence our word choices when we're trying to get something across, something she has never denied. In that minimal sense, we do make decisions about how to use words. However, we do *not* make "decisions" about how to apply our *concepts*. What would that even mean? We would already need to have the concepts to have the decision-thought.

The second and more fundamental mistake is that, even supposing that we do make decisions about how to use both words and concepts, this would not be relevant to determining their reference. The function of both words and concepts is to identify, not to classify. Identifying a natural clump or bump in nature (including the human parts of nature) does not require clear demarcation of that clump or bump from its surround, unlike classification, which does. To be anxious about how to apply a word or concept in unfamiliar cases is to make a mistake about what words and concepts are for.

Cynthia and Graham Macdonald take up one of the strands in Millikan's work identified by Prinz: her rejection of Fregean senses. Millikan's externalist view of linguistic meaning has the consequence that, in an important sense, language-users may not know the meanings of their own words. The Macdonalds suggest that in fact teleosemantics and the existence of something like Fregean senses can be reconciled, and so this counterintuitive consequence can be avoided.

The slimmed-down version of Fregean sense that the Macdonalds favor comes from John McDowell. The Macdonalds argue that McDowell's *de re* senses have none of the features that Millikan finds objectionable in other Fregean accounts of conceptual knowledge. In particular, *de re* senses are externalist and they do not include conceptions (which the Macdonalds agree vary among speakers and therefore cannot serve as senses). Instead, they are akin to informational channels that put a thinker in direct contact with objects. The benefit of such an account is that speakers count as knowing the meaning of their words in a way that makes their linguistic behavior subject to rational explanation, a desideratum that, according to the Macdonalds, Millikan's account puts in jeopardy.

Millikan disagrees. There is no need for any kind of Fregean senses to account for successful psychological explanations of linguistic behavior, since conceptions can do the job. Better or worse knowledge of what one is thinking of or speaking of corresponds to more or less adequate conceptions of the real humps and clumps found in nature, and so better or worse capacities to reidentify them, but there is no guarantee our concepts are adequate. (And to suppose that we know what we are thinking of when our concepts are confused is itself a confusion.) In addition, Millikan has non-Fregean solutions to the other problems traditionally addressed by appeal to senses, namely indirect discourse and opacity more generally, informative identity claims, and apparent reference to non-existents. Finally, Millikan clarifies her view of "direct" perception through the sense modalities and language, distancing herself from McDowell where the Macdonalds saw an overlap.

Millikan argues in her 2010 paper "On Knowing the Meaning: With a Coda on Swampman" that purely *a priori* analysis is not the right way to go about examining the meanings of empirical terms: empirical meaning is immutably embedded in the actual world. In "Weasels and the *A Priori*," David Braddon-Mitchell tries to reconcile this view with his own commitment to there being analytic truths about the content of concepts/terms.

According to two-dimensional semantics, associated with a term in a language there is a description that is analytic: e.g., the A-intension of "water" is something like "that actual substance which is drinkable and which forms the dominant fluid found in the lakes and rivers on earth." The speakers of Twin-English have a word "water" which has the same A-intension but a different referent.

Millikan points out that there is no shared introspectable A-intension, and Braddon-Mitchell concedes that she is right. There are lots of different ways of identifying water. Braddon-Mitchell's way of putting this point is that if two-dimensional semantics and neo-descriptivism are right (considering concepts rather than words, for the moment), we really do have massively varied concepts: in one dimension of content, the contents of our water thoughts vary dramatically.

Braddon-Mitchell backs off from his concession, however: he thinks that because we have dispositions to coordinate with others and defer to experts, our mental idiolects will not be that diverse after all. But, as Millikan points out, we are not disposed to coordinate our *concepts* in this sense (our identifying descriptions), or in Millikan's terminology our conceptions, with others – how would we even know what others' conceptions are? What we coordinate are referents.

The coordination story is more plausible for words. Braddon-Mitchell says: "… the meaning of words in languages supervenes on what concepts the words are typically used to communicate, and if most speakers use 'water' to communicate that something is the same substance as the potable stuff that fills the rivers, then that's what the word means." So long as the extension of all of our concepts of water overlaps, the word "water" does its job. However, "… perhaps the community average that we would defer to after equilibration is that the extension of the word 'water' is one which is given by the same-substance relation and a wide set of identifying descriptions."

Braddon-Mitchell accepts a large part of the Millikan view, and wants to integrate it with the two-dimensionalist view. There are real substances, such as water, and our words and concepts pick them out. "Water" refers to that substance no matter what idiosyncratic handle we may have on it, or what recognitional cues we have for detecting it. And, Braddon-Mitchell suggests, the two-dimensionalist could take Millikan's story about the "same substance" relation and use it to fill out his A-intensions: the story is something we're disposed to accept upon critical reflection (at least for many words and concepts), and so helps determine their extensions.

The disagreements that remain are about whether once you have accepted this much of the Millikan view there is any reason not to simply adopt it (rather than using it as an adjunct to your two-dimensional semantics), and, relatedly, about whether or not there is any commonality in content between agents in states which are narrowly the same but in different environments (Earth and Twin-Earth). The two-dimensionalist thinks the answer to the second question is "Obviously yes," and that this is a reason to prefer the two-dimensional view over Millikan's. Millikan replies that this would be to misunderstand concepts and the words that express them as classifiers rather than reidentifiers. Since they aren't classifiers, the dispositions alluded to by the two-dimensionalist are an idle fancy with no theoretical relevance. Perhaps they reveal potential trajectories in the development of semantic mapping functions in the language if certain changes were in fact to take place, either locally or in our broader understanding of the world. But current usage cannot determine this, and such potential trajectories have nothing to do with current linguistic meaning. (This is exactly why Millikan claims the case of Swampman is not helpful for understanding the nature of meaning, as explained in her 2010 paper.)

Braddon-Mitchell's concluding suggestion is that there are two different projects that need to be distinguished: Millikan's, which is "… explaining the existence and persistence of mental or linguistic phenomena" and a second project which is about "… the dispositions we have to respond to new information and new environments." He concedes that Millikan's project is well motivated and the other needs defending, not so much as regards its coherence, but as regards its explanatory utility. Millikan agrees. The first project is the one that she and everyone else engaged in the discussion is (or should be) interested in.

Millikan's 2005 paper "The Son and the Daughter: On Sellars, Brandom and Millikan" invites reflection on her philosophical ancestry and the family resemblances (or lack thereof) between her and her surprising philosophical sibling, Robert Brandom. Two of the authors in this collection take up this theme. Charles Nussbaum suggests that Hegel is Millikan's unacknowledged philosophical grandfather, at least in her view of kinds and of the law of noncontradiction, whereas Willem deVries, in "All in the Family," focuses on Sellars the father. Both have in mind the contrast between the causal order and the rational order, threads that are combined in Hegel and Sellars, but split apart in Millikan and Brandom. Millikan has pursued the project entirely within the causal order, whereas Brandom has focused on the rational order; Nussbaum and deVries are united in their concern that the son and the daughter are both neglecting half their inheritance.

DeVries makes this issue the centerpiece of his contribution. There are two ways of looking at the patterns to be found in linguistic behavior: as selected, where the pattern *does not* occur "because of the rules"; and as endorsed by the community, where the pattern *does* occur "because of the rules." In her account of language, Millikan in particular has focused exclusively on the first, thereby neglecting the importance of a speaker's *grasping* the rules of the language, rules or community norms that cannot be defined in causal-order terms. He claims that it is only through this grasp – "reflexivity" – that linguistic behavior can be truly rule-governed, allowing for the possibility of real inference and semantics (constituted by inferential role). There is no intentionality without reasoning, and no reasoning without acts governed by community norms.

Millikan replies that community norms do not play the central role that Sellars and deVries take them to. Much of language learning proceeds without correction, and linguistic mistakes are best seen as departures from one's own purposes (which may include purposes to speak as others do), rather than from community norms. The role Millikan envisages for the language community is not as a source of norms, but rather as a medium through which we perceive the world. Since language is just another medium for reidentification, the thesis of the diversity of reidentification methods and its consequences applies. For most language, there are no stable "rules of language use" to be learned, other than rules for how words map onto the world: these are simple correspondence rules, not complex rules of inference. As a result, the normative structures that Sellars takes to be constitutive of the semantic (community norms of inference and language entry/exit) play only a supportive causal role for Millikan in the service of the fundamental conceptual function of reidentification. The simplicity of that function ensures that the semantic is *not* holistic, *contra* Sellars. Millikan also argues that going metalinguistic should not be seen as moving out of the causal order and into the "rational order": she has grave doubts about the "autonomy of reason," so dear to Sellars. Millikan is content to stay within the causal order in explaining semantic facts.

This volume will hopefully serve many functions. (If it goes through enough printings, those functions may even become direct proper functions, not merely derived from the editors' intentions.) Some of those new to Millikan will hopefully find one or two of the contributions to be just the right introduction for them, given their background; we also believe that there is much here to enrich the knowledge of Millikan experts. Other readers, familiar only with Millikan's widely known work on biological function and teleosemantics, will hopefully discover her insights in other areas, like epistemology and philosophy of language. We hope that still others will come to realize that in dismissing Millikan's teleosemantics on the basis of intuitions about Swampman, they have inadvertently been begging the question against her. It is impossible to subject the full compass of Millikan's work to scrutiny in the span of a single volume, but we hope the following essays and replies will help bring new appreciation and understanding for the work of this remarkable philosopher.

Finally, the editors would like to express our profound appreciation to all of our contributing authors and the Wiley-Blackwell editorial team for the care and attention they brought to this project, and for their patience with us through the long process from inception to completion. Most of all, we owe a great debt to Ruth Millikan, not only for the enormous amount of work she put in (with an alacrity that put us to shame), but for her generosity to us and the many other younger scholars who, whatever their philosophical orientation, have benefited immeasurably from knowing her. (Many of our contributors know exactly what we're talking about.) Here's to you, Ruth!

1

Toward an Informational Teleosemantics

KAREN NEANDER

One might think that a natural place to look for mental content is in the information carrying functions of mental representations. One might even think that the proponents of teleosemantics would especially like to look there. Fred Dretske (1986) does look there. However, Ruth Millikan and David Papineau, who have most extensively developed and defended teleosemantics, are critics of this approach. They deny that representations have the function to carry information or that representation producers have the function to produce representations that carry information. In my view, they are wrong, and once we see why they are wrong we can develop an alternative approach, an informational yet teleosemantic theory of content. This chapter makes a start with the contents of sensory representations.

Introduction

Informational theories use a "natural" (as opposed to intentional) notion of information.[1] The starting idea is that sensory representations and perhaps other perceptual, conceptual, and belief-like states, carry information *about* things in the world insofar as they are caused by them, are correlated with them, or make their presence more probable. On this type of view, the "aboutness" of content is thought to derive from the "aboutness" of natural information. Teleological theories of mental content instead (or possibly, as well) appeal to functions that are given an etiological analysis. A function of something is what it was selected for.[2] Content is said to be "normative" because representations can

---

[1] See esp. Dretske (1981) and Fodor (1990).
[2] See esp. Millikan (LTOBC, 1989a); Papineau (1987).

misrepresent. And these functions are said to be "normative" too because items with such functions can malfunction. For teleosemantics, the core idea is that the "normativity" of content derives from the "normativity" of functions.

Each idea is further refined in different ways by different people but, at a first pass, teleosemantic theories, by invoking functions, are said to focus on the effects or results of representations, whereas informational theories, by invoking natural information, are said to focus on the causes of representations or on conditions that co-occur with their causes. The difference between the two approaches can thus seem to be one of output-based versus input-based views of the constitutive conditions for mental content. One might nonetheless think that the two approaches should be combined. A hybrid idea is that representation-producing systems have the function of producing states that carry natural information. Then the "aboutness" of content could be explained as originating with the "aboutness" of information *and* the "normativity" of content could be explained as originating with the "normativity" of functions. Here is the Holy Grail! A source of naturalistic normative aboutness! Or so one might think.

Yet Millikan scotches this idea. In her early criticism of informational theories,[3] she says that "... [i]f 'detecting' is a function of a representational state it must be something that the state effects or produces. For example, it cannot be the function of a state to have been produced in response to something" (Millikan 1989a, 283).[4] She then argues that a theory of content should ignore the production of representations altogether, in favor of their "consumption." More recently, this emphasis on consumption versus production has been softened in her theory. Now Millikan speaks of co-adaptation between producing and consuming systems. However, her view is still that "... intentional signs often do carry natural information ... [but] ... it is not the purpose of an intentional sign to carry natural information" (VOM, 31). And, when discussing Nicholas Shea's (2007b) argument for an input condition for teleosemantics, she approves his observation that, on her view, "... carrying information is not a function of a representation" and "... it is not a purpose of the producer system to produce items that carry correlational [or I would add [she adds] any other kind of natural] information" (Millikan 2007, 445).

David Papineau (1998, 3) agrees with Millikan about this and chides me for ignoring her point:

> I would argue that Neander has taken insufficient note of Millikan's point that representational content hinges on how the representation is used, not on what causes it. In her general discussion of teleology, Neander focuses, quite rightly, on the effects of biological traits. But as soon as she turns to representation she shifts to the question of what ... it is supposed to detect.... From the teleological point of view this is to start at the wrong end. The teleosemantic strategy requires us first to identify which result the state is supposed to produce and then use this to tell us what it is representing. (Papineau 1998, 7)

I am not about to repent. This chapter explains why, despite being in broad sympathy with Millikan and Papineau, I believe that asking what a system is supposed to detect is to start

---

[3] Including the teleonomically inclined, such as Stampe (1979), Dretske (1986), and Matthen (1988).
[4] See also Millikan (WQ, ch. 11, OCCI, VOM chs. 3–6, 2007).

in the right place.⁵ My plan is first to explain why there are what I call "response functions," which are functions to respond to something by doing something (in section 2). Sensory systems, I claim, have response functions. Then (in section 3) I argue that sensory systems have the function of producing representations that carry natural information, on a certain (singular, causal) understanding of information that I recommend. Section 4 briefly considers the implications of this for what we can say about the informational functions of the sensory representations themselves (the argument in section 4 is not crucial). Then I give a two-part proposal toward an informational teleosemantic (IT) theory, which is for the indicative content of sensory representations. Section 5 explains how, on this version of IT, the problem of error is handled and section 6 explains how the problem of distal content is handled.

Response Functions

A system has a response function if it has the function to respond to something by doing something else. The term 'response' is used in a purely causal and non-intentional sense here; a system counts as producing a RED-state in response to red if some red (the instantiating of red) caused the system to produce a RED-state.

The first question I want to consider is whether there can be response functions. Against this, Millikan (2007, 447) claims that "The function of an item, in the teleologist's sense, is always something it effects." But, before hastening to agree with this apparent truism, bear in mind that the issue is not whether functions must involve effects, which they must, but whether they must be effects to the exclusion of input causes. Might not a sensory system have the function, in part, to produce certain inner states in response to certain features of or occurrences in the environment?

A philosophical problem about biological functions is often posed as the problem of elucidating which effect(s) of an item count as its function(s). Why is it the function of the heart to pump blood and to circulate it through the body? Why are these effects its functions, not its making a whooshing noise? It is assumed at the outset that functions are certain effects of the items that have them, or anyway that functions are certain effects that they have in appropriate circumstances when they are functioning properly and not malfunctioning. But let's look beyond the initial posing of the problem of functions to consult our best solution to it.

The proponents of teleosemantics invariably agree that the best solution is the etiological theory, and this includes Millikan and Papineau⁶ and me. Leaving aside details that need not concern us now, the etiological theory says that functions are what things were selected for.

The details are variably elucidated, but note that phenotypic traits can be selected for the adaptive things that they *do*, and *doings* can involve input causes. Consider an analogous case first.

---

⁵ And "[a natural sign] carries information if it is true and has been normally produced, that is all" (Millikan, 2007, 445–448).

⁶ See Millikan (LTOBC, 1989b). Papineau (1987) approves of the etiological theory that I give in my (1983) PhD dissertation and later defend in Neander (1991).

If I am waiting for my friends to phone to tell me if they're coming to dinner, there are different things that I can do. I can pick up the phone at random or whenever I remember that my friends will call, or I can wait until it rings and pick it up only then. If I pick it up at random or whenever I remember that my friends will call, I'll waste my time and possibly frustrate my friends' efforts to get through. So doing one thing rather than another will serve my purposes better. And the crucial difference in these different things that I could do are differences in the input cause, or in other words in what I am responding to.

Now think about those biological functions that depend on natural selection. One type of mechanism might secrete melatonin in response to the dimming of light,[7] while another type of mechanism might secrete melatonin in response to light brightening and a third more or less randomly. Since melatonin makes us sleepy, the first mechanism might be more adaptive in creatures like us who have poor night vision. So the first type of mechanism might be selected in preference to the second and third because the first differs in its input cause, in what triggers its production of melatonin.

It is a function of our pineal gland to secrete melatonin in response to the dimming of light. It is also a function of B-lymphocytes to produce antibodies in response to antigens, a function of certain mechanisms to produce shivering in response to the early stages of hypothermia, and a function of the pancreas to secrete insulin in response to elevated levels of glucose in the blood.

A visual system can have the function, in part, to produce REDs in response to red being instantiated. Or, more generally, there can be a range of determinate values, $e_1 \ldots e_n$, of a determinable (E), and a range of determinate values, $r_1 \ldots r_n$, of another determinable (R), and a sensory system can have the function, in part, to produce certain values of R in response to certain values of E. It can have the function to produce r_1 in response to e_1, r_2 in response to e_2, r_3 in response to e_3, and so on.[8]

To be sure, functions *must* also involve effects. For natural selection to occur, there must be downstream replication. But, to repeat, the issue is not whether functions involve effects but whether they can also involve input causes. And they can. Biological mechanisms are selected for their causal roles, which can include dispositions to respond to specific types of causes. Someone could *stipulate* that by the term 'function' she will refer to *effects* exclusively. But denying that there are response functions is meant to have significant consequences for how teleosemantics should be developed and mere stipulation is no motivation.

I am anyway not being revisionist. Let us take a look at some statements of the etiological theory. Larry Wright (1976, 81), for instance, says that "The function of X is Z iff: (i) Z is a consequence (result) of X's being there, and (ii) X is there because it does (results in) Z." A result of a sensory system "being there" can be its producing inner states in response to outer states and this can be, in part, why it is there (once we appropriately disentangle talk of types versus tokens, which we would anyway need to do here).

---

[7] More fully, in response to the norepinephrine produced in retinal photoreceptors in response to the dimming of light.

[8] It could be that the inner states are caused by the outer states in a way that involves second-order similarity or some other mapping, a complication that I don't explore but intend to allow here.

If we change the symbols for the sake of conformity, Millikan (1989b, 288) says that X has the proper function, Z, "... if X originated as a reproduction ... of some prior item or items that, due in part to possession of the properties reproduced, have actually performed Z in the past, and X exists because (causally, historically because) of this or these performances." Presumably, a performance could be a simple effecting. However, without an explicit ruling against this, it could also be an effecting of certain inner states in response to certain conditions obtaining.

In a paper titled "Functions as Selected Effects" I argue that functions are selected effects, thus apparently committing myself to the claim that I here deny. But my point there was that functions are effects of phenotypic traits for which the traits were selected rather than effects of traits that are currently adaptive. That is, the emphasis was on functions being *selected* effects, not selected *effects*. Of ordinary physiological functions, I say, "[i]t is the/a proper function of an item (X) of an organism (O) to do that which items of X's type did to contribute to the inclusive fitness of O's ancestors, and which caused the genotype, of which X is the phenotypic expression, to be selected by natural selection" (Neander 1991, 174). Again, response functions are not excluded. Something that a sensory system can do and be selected for doing is produce an inner state in response to a certain type of cause.

In short, that functions are selected effects is true enough for most dialectical situations, but don't mistake the slogan for the theory. The etiological theory, as standardly formulated, and as it should be formulated in my view, does not rule out response functions.

That functions are effects and not causes is often taken to be too obvious to need arguing (Millikan's and Papineau's claims are often repeated by others). However, Shea (2007b) offers some motivation for the claim. He says:

> Teleosemantics is sometimes mistakenly taken to be a refinement of informational semantics according to which items represent what they were designed by evolution to carry information about. The trouble with this gloss is that evolutionary functions are a matter of effects. Amongst the various effects that an evolved system can produce, those which have contributed systematically to the system's survival and reproduction in the past are its evolutionary functions. Evolution acts only on effects. It is blind to the mechanism by which those effects are produced. An effect produced by a system at random will be an evolutionary function if its production has contributed systematically to the survival and reproduction of the system. It is not part of any evolutionary function for the effect to have been caused in a particular way. (Shea 2007b, 409; italics added)

Shea goes on to argue that we could nonetheless still *supplement* teleosemantics with an input condition, and this is an important point. Teleosemantic theories can use other ingredients, besides functions, when they give a recipe for thought. But in defending the possibility of an input-based teleosemantics, Shea concedes more than he needs to.

It is true that natural selection cannot discriminate between two mechanisms that have the same impact on fitness. If one type of pineal gland with associated structures had randomly secreted melatonin *and* had contributed to survival and reproduction just as well as a system that secreted melatonin in response to the dimming of light, then natural selection could not have discriminated between them. More generally, something that gets something done one way and something that gets exactly the same thing done another way are equal in the eyes of selection (all else being equal).[9]

However, there are cases where different things are done that have different impacts on fitness because they involve responses to different causes. It is safe to assume that a system that secreted melatonin randomly did *not* contribute to the survival and reproduction of our ancestors as effectively as one that secreted melatonin in response to the dimming of light. And it is safe to assume that a human visual system that produced REDs randomly did *not* contribute to the survival and reproduction of our ancestors as effectively as one that produced REDs in response to red. Given the complexity of the neurobiological systems involved, the inference to the best explanation is that natural selection selected the relevant pathways in our visual system, in part, for their disposition to respond to red by producing REDs (and to respond to green by producing GREENs and so on).

A difference in the triggering causes of a process cannot make a difference to fitness unless the difference in triggering causes *makes a difference* and impacts fitness. However, it does not follow that natural selection is blind to triggering causes (or mechanisms) and only sensitive to effects. Rather, natural selection is sensitive to triggering causes (and mechanisms) because it is sensitive to the different effects that different triggering causes (and different mechanisms) can bring about.

Information and Singular Causation

How do response functions bear on information carrying functions? In a while, I'll argue for a different understanding of natural information, but it helps to appreciate a couple of points that Millikan makes about natural information construed as correlation first.

If a sensory system were selected, in part, for producing REDs in response to instantiations of red, this could increase the probability of a RED given red and the probability of some red given RED. But, as Millikan (2007) reminds us, a result of a system's normal functioning is not necessarily a result that it has the function to effect (not all that it normally does is something that it has the function to do). The normal functioning of the heart results in a whooshing sound, but the heart does not have the function to make a whooshing sound.

Further, correlations hold between *types* and whether two types of things are correlated and to what extent they are correlated depends on what is counted. As Millikan explains, all sorts of troubles follow on this score. An informational teleosemantics that interprets information as correlation will have to pay careful attention to specifying the relevant reference classes and it is hard to see how to appropriately circumscribe them.

In any event, Millikan argues, it cannot be the function of any one visual system to bring about such a correlation among REDs and red instantiations and thus ensure that

---

[9] See also Rosenberg and McShea (2008, 96–126).

its REDs carry information about red, no matter how the reference class is specified. My visual system does not have the function of ensuring that your visual system tends to produce REDs in (or only in) the presence of red.

Even if the correlation need only hold on an individual basis (i.e., even if the function of my visual system is to ensure that my REDs correlate with red in my vicinity), I would argue that there is a related problem. Johnny's visual system might function perfectly well until he is six, producing REDs when and only when he sees red. But his visual system could malfunction in the future, with the result that, for Johnny, his lifelong record of REDs in and only in the presence of red is very poor. By the time that he is celebrating his ninety-sixth birthday, his REDs might have correlated far more highly with green instantiations instead. If the function of his visual system is to ensure an individual lifelong correlation between REDs and red, it seems that in this scenario it never performed its function properly. The trouble with this is that, for his first six years, his visual system was functioning properly, and it was producing REDs that carry information about red, despite its later failure.

At this juncture we could, with Millikan, abandon the idea that sensory systems have the function of producing representations that carry natural information. Or we could explore other ways to understand natural information. Millikan explores other ways with no better outcome but there is one that she does not explore.

I suggest that r carries the natural indicative information that e if e is a cause of r, where this is singular causation and r and e are particulars.[10] Thus a particular RED carries the information that there is red if red being instantiated caused RED.[11] Sensory systems have the function to produce inner states in response to outer states (as I've argued). So they will have the function to produce inner states that carry information about outer states, on this proposal.

On this proposal, there is no difficulty of the kind that Millikan has raised for other informational proposals. On this proposal the information relation holds primarily between tokens and only secondarily between types.[12] In what follows, I will sometimes refer to this singular causal indicative information as 'information' (or 'natural information') without further qualification.

---

[10] In the end (elsewhere) I will say that r carries the information that e when e causes r or r causes e, which lets motor instructions carry information about the movements they cause. Thus I say that this is for *indicative* information.

[11] Fodor (1990) sometimes seems to have this in mind. Millikan (2001) thinks of information as "grounded" correlation. Millikan (VOM, 36) seems sympathetic to the idea that information is a matter of singular causation. But then she introduces a notion of a recurrent local sign (which reintroduces statistical considerations) and focuses on that. Of the latter, she says, a causal connection is not required (VOM, 44). Her concern there (in chapter 3 of VOM) is different from my concern here. Millikan is there concerned with one feature of the environment being a sign of another feature of the environment (e.g., magnetic north being a sign of the direction of oxygen-free water) and with a creature being able to use one as a sign of the other. Hence the connection between sign and signified, she says, needs to be at least locally recurrent. This epistemic issue is not my concern here.

[12] It holds between types only insofar as we aggregate what holds between tokens. For example, most REDs will carry information about red if most RED-tokens are caused by red instantiations.

It is often said that natural information must be "grounded." On the present proposal, if *e* causes *r* and so *r* carries information about *e*, this will be in accord with natural laws and not a miracle. But singular causal information does not require that, if some REDs carry the information that there is red, there is a law to the effect that red and RED are correlated, or that there be reliable head-to-world indication, or that red instantiations always cause REDs or that REDs are only caused by red instantiations. A sensory system of a certain type can be selected for producing *r*s in response to *e*s in one species, one part of the world, or one period of time.

Natural information, so understood, has a key feature that natural information is supposed to have. Crucially, misinformation is impossible. A RED carries the information that there is red at a certain location only if red at that location caused that RED tokening. Thus, if a RED carries the information that there is red, there is red.[13]

For what it is worth, this understanding of natural information is consistent with many of our intuitions about natural signs. Echoing, with modifications, Grice and Dretske, we take six rings in a cross-section of a tree to be a natural sign that the tree is six years old, but this is defeasible if the tree isn't six years old. If an unseasonable cold spell one spring added an extra dark ring and the tree is only five years old instead, the fact of there being six rings is not a natural sign that the tree is six years old. Six years of growth must cause the six rings, if they are to be a natural sign that the tree has been growing for six years. The red spots on Johnny's face signify that he has the measles only if he has the measles. If Johnny does not in fact have the measles, then the presence of such spots is not a natural sign that he does.

Even if Johnny has the measles, I would add, his spots do not carry the information that he has them *unless his measles cause his spots*. Imagine that Johnny has a mutation that prevents his developing spots in the course of the measles. If his spots were coincidentally caused during a bout of the measles by an allergic reaction to strawberries, so that his having the measles were a mere coincidence, his spots would not carry the information that he has the measles, no matter how similar his spots otherwise were to those usually caused by the measles.

Johnny's spots would not carry the information that he has the measles if his having the measles were a mere coincidence, even if the correlation (speaking in terms of mere frequency) between measles and the presence of such spots were perfect.

Perhaps Johnny's allergic reaction to strawberries is rare, or even also unique. Someone might balk at the implication of my proposal that Johnny's spots carry information about his allergy to strawberries. Such a rare causal relation will not be informat*ive* to us. But take care not to conflate an epistemic sense of information with the natural notion that we want here. Natural selection will bring it about that our sensory systems are adapted to produce inner states that carry information that our brains can exploit and it will bring it about that our brains are adapted to exploit the information that our sensory systems are adapted to provide. But that natural information be exploitable (let alone for adaptive

---

[13] Does this preserve the informational entailments of concern to Dretske (1981)? Perhaps. If a RED carries the information that there is red and a LINE$_/$ carries the information that there is a line with a certain orientation, the conjunction of that RED and that LINE$_/$ carries the information that there is red and a line so orientated.

behavioral responses or for the acquisition of knowledge) is not something that our notion of natural information needs to guarantee. A great deal could be useless, epistemically. Natural information *as such* need not be even locally recurrent. Natural selection can be left to sort out what is usable. Our analysis of natural information need not do that. Sensory systems will not be adapted to pick up on information unless it is at least locally recurrent, but this is not something that must be built into our analysis of natural information.

We could, if we want, skip talk of information and go straight to a causal theory of content. However, on this proposal, we can make good sense of the claim that sensory systems have the function of producing inner states that carry information about outer states. Their function is, in part, to causally mediate between the outer and inner states and hence to make it the case that certain inner states carry natural information about certain outer states. They have, in other words, information processing functions. Making sense of this way of speaking is salutary.

Singular causal information is strict enough without being overly strict. To use one of Dretske's well-known examples, suppose that his neighbors begin making their doorbells out of nuts and squirrels start chewing on them and making them ring. Doorbell systems could still be useful in his neighborhood and people could still install them and use them if *often enough* when a doorbell rings someone is causing it to ring. A parallel point can be made for biological functions and natural selection. If our ancestors experienced color contrast illusions, or the light wasn't always white light, with the result that REDs were sometimes produced in the absence of red, certain pathways in the human visual system could nonetheless have been adaptive and could have been selected because *often enough* the REDs they produced carried information about red.[14]

The Functions of Sensory Representations

Until now, I have discussed the functions of sensory *systems* and not, or not directly, the functions of the *states* that it is the function of these systems to produce. There are two terminological issues to be decided before we conclude (if we do) that the sensory representations themselves have response functions, or functions to carry information.

One is whether only systems, structures, mechanisms, and so on can count as having functions or whether their states have functions too. Someone might allow that the left ventricle of the heart has the function of relaxing after it contracts in order to let in blood, and yet deny that the relaxation of the left ventricle, after its contraction, has the function of letting in blood. I see no good reason why we should stick at this.

A second matter is this. Even if we grant that a sensory system has the function of producing REDs in response to red, and that the RED states can count as having a function of effecting something, one might still protest at the idea that these RED-states have the function of being produced in response to instantiations of red. As Millikan (2007, 447)

---

[14] Thus this proposal, though it is on Dretske's side of the teleosemantic family feud, avoids the worries that Millikan (OCCI, VOM) and others raise for Dretske's (1981) strict definition of 'indication'.

says, "... a thing cannot effect its own history." Nothing can, so to speak, have the responsibility to bring itself into existence in a certain way. Functions are things that were done, that were adaptive, and that caused things of a type (the type of thing that did them) to be selected. Coming into existence in a certain way isn't something that anything can do. Or at any rate it isn't something that anything can be "responsible" for doing.

However, it is of little consequence what we say about this for present purposes, because even if it is infelicitous to ascribe a response function to a sensory *representation*, we can always translate such an infelicitous ascription into an ascription of a response function to the sensory *system* that produces the representation. Crucially, to say that a sensory system has a response function is not to say that *the system* is responsible for bringing itself about. It is to say that it is responsible for bringing about a state change in itself in response to a certain type of triggering cause.

We may also say that REDs are "supposed" to be caused by red instantiations (unless it is objectionable to speak in this colloquial style) if we read this as saying that the RED-producing system has the function of producing REDs in response to instantiations of red.

The Contents of Sensory Representations: The Problem of Error

Finally, we come to the content of sensory representations. How does any of this bear on their content? The now all too obvious suggestion, or so I hope, is that a sensory representation, RED, has the indicative content that there is red if the RED-producing system has the function of producing REDs in response to red, or (synonymously) if REDs are supposed to carry the information that there is red.[15] Similarly, a sensory representation, $LINE_i$, will have the indicative content that there is a line with a certain orientation if the $LINE_i$-producing system has the function of producing $LINE_i$s in response to lines with that orientation, or (synonymously) if $LINE_i$s are supposed to carry the information that there is a line with that orientation.

Let's look at a third example.[16] Simplifying somewhat, this involves representation by certain tectal (T) cells in toads. These cells are referred to as the T(5)2 cells, and they normally fire strongly in response to any stimulus that is within certain size parameters and elongated and moving in a direction that parallels its longest axis.[17] Neuroethologists refer to this configuration of features, for short, as "worm-like motion." Let 'r' stand for the relevant sensory representation in the toad. Often enough, the stimulus to which r is a response is something nutritious, such as a worm, a cricket, a millipede, or some other nutritious tidbit (even a bird, frog, or a smaller toad, for toads are not fussy eaters). However, a toad cannot distinguish something that is nutritious from something that is not nutritious.

---

[15] I'm not implying that sensory representations have a syntactic structure isomorphic to the grammatical structure of the sentence, "There is red."

[16] See Neander (2006) for more detailed discussion of the example and the points made in this section.

[17] The content isn't necessarily the property that triggers the strongest reaction. Super stimuli might never have occurred in the past, let alone been systematically involved in past selection.

When r is tokened, the toad orients toward the stimulus, moves around obstacles and approaches if necessary, and then it tries to catch and swallow what is, if it is lucky, some nutritious prey. The tokening of r initiates the orienting and prey-capture behavior, but more information processing concerning localization and motor coordination is involved before the prey is caught.

The information processing that produces r is not mere transduction, as has sometimes been suggested in philosophical papers. It is not completed in the retina or even in the retinal ganglion cells. The processing that produces r involves mid-brain structures and it is complicated enough to remain not yet fully understood even after intensive neuroethological study.

The relevant pathways in the toad's brain normally produce this representation in response to the configuration of features characterized as worm-like motion, and it is safe to assume that they were adapted in part for doing so and hence have the function of doing so. Thus, on the present proposal, the r-states that it is their function to produce are supposed to carry information about worm-like motion.

A complication is that the activity in the toad's tectal cells normally carries both *what* and *where* information and they were adapted for doing so. So they have both *what* and *where* content on IT. There are different T(5)2 cells, and the different T(5)2 cells respond to moving stimuli in different parts of the toad's visual field (by which I mean different parts of the outside world, as it comes within the toad's field of view). Neighboring regions of the toad's retinas feed information forward to (causally feed forward to) neighboring regions in the tectum. So which T(5)2 cells fire tells the toad's brain the (rough) location of the moving stimuli. They tell it that *there* is worm-like motion *there*, and where *there* is depends on which T(5)2 cells fire. Thus, there is a range of relevant representations ($r_1 \ldots r_n$) with different localization contents.

Elsewhere (Neander 2006) I argue that this is the right content to attribute to $r_1 \ldots r_n$ if they are to play a role in the information-processing explanation of the toad's capacities that the neuroethologist seeks to provide. My argument there leans on the fact that the perceptual capacity to be explained is a capacity to distinguish between items in worm-like motion and items that are not in worm-like motion. This is the capacity to be explained because it is the capacity that the normal toad has. In contrast, the normal toad has *no* capacity to distinguish nutritional from non-nutritional substances and so this is not the capacity to be explained.

If we want to explain why the toad's capacity to detect worm-like motion *evolved*, then the relationship between worm-like motion and nutritional substances will be paramount. But, if we want to understand the toad's perceptual capacity and its prey-capture behavior and the role of its representations in this capacity we cannot afford to ignore the information processing that produces the representations. It is a mistake to think that talk of proper or normal functions has as its main or anyway its sole aim the explaining of the evolutionary origins of things and their capacities. Talk of normal or proper functions also has an important role in physiology and in neurophysiology, in explaining how the capacities are realized, or in other words in explaining how the *normal* system works. Informational teleosemantics (mine, anyway) has its eye on explanations like these and the role of representations in them.

The normal toad's brain processes information about the configuration of visible features that constitutes what we're calling "worm-like motion." That is, its perceptual processing in the detection of prey concerns the size, shape, motion, and motion relative to shape of the stimulus. There is no justification from a neurobiological point of view for the assumption that there is any inference on the part of the toad from this configuration of visible features to the presence of invisible nutrients. Such an inference surely does not occur. And even if there were a further inference to the presence of nutrients, it is not the content of the conclusion of any further inference that is under discussion. What is under discussion is the content of a sensory representation. To ignore the underlying information processing that produces this representation is to risk making the content that we ascribe to the sensory representation irrelevant to the explanation of the toad's perceptual capacity.

The pathways that produce r have the function to do so in response to a certain type of cause (worm-like motion) but they also have the function to effect certain things thereby. Most directly they have the function to initiate certain behavioral responses, starting with the orientation of the toad toward the worm-like motion that has been detected. Notice that when we focus on the further effects of the representation (on its "consumption") we are not anyway forced to start talking about nutritional substances right away. The components in the toad's brain that go on to more precisely locate the stimulus, and that go on to control the orienting, approaching, and catching behavior of the toad, have functions that can be described in terms of the item in worm-like motion that has been detected, rather than in terms of any nutritive potential it has.[18]

Millikan (2001, 105; cf. OCCI, 217) worries that informational semantics (including informational *teleo*semantics) implies a kind of verificationism but, to be clear, I am only making a claim about sensory representations here. I agree that we have what Millikan (OCCI) calls "substance concepts." I agree that we can have a concept of X but not Y, even if we cannot distinguish between Xs and Ys (extensionally speaking, we can have a concept of H_2O but not of X_YZ, even if we cannot distinguish between them). But a toad has no substance concepts. Nor does a sensory system. Substance concepts will be involved in sophisticated acts of perceptual recognition or post-perceptual recognition, such as perceiving or judging that something is gold or a lion or recognizing Ruth when she walks into the room. The contents of sensory representations, in contrast, are constrained by what can be sensed and so by the distinctions to which our sensory systems are causally sensitive. Though some of our concepts must be able to refer to things with hidden structural or historical "essences," sensory representations do not do so. Sensory

---

[18] In Millikan's theory of content, it is not the focus on consuming systems *as such* that dictates that the indicative content of r is something nutritious (though her theory does, she says, entail this). What principally does that job in her theory is the notion of a Normal condition. As I understand it, the Normal condition is an environmental condition that corresponds (more or less, but well enough) with past uses of a representation and that crucially explains why its use was beneficial. The latter is cashed out in terms of the condition allowing the consuming system to perform its function. The Normal condition need not be the condition to which a sensory system is causally sensitive. It need not be the condition to which it is responding. Hence Pietroski's (1992) worry about the kimu who cannot tell a snorf from a non-snorf though, on Millikan's theory, the kimu love snorf-free spaces and seek them out.

representations represent the visible properties of things. How we get from them to substance concepts is a separate question.

In my view, the relevant tectal firings in the toad represent the visible properties of a toad's prey, not its invisible properties (such as its ability to nourish the toad). The version of IT that I am offering gives that result. IT is inconsistent with the possibility that rs refer to toad food (or at least with the possibility that this is their indicative content) because the relevant pathways in the toad were not selected for causal sensitivity to toad food as such.

In some neurologically damaged toads, "anti-worm motion" by something that is elongated but moving perpendicular to its longest axis can trigger a spike in T(5)2 firing. In some neurologically damaged toads, large predator-like objects (e.g., the laboratory technician's hand) can trigger it too. IT counts the resulting r-states as erroneous because the relevant pathways in the toad were not selected for responding to such stimuli as these by producing r-states. These inappropriate responses on the part of a neurologically damaged toad also count as *abnormal* for the neuroethologist. That is, strong firing in T(5)2 cells in response to a worm that is hung by its tail and moved sideways (in anti-worm motion) is considered abnormal, no matter how nutritious the worm is. So is strong firing in response to a large (predator-like) chunk of processed toad food, even if a good bite out of that chunk would feed a hungry toad. The notion of normality invoked here is teleonomic and not statistical. It is the notion of normal invoked in talk of normal functioning that is synonymous with talk of proper functioning.

IT does not equate correct representation with representations produced as a result of normal functioning, even in the case of sensory systems, but it brings the two into closer alignment, as just noted. In neurologically normal toads, an electrode in the right place could cause rs to be produced and that would also be erroneous although no malfunctioning would be involved. But to explain this further, we need the second part of IT, given in the next section.

The Contents of Sensory Representation: The Distality Problem

In her "What has Natural Information to do with Intentional Representation" (2001, 118; cf. OCCI, 231), Millikan says that Neander (1995) "seriously claims that all representations must only be of proximal stimuli." As it happens, Millikan and I agree that there is distal content. Toward the end of my "Misrepresenting and Malfunctioning" (1995, hereafter M&M) I raised the problem of distal content as an, as yet, unsolved problem for my proposal in that paper. I had argued that several content indeterminacy problems had been conflated and that we needed to adopt a divide and conquer approach to them. And, while I did not tackle the problem of distal content, I did not deny that it needed a solution (on the contrary, see esp. Neander 1995, 136). I did not try to solve it then because I did not know how. But let me try to solve it now.

Suppose that you agree with me so far. (Just suppose.) So you agree, if only for the sake of the argument, that $r_1 \ldots r_n$ are, so to speak, supposed to carry information about worm-like motion in various locations and that this is therefore what they represent. Now consider a specific representation in the range (r_i) and the singular causal chain represented on the following line:

Wormy motion at L ⇨ light ⇨ *retinal firings* ⇨ *other stuff* ⇨ r_i

The problem of distal content in this case is to say why r_i represents worm-like motion at location L rather than the more proximal things that carry information about the worm-like motion to r_i. The more proximal things include the light reflected from the item in worm-like motion and the retinal firings that result when this light hits the retina and various other neurophysiological events that occur in brain structures *en route* to the tectum and the tokening of r_i.

It is an interesting question just how the problem of distal content relates to the contents of representations of depth. Just briefly (as explained in Neander 2006) the toad does have some depth perception. But r_i itself is not thought to carry information about depth. In part, this is because it is only when the toad subsequently orients and fixates on the worm-like motion that it can see it with both eyes. Anyway, as is usual, I'll treat the problem of distal content as in need of a solution that does not depend on the representation of depth relations, although these might not be wholly separate matters. 'Location L' refers to a region within the toad's field of view and this region will include both near and far. Our question is not how to account for the more precise localization content on the near–far dimension that later representations possess. Our question is how we can count r_i itself as representing the distal worm-like motion rather than the light waves reflected from the item whose motion it is, or the retinal firings that result, and so on.

The intuition behind my idea for a solution is that a sensory system is only adapted to respond to the more proximal items because doing so is the means by which it responds to the more distal ones, and not vice versa. This is a principle I wish to apply only to the determination of content on the distal–proximal dimension. Some of the more proximal items are states of the system itself, and the sensory system was not selected for carrying information about itself. So those states are ruled out anyway. But there are also proximal items outside of the system, namely the patterns of light. So, here goes.

Let there be, as before, a range of determinate values, $e_1 \ldots e_n$, of a determinable (E), and a range of determinate values, $r_1 \ldots r_n$, of a determinable (R), and a sensory system (S), which has the function of producing instances of certain determinates of R in response to instances of certain determinates of E. Thus, S will have the function to produce r_1 in response to e_1, r_2 in response to e_2, r_3 in response to e_3, and so on. Now, in addition, let there be more proximal candidate contents (prox-e_1 … prox-e_n), which are values of the determinable Prox-E. And let 'r_i', 'e_i' and 'prox-e_i' stand for arbitrary values in their respective ranges. Then the addition to IT is this: r_i refers to e_i, and not prox-e_i, if S was selected for producing determinates of R in response to determinates of E *by* producing determinates of R in response to determinates of Prox-E, and not vice versa (i.e., S was not selected for producing determinates of R in response to determinates of Prox-E *by* producing determinates of R in response to determinates of E).

Back to the toad. The pathways in the toad's brain were selected for responding to both the distal worm-like motion and the more proximal patterns of light that carry information about the distal worm-like motion to the toad. But there is an important asymmetry here.[19] These pathways in the toad's visual system were selected for responding to the light by producing certain tectal firings because *by that means* they responded to the distal worm-like motion, and not vice versa. That is, they were not selected for responding to the distal worm-like motion by producing certain tectal firings because *by that means* they responded to the more proximal patterns of light. That just isn't how the means–end analysis pans out. I believe that this solves the problem of distal content.

Millikan's (2001, 119 n7; cf. OCCI, 231 n7) also attributes to Neander (1995) the view that "…things really can't have distal functions," and so perhaps I should say something about this too. Actually, I did not mean to make this claim either. M&M is mostly about how, because things are selected for complex causal roles, there is a functional indeterminacy problem that needs to be distinguished from other indeterminacy problems for teleosemantics,[20] and about how this particular functional indeterminacy problem might be met. Things are selected for complex causal roles, I said, and so we can describe functions in complementary but different ways by focusing on different aspects of these roles. But we can, in a principled way, focus on different aspects of these roles in certain contexts, I said. For instance, it is the function of ovaries not only to produce ova, but also to contribute to conception and to live births. *All* of this is their function. However, ovaries do not *mal*function if they cannot contribute to conception or a live birth because the fallopian tubes are blocked. Ovaries only *mal*function if they cannot make their own individual contribution, I said. It was my attempt to characterize this "individual contribution" that led to so much misunderstanding.[21]

Of course, producing ova is not something that ovaries can do all by themselves, without assistance. The ovaries must be fed oxygen by the circulatory system, for instance.

I also spoke of the more "peculiar" function of something as that which it does "more immediately," and this is easily conflated with talk of the "more proximal," in discussions of distal content. I meant these contrasts to be orthogonal, but that wasn't made completely clear. In fact, I assumed that certain pathways in the frog's (or toad's) brain had the more "immediate" function, so to speak, of detecting something small, dark, and moving (or something in worm-like motion) that was in fact distal (though I had no solution to the problem of distal content then to offer). The less "immediate" function of these pathways, as I saw it, was to help the creature catch and eat prey and be better nourished and so on. That is in part to say that the relevant *sensory* pathways do not *mal*function if the frog (or toad) cannot catch, eat, and benefit from its prey because, say, it cannot swallow, or its tongue-snapping mechanism misfires, or it is missing necessary digestive enzymes. The relevant *sensory* pathways count as *mal*functioning if they cannot detect the type of stimuli to which it is their function to respond. This was not (and is not) by itself meant to be conclusive on its own for how we should approach content, but it seemed suggestive

---

[19] Shades of Fodor's asymmetric dependency theory here, but this is not at all like his attempt to solve the problem of distal content.
[20] Such as Fodor's (1990) problem, which he understands to be due, as he puts it, to natural selection being "extensional."
[21] Others besides Millikan were also misled (see Price 1998). Happily, Papineau (1998) understood what I said.

to me at the time. The idea was that the aspect of the complex causal role which is something's function that is most relevant to judgments of malfunction might also, in the case of representational systems, be most relevant to judgments about misrepresentation too.

To sum up. One phenotypic characteristic can be more adaptive than another and can be selected in preference to another because it brings about a certain effect in response to a certain cause. Sensory systems, more specifically, have the function of producing certain inner states in response to certain environmental conditions obtaining. They therefore have the function of producing states that carry natural indicative information if we understand natural indicative information in terms of singular causation, such that one thing carries natural indicative information about another if it is caused by it. We can then entertain the rather obvious suggestion that sensory representations represent that which they are supposed to carry information about. It is wrong to reject such a suggestion on the grounds that functions must be effects. I have tried to show that this idea can be formulated in a respectable way, and that as formulated here it generates suitably determinate content for at least sensory representations.

Hard questions remain. The hardest is how to get from here to where we want to go in the end, which is to much more sophisticated representational contents. Millikan and Papineau believe that their output-oriented approach will succeed, and that this more input-oriented approach will fail. They seem to think that the input-oriented approach is stillborn. That it is not. And I am betting that they are wrong about its future prospects too. It seems to me that they have begun in the wrong place, assuming (as I do) that we want a treatment of sensory representations that is sensitive to the actual capacities of sensory systems and to the actual information processing that is responsible for them. In Millikan's view, and in Papineau's view too, I turn teleosemantics on its head. But I think I've turned it the right way up and now, I hope, IT can get up and walk. I am in sympathy with much of what Millikan (OCCI) says in her important discussion of substance concepts. Much of it is, I think, compatible with the informational teleosemantics supported here. However, this is where I take my departure with the customary parting comment that a careful discussion of this must wait for another time.

Reply to Neander

Karen Neander's and my own positions on intentional representation are very close in broad outline but different in detail. The details do make important differences, however, and Neander offers me an opportunity to clarify my own stance for which I am very grateful. I will leave aside the question whether there are, in general, things that might reasonably be called "response functions," because I don't think that either of us cares that much about how the ordinary word "function" is to be used so long as it is made clear. In any event, I believe that both Neander's "functions" and my "direct proper functions" are intended to pick up only functions that have been selected for, by evolution or learning or in other ways. And Neander and I seem to agree that the basic question to be asked concerns what the mechanisms that produce mental representations are selected for doing. So let me go directly to the question whether, as Neander claims, mental representation producers are selected for producing representations in response to the things these representations represent. (Mental representations themselves have not been selected for at all, of course. They have been produced by mechanisms that were selected for producing them. In my terminology, they have "derived proper functions.")

I also am sure there is agreement between Neander and myself that it is important to draw a distinction between what has merely been selected and what has been selected by being selected *for*. Elliott Sober's classic example was of his niece's toy that sorts little balls by size when you turn it upside down, only the smallest balls, which are also the green ones, getting through all the size filters to the bottom. The toy selects small green balls to go to the bottom, but it only selects *for* small balls (Sober 1984). The basic question that divides Neander and me, then, is do natural selection processes (or perceptual and/or cognitive learning in more advanced species) merely select representation-producing systems that have been producing representations in response to what these representations represent, or do they, further, select these systems *for* producing representations in response to what they represent? Better, since before the selection process there are no intentional representations but only, maybe, some natural signs, we should mark this difference in our question. Labeling the kinds of physical mechanisms that, if selected for doing the right things, will become representation producers as *representation producers*, the things that will then become representations as *representations*, and so forth, the question becomes, do the relevant selection processes merely select *representation-producing* systems that have been producing *representations* in response to what these *representations* *represent*, or do they, further, select these systems *for* producing *representations* in response to what they *represent*? The words between stars then refer to certain physical types only, and it is clear that no questions are being begged.

The first key to an answer to our question is to notice, what is obvious once pointed out, that in order to be selected, these *representation producers* will have to produce, enough of the time, not just *representations*, but *representations* that actually coincide, or "fit" with their *representeds*. They must effect, enough of the time, that there be relations between the *representations* that they produce and the world of the sort that will come, later, to constitute the truth – the being true – of what will later be representations. In the case of *descriptive representation producers*, they will do this, of course, not by changing the world but by producing *representations* to fit the world in accordance with some given rule of "fit." Since, in order to be selected, they must do this (as the balls, to get selected, must be small enough though not necessarily green), they are selected, in part at least, *for* doing this. They are selected, at least in part, for making representations that fit the world in accordance with some given rule of correspondence. (A rule such as, for example, *neuron n firing in frog F at time t* [the *representation*] is to correspond to *an edible bug* (or a worm-like shape – whatever) *moving in the visual field of F at t* [the *represented*]. See next paragraph.)

What "given rule?" The second key is to be sure we address this question. Obviously the *representation producers* won't be selected, merely, for often enough making *representations* that "fit" the world by any old arbitrary rule. That requirement would be empty (see Shea, this volume, and my reply). They have to make *representations* that fit according to a correspondence rule that the consumers, the interpreters, of these *representations* will be able to "understand," that is, to make use of in some way in going about whatever business they are designed to go about. Being a little more careful, since producers and consumers are often designed together, the *producers* have to make *representations* that fit the world by rules that their *interpreters* often enough make use of in ways that help cause them – the *interpreters*, but consequently also the *representation producers* – to be selected.

Thus a reference to the interpreters of representations, a reference to the consumer side of the representation making and using systems, is required for a full analysis of what the *representation producers* are selected for doing. This is why teleosemantics needs to be, at least in part, consumer semantics. The correspondence rule between representation and world has to be one the consumer relies on in performing some activity or activities. Otherwise there would be no selection forces acting on the *representation producers* at all, or rather, none that would tend to turn them into representation producers. (The pigment rearrangers in a chameleon's skin are selected for making its skin color correspond to what the chameleon is sitting on but they are not representation producers because they lack co-evolved interpreters.)

Returning, *representation producers* are selected to become representation producers, at least in part, *for* making *representations* that fit the world in accordance with some given rule of correspondence. Are they also selected *for* being caused to produce this correspondence by their *representeds* – for being caused by what they will come to represent? *Representation producers* that do effect correspondence of their *representations* to the world in accordance with useful correspondence rules by means of reacting to their *representeds*, or by means of reacting to other things that have reacted to their *representeds*, and so forth, will of course be selected. But if the appropriate correspondences were to be produced by some different means, they would be selected as well. This shows that they are not selected *for* being caused to produce *representations* by responding to their *representeds*. (What makes the issue confusing is that they are, of course, selected *because* they react to their *representeds*, or to other things that have reacted to their *representeds*. This is why, *in these particular cases*, they are able to make representations that correspond, by appropriate rules, to *representeds*. But in other cases this may not be why the producers are able to produce the right correspondences.) How the producer produces the right correspondences is not selected for. Only *that* the producer produces these correspondences.

There are a good number of ways that representation producers manage to produce the right correspondences between their representations and their representeds besides being caused to do so by these representeds. An obvious way, explicitly recognized by Dretske (also alluded to by Neander at the start of her essay), is that the representations and the representeds have a common cause. This is a common way that future things may get represented. It is how the outfielder perceives where the ball is going to land so as to get there first and catch it. But there are also many natural signs used by representation producers that bear no causal relations to the representations they produce. The compass needle tells us the rough direction of geographic North, but it is not causally related to it. The North Star also tells us the direction of geographic North, but is not causally related to it. Similarly, the directions that the magnetosomes of certain anaerobic bacteria point to tell the direction of lesser oxygen, but there is no causal connection between these. Leaves turning red and yellow and dropping off the trees tell us that the weather will soon become colder, but there is no causal connection. The minute minder's ding tells me that the roast is done, but there is no causal connection. That I am passing the cow barn with the "Random Farm" sign tells me that I am about a mile from home, but there is no causal connection. And most important, the shape, size, and color of the apple are what tell me that it is an apple, but the connection is not causal but one of property possession. Similarly, the place of the shape, size, color, and movement of my husband often tells where he is, but he is not their cause; indeed, individuals, as such, do not figure in causal laws. Representations of individuals would be impossible on a pure causal theory.

It is the proper function, the selected-for function, of a representation producer to produce true representations, those that correspond to the world in the way its consumers need them to, but not to accomplish this in any particular way. *How* a producer has usually accomplished this corresponds to what I have called the "Normal explanation" for the producer's proper performance. (Distinguishing proper functions from Normal explanations has many theoretically important uses, and I think this is one of them.) Proper functions are sometimes performed yet performed abNormally; indeed, sometimes performed quite accidentally. For example, true belief is not always knowledge. Only true belief is selected for, but of course it will not be produced enough of the time to cause systematic selection unless it is produced by a mechanism having some Normal explanation for its production. As Socrates told us, true belief is just as good as knowledge, though only so long as it stays put.

Notice that when a representation is true but has not been produced in a Normal way, it is unlikely to have been caused by any natural sign of its represented state of affairs, and is unlikely to be, itself, any kind of natural sign of that state of affairs. It follows that if producing *true representations* is what *representation producing systems* are selected for, informational teleosemantics can't be quite right. The problem to be solved by teleosemantics is not just how false representations are possible, but also how representations can sometimes be representation and be true despite having been arrived at through accident or malfunction.

Neander has another concern – about what she calls "the norms of malfunction." I think that this concern can also be met by carefully distinguishing between a thing's proper functions and the Normal explanations for its accomplishing those functions. The Normal explanation for the accomplishment of a function by a device always includes reference to background conditions that are required to be in place but that are not caused to be in place by that device itself. Failure of a device to serve some proper function is usually the result of the absence of some of these background conditions. For example, I have argued that failures of the cognitive systems to produce true beliefs generally result from absence of the right external-world background conditions (Millikan 1998b). When a device does not perform some proper function only because the necessary background conditions are absent, we do not consider that to be a malfunction – though it might, of course, be due to malfunction of some other device whose job was to put the necessary conditions in place. Malfunction results only from abnormalities in the constitution of the device itself.

2

Signals, Icons, and Beliefs

PETER GODFREY-SMITH

Introduction

One way to approach the explanation of meaning and semantic content is to hold that any object which is a sign or a representation – anything that has semantic properties – has this status because of the way it is used in an interaction between two other kinds of things, a *sender* and a *receiver*. "Sending" is understood here in a very broad way, to include any kind of creation, display, or inscription of a sign. "Receiving" involves the sign's interpretation and use.

A defender of this approach has to say something about how semantic phenomena that do not apparently involve senders and receivers can be understood. If our thoughts have content, for example, that does not seem to be because they are being communicated from one place to another. The defender of the sender–receiver view might respond by retreating from the attempt to cover all cases, or they might claim that their framework is more general than it looks.

There are also very different approaches to explaining content. We might start with an explanation of the content of thoughts and explain other semantic phenomena in terms of the *expression* of these thoughts. Such an approach might be motivated by the view that anything that deserves to be called a "sender," "receiver," or other user of a message must itself have states with semantic or intentional properties. Perhaps only thoughts have "intrinsic intentionality," although various other signs (maps, spoken sentences, etc.) have "derived intentionality" as a result of their use by thinkers (Searle 1992).

Ruth Millikan's theory is the most carefully worked-out version of the sender–receiver approach to semantic phenomena within naturalistic philosophy – perhaps within any kind of philosophy. The element of Millikan's view that has caught the most attention is the "teleological" element, the use of a biological notion of function, grounded in history, to explain meaning. Less discussed is the fact that hers is an ambitious version of the sender–receiver approach; any entity that has semantic content, according to Millikan's analysis, has it as a consequence of its relations to a "producer" on one side and an "interpreter" or "consumer" on the other. So if our thoughts have content, there must be "producer and consumer" arrangements inside us, as well as between us. The same schema is used when explaining the semantic properties of external signs and internal states, and neither has semantic properties that are prior to or reducible to the other.

An additional perspective on sender–receiver systems has recently been provided by Brian Skyrms, in *Signals* (2009). Skyrms takes a modeling approach, looking at how sender–receiver configurations come into existence and are maintained, especially in situations where foresight and deliberation are not involved. Recent years have also seen much new work on this topic within evolutionary biology.[1] Bringing these ideas together opens up new avenues for the development of the sender–receiver approach. This chapter pursues a synthetic project of that kind (as does Harms 2004). The next section describes sender–receiver configurations. I then compare some explanations of semantic content within that framework. The final section looks at the application of the sender–receiver approach to mental states and internal representations.

Senders and Receivers

Vervet monkeys give alarm calls that alert their troop when a predator is spotted (Seyfarth et al. 1980). The hearers use the calls as a guide to the state of their environment, and the vervets give different alarms when they see different predators (leopards, snakes, eagles). Paul Revere gave an alarm signal during the American Revolution ("one if by land, two if by sea"). Actions directed on the British army were guided by attending to a lantern in a church. Obvious features present in these cases include a sender, a signal, and one or more receivers. The signal is produced by the sender, and this is done in a way that is responsive to the state of something in the world beyond the signal itself. A receiver produces behavior according to what the signal is like. These behaviors are also directed on something beyond the signal – often, on the state of the world that the sender was responding to when producing the signal.

Sender–receiver configurations have been treated as a starting point in discussions of meaning within very different traditions.[2] This is also the set-up that provides the starting-point for the mathematical theory of information and communication, and was pictured by Claude Shannon (1948) in a famous diagram (Figure 2.1). Shannon's diagram adds some elements beyond the basic sender–receiver structure described above, and omits others. He includes a source of "noise," for example, and aside from the signal itself there are extra "messages" on each side. Shannon says that the "destination" on the far right is "the person (or thing) for whom the message is intended."

[1] See Maynard Smith and Harper (2003), Lachman et al. (2001), Rendall et al. (2009).
[2] See, for example, Eco (1976).

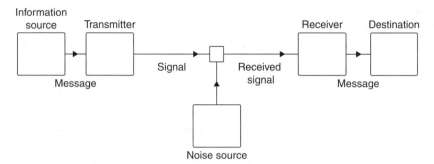

Figure 2.1 Shannon's diagram of a "general communication system."
Source: Shannon (1948). Reprinted with permission of Alcatel-Lucent USA Inc.

Many different terminologies are used. Shannon's picture has a "transmitter" and a "receiver." The game-theoretic literature tends to say "sender" and "receiver." In Millikan's framework we have "producers" on one side and "interpreters" (LTOBC) or "consumers" (1989a, VOM) on the other. Between the two we may find "signals," "symbols," "representations," or "signs." Millikan's general term for a sign-like object is "icon." She reserves "representation" and "signal" for narrower categories. In this chapter I will move between different terminologies in a way that is not intended to reflect theoretical differences, and the aim is to discuss a model that can be applied very broadly. A sender creates or emits things that can be called "signs" or "signals," and they are read or interpreted and put to use by a receiver. The term "reader" will also be used, as it emphasizes the idea that what a receiver does is apply an interpretation rule to the sign. When discussing Millikan's work I will mostly use her preferred terms. Millikan's term "producer" is in some ways better than "sender," as it emphasizes that a sign is being created as well as transmitted or transported. Her "consumer" is perhaps *too* broad in its connotations, and that will become an issue in the last section of this chapter. But I will not impose a standardized terminology, and will use whichever terms have the most useful connotations for the case being discussed.

As the sender–receiver approach sees things, just about anything can function as a sign, and can have any semantic content, if it has an appropriate sender and receiver. A receiver should not be mistaken here for an interpreter in the sense of a mere onlooker, someone trying to "make sense" of the sign. Senders and receivers have to be tied together more closely than that, as will be discussed below. A derivative case may be recognized, where an interpreter makes use of "natural signs" or "cues" to guide action. (Smoke is a natural sign of fire.) These are produced in a way that involves no coordination with the interpreter. These signs can carry information in the sense described by information theory – anything that changes the probability of some other event in the world carries information about it. They might also be described in other quasi-semantic terms, perhaps through a pretense that they have an appropriate sender. But they are not taken to have content in the same way that a sign within a sender–receiver configuration does.

When we fix the sender–receiver configuration in our minds and go looking through nature, we find a great diversity of things that fit the pattern. Some are very simple, some complex. Sometimes there is a single receiver, sometimes several. Many are found between whole organisms: alarm calls, mating calls, chemical signals that mark trails and territories, and signals used by bacteria to work out how many of them are in the same area ("quorum sensing"). We also find sender–receiver configurations within single organisms; hormones (like adrenalin, thyroxin, and testosterone) are examples of such internal signals.

Signaling within certain tightly integrated colonies and societies is a special case. Bee dances, which Millikan has discussed in detail, provide an ideal example. When a honeybee finds a source of nectar that is fairly far from the hive, she will return to the hive and perform a looping dance, near a vertical surface, which includes a "waggle" in the middle. The dance signals the location of the nectar source. The dance's angle off the vertical corresponds to the angle between the nectar and the point where a line drawn down from the sun hits the horizon, in relation to the hive. The length of the "waggle" segment of the dance represents the distance. Worker bees, which are normally sterile, have tightly connected evolutionary interests. Some biologists see honeybee colonies as "superorganisms," with individual bees akin to cells (Holldobler and Wilson 2008). Then signaling between individual bees is more like signaling between parts of a single body. Maybe the bee dance is best seen as intermediate between a case of signaling between multiple individuals and signaling between the parts of one. And when we look closer at some "organisms," including familiar trees such as an oak, we find that they are in some ways like collections of individuals, not single ones (White 1979). When there is signaling within a plant (for example, using the hormone auxin), this takes place within a collection that is more physically integrated than the bees in a colony, but with less reproductive specialization – all the branches of an oak tree have independent reproductive capacities. So there is a continuum between cases with separate agents who come together and signal in their own interests, and cases involving signaling within one agent. Here it is interesting to note that some neurotransmitters are thought to have evolved from chemicals whose initial role was inter-organism signaling.

There are also unclear and questionable cases of the sender–receiver configuration. DNA looks very much like it was designed to send information (Shea 2007a, Bergstrom and Rosvall 2011). But who is sender and who is receiver? "Messenger RNA" looks like a message sent *from* the genes to the machinery of protein synthesis, but who is the sender in the case of genes themselves? Even with the questionable cases set aside, we have a large and diverse collection: human language, subway maps, bee dances, alarm calls, birdsong, hormones, pheromones.... The sender–receiver configuration is found at many different scales and has arisen independently many times. So now we can ask: Why should we find things in this arrangement? Why is it common? What makes it come about, and what keeps it from falling apart?

David Lewis (1969) gave a game-theoretic model of sender–receiver coordination. In a ground-floor version, assume there is a sender who can see the state of the world and produce signals. The receiver can see only the signal but can act according to the signal he sees. There are two states of the world and two actions available. The sender can send the same signal all the time or adjust the signal to the state of world. The receiver can act the same way all the time or adjust his behavior to the state of the signal. A situation of *common interest* is assumed; for each state of the world there is a right and wrong action, and both agents benefit if and only if the receiver produces the right action. A *signaling system* is a combination of policies by sender and receiver such that both of them do reliably get the payoff. That means that the sender is adjusting its signals to the world's state, and the receiver is interpreting these in the right way. A signaling system in this scenario is a *Nash equilibrium* – a combination of behaviors such that neither agent does better by unilateral change. Each behavior is the "best response" to the other. Signaling can be maintained by common knowledge of this fact.

Skyrms (1995, 2009) takes an evolutionary approach. His model does not assume "smart" senders and receivers, and applies to bees and bacteria as well as to more complex agents. In one of his models, we assume a population of individuals who take both sender and receiver roles on different occasions, occupying each role half the time. As in the Lewis model, only senders can see the state of the world but only receivers can act on it. Each individual has a behavioral profile that combines a sending policy and a receiving policy. With two states of the world, two possible signals, and two possible actions, there are 16 such profiles. The agents interact randomly, receive payoffs, and reproduce asexually according to those payoffs. Both agents receive a payoff whenever the receiver does the right act for the state of the world, and otherwise no one is paid. The states of the world are equally likely. Agents reproduce asexually, in a way reflecting their payoff. Then it can be shown that evolution favors behavioral profiles that give rise to signaling systems. Those behaviors are *evolutionarily stable* – they cannot be invaded by rival strategies once they are common. Other behavioral profiles are not evolutionarily stable. Skyrms found in simulations that his idealized populations did make their way from a variety of initial states to signaling systems. In other models, the sender and receiver roles are separate and fixed, there may be multiple receivers, and so on. Adaptive change may also be due to learning dynamics rather than evolution; there are several different processes by which sender–receiver configurations can arise and be maintained.

In the models sketched so far, "common interest" between sender and receiver is assumed. Similarly, Millikan's discussions of "producer–consumer" systems include the requirement that these two devices "cooperate" with each other, as a consequence of evolutionary design. Millikan does not say a lot about what degree of cooperation is required here, and she views the problem of explaining the maintenance of cooperation as overemphasized in some discussions of social evolution, at least in the case of humans.[3] But clearly it is worth asking how sensitive models of signaling are to assumptions about common interest.

In Skyrms' (2009) models, a total breakdown of common interest dooms signaling – this is what we have when sender and receiver benefit from completely different acts being performed in a given state of the world. But there are also cases of *partially* aligned interests. In Table 2.1, the payoffs are given for sender (first) and receiver (second) in each

---

[3] "Most aspects of social living involve cooperation in ways that benefit everyone" (VOM, 21). Even if this is true of present-day interactions that maintain some central practices of public language use, especially in Connecticut, it is not something that can be generally assumed in an evolutionary context. In those contexts it is also relative advantage that usually matters.

Table 2.1 Partially aligned interests (from Skyrms 2009, 81).

	Act 1	Act 2	Act 3
State 1	2, 10	0, 0	10, 8
State 2	0, 0	2, 10	10, 8
State 3	0, 0	10, 10	0, 0

combination of a state of the world and an act performed by the receiver. For example, if the receiver produces act 1 in state 1, the sender gets 2 units and the receiver gets 10. This is a case where in states 1 and 2 there is a partial conflict of interest: the sender would rather the receiver performed act 3 in states 1 and 2, while the receiver would rather perform act 1 in state 1 and act 2 in state 2. In state 3, their interests align.

In a system of this kind, there is an equilibrium if everyone uses the following strategy: *if sender, send signal 1 in states 1 and 2, and send signal 2 in state 3; if receiver, do act 3 in response to signal 1, and do act 2 in response to signal 2*. Effectively, senders refuse to discriminate between states 1 and 2, making it rational for the receiver to choose their second-best option. This *second*-best option for the receiver gives the *best* outcome for the sender. The receiver would benefit from getting perfect information about states 1 and 2, but there is no reason for the sender to send it.[4]

There are also situations where full and honest signaling can be maintained even though it has an "altruistic" character. Alarm calls are good cases to think about. Suppose there are two types, a signaler and a "free-rider" which responds to alarm calls but keeps quiet when they are the one who sees a predator. If these individuals spend time foraging in pairs (not mating pairs), the best thing to be is a free-rider in the company of a signaler: they will alert you to predators, incurring some danger by making the call, and you get the benefit of their calls without making any of your own. The next best thing is to be a signaler with another signaler. The next best is to be a free-rider with a free-rider, getting no warnings but taking no special risks, and the worst is to be a signaler with a free-rider. This is a "prisoner's dilemma."[5] In a population containing both types, signalers are always exploited when in mixed pairs, but signaling can prevail provided that interaction is sufficiently *correlated* (Hamilton 1975), so that most signals are sent to other signalers and free-riders are stuck with their own kind.[6]

Above, I noted that sender–receiver configurations are found in nature both within and across organisms, and also in cases where the boundary of an organism is not clear. In general, as we move further from the within-organism case, breakdowns of common interest become more likely. They are not impossible even within organisms like ourselves, however.

---

[4] See Stegmann (2009) for discussion of a different kind of case involving conflict of interest, featuring exploitation of a sender–receiver system by a malicious (predatory) alternative sender.
[5] It is sometimes also required that the payoff from mutual cooperation be less than half the average of the payoff from defecting on a cooperator and the payoff from being a cooperator who is defected on (Axelrod 1984). I assume here that there is no possibility of conditional behavior in response to the behavior of a partner.
[6] The biological literature has also looked extensively at the idea that the *costliness* of certain signals can maintain their honest use in a population. See Maynard Smith and Harper (2003).

David Haig argues that within our bodies there is the potential for competition between *paternally derived* and *maternally derived* genes. Roughly speaking, the genes an animal gets from its male parent have less interest in the well-being of the animal's siblings than its maternally derived genes do, as long as there is any uncertainty about the paternity of those siblings (Haig 1996). So our paternally derived genes would have us monopolize resources in a family situation more than our maternally derived genes would. This sort of conflict has been well documented in pregnancy, where the "resources" in question are supplied directly through the mother's own body, and the conflict is played out in a dynamic between senders and receivers in a hormonal signaling system. The extension of the hypothesized conflict beyond pregnancy is, I understand, still more speculative (Haig 2010).

In other within-organism cases, the idea of common "interest" between sender and receiver becomes tenuous and metaphorical. But the parts of an organism can have a "cooperative" relationship if they have evolved to interact with each other in a way that furthers the interests of the whole organism or some other reproducing entity. For example, the most physically tiny signaling systems are probably those within individual bacteria (Camilli and Bassler 2006). A receptor within a cell may function as a receiver of messages sent from its genome, or a relay-like "sender" which transduces an external cue. The receptor itself cannot reproduce or help other receptors similar to it to reproduce, but the bacteria have evolved to make receptors that interact with other parts of the cell in a way that furthers bacterial reproduction.

Very simple cases like this are also significant because they probe the boundaries between sender–receiver systems and set-ups that should be understood in different terms. There is a tendency in cell biology to use the language of "programs" and "computation" in a very broad way about intricate arrangements that are the products of evolutionary design, and this leads also to descriptions in terms of meaning and intention. But being a sender–receiver system is a quite different matter from (for example) being a "programmed device." In the paradigm cases of sender–receiver configurations, receivers are physically distinct from the signs they interpret, and receivers can act in ways that have consequences for senders and signs as well as for themselves. Sender, sign, and receiver are not merely three things connected in a causal chain. In some discussions, it is also said that an essential feature of sign use is a potential flexibility in the interpretation rule applied by a receiver. This is referred to as "arbitrariness" (Maynard Smith 2000) or "conventionality" (Skyrms 2009). Something like this is certainly a feature of familiar cases of public representations used by humans. It is not so clear in some of the biological cases above, where we find mechanically simple receivers applying "hard-wired" rules. Skyrms says that at least "contingency" in the interpretation rule is seen in those cases. I think this intuition is linked to the basic requirement of the separability of sign and reader: each could be independently modified in principle. It also seems that when a sign's effect on a reader is too physically constrained – when the sign's physical impact leaves "no room for interpretation," as will often be the case with the parts within a cell – then we have something that is not a clear case of the sender–receiver configuration.[7]

---

[7] Chapter 7 of Millikan's LTOBC, "Kinds of Signs," surveys other kinds of simplicity and sophistication in sign systems, emphasizing further distinctions between cases that clearly fit the basic requirements described here.

Content

Sender–receiver configurations arise often in nature and at many scales. They are also prominent everyday features of our own social lives. As a result, we have culturally entrenched habits of dealing with them and talking about them – habits of description and ascription of content. Using language (a sign system) we constantly talk about what messages in other sign systems "mean" and "say." So there are two sets of facts here. First, there are facts about the relations between signs, their users, and the world that affect how signaling systems work. Second, there are facts about our habitual ways of talking about signs within social life. This second dimension might be seen as a kind of "folk theory" of signs, though it probably has roles in social coordination that are poorly described with the term "theory."

When investigating sender–receiver systems in nature we can expect to be guided by those habits. This might have both good and bad effects: good, as we are sensitive to the presence of signaling systems, perhaps bad, as we may describe naturally occurring sign systems in ways that are guided too much by everyday cases. For example, we might describe the meanings of simple signs in a way that is affected by our experience with the very sophisticated semantic properties found in language.

In this section I will look at the content of signs in sender–receiver systems. People can agree that sender–receiver configurations are the natural home of meaning, but disagree about how the content of messages in such systems should be described. I will start by looking at an approach making use of information theory. This approach was pioneered by Dretske (1981), but here I discuss a version due to Skyrms (2009). I then compare this approach to Millikan's.

On an information-theoretic view, a signal contains information when it changes the probabilities of states of the world. Skyrms' suggestion is that the *amount* of information in a signal is measured by the amount of change the signal makes, on average, to the probabilities of each of the possible states of the world. The *content* of a signal – what it says – is represented with a list of *all* the differences made to the possible states. "Propositional" content – content expressible with a "that ..." clause – is for Skyrms a special case. When some states of the world are not just made less likely by a signal but are ruled out, then the content of the signal can be expressed as a proposition that disjoins the states that remain.

Clearly these are real features of any sender–receiver configuration (bracketing foundational questions about probabilities), and they figure in explanations of how the system works. If a signal did not change the probabilities of relevant states of the world there would be no reason to attend to it.[8] The case of partially aligned interests in sender and receiver discussed in the previous section illustrates the possibility of more fine-grained analysis in these terms. When that system is at the equilibrium described earlier, the

---

[8] Here I set aside a very special case, where an uninformative signal might act as a useful randomizing device in a situation where mixing of behaviors reduces variance in payoff in a way that is evolutionarily advantageous (Godfrey-Smith 1996, ch. 7). I should also note that Millikan thinks that the foundational questions about probabilities set aside here do raise problems for information-theoretic views.

mixture of matched and mismatched interests across sender and receiver is reflected in the amount of information contained in different signals. The signal sent in states 1 and 2, when there is some conflict, contains less information than the signal sent in state 3, when there is none.

Not all of the properties that might be recognized by an information-theoretic analysis have this role, however. Skyrms also says that a signal in a sender–receiver system has a second kind of content. As well as changing the probabilities of states of the world, a signal changes the probabilities of acts by the receiver. Skyrms says a signal contains information about acts in the same sense in which it contains information about the states of the world. But this sort of "content" does not have anything like the same role as the first. In a sender–receiver configuration, the role played by a signal is to tell the receiver something about the state of the world, not to tell the receiver how the receiver is likely to act. We as onlookers can learn about actions likely to occur by looking at the signals, but this is surely not something the receiver is being told. This is an illustration of a more general fact: any signal will have effects on the probabilities of a huge range of goings-on, but almost all of these are irrelevant to a description of how the signaling system works and why it exists.

Information-theoretic views have often had problems dealing with phenomena involving error and false content.[9] Here Skyrms attempts a very simple treatment. He says that when a signal raises the probability of a state that is not actual, that is "misleading information, or *misinformation*" (Skyrms 2009, 80). This only provides a very partial analysis, however. To explain the possibility of error or false content is usually seen as a matter of saying how a representation can have a propositional content that is false: How can a signal say that p is the case when it is not? For Skyrms, a signal only has propositional content when it rules out one or more states of the world completely. So for a propositional content to be false, a signal must say that some state has been ruled out when in fact it is actual. That is impossible within Skyrms' model. When a non-actual state's probability is raised and the actual state's probability is lowered by a signal, that does give the signal the potential to mislead an interpreter. But in Skyrms' model a signal cannot *say that* something is the case when it is not.[10]

As Dretske found in 1981, there is a certain amount about semantic phenomena that can be explained very readily in an information-theoretic framework, and a residue that is a persistent problem. One response is to add extra elements to the information-theoretic framework, as Dretske (1988) did. Another is to set information theory aside and start again, as Millikan does. Yet another would be to say that some of the recalcitrant "semantic phenomena" are just features of the way we *talk* about representations, features of our

---

[9] See Fodor (1984), Godfrey-Smith (1989, 1992).
[10] Skyrms also says that a "misinformation" has been transmitted when a receiver acts differently from how he would if he had "direct knowledge" of the state of the world. If what really matters here is a counterfactual about rational behavior, this compromises the naturalistic status of the analysis and its applicability to simple systems. The idea might be expressed in a more direct way, by saying that when the payoff is low then misinformation has been transmitted. But why, within a non-historical analysis of content like Skyrms', does the size of a payoff occurring *after* the signal has been received matter to what the signal *said*?

interpretive habits. Or these features might be real but found only in the most sophisticated sign systems, and should not be read back into simple sender–receiver configurations.

I now turn to Millikan's view. For Millikan, the content of a sign or "icon" derives from the biological functions of its producers (senders) and consumers (receivers), and from the normal conditions under which these functions are performed. Both the functions and the "normal conditions" depend on the system's history. "History" sounds like a long time ago, but the relevant history is made up of the processes of maintenance and stabilization discussed throughout this chapter.

In simple cases like animal danger signals and bee dances, Millikan thinks that an icon has two kinds of content – or, more accurately, a kind of content that has a dual character when we express it in language. A bee dance says both "There is nectar at location X" and "Go to location X." There is both a descriptive element (indicative) and a command (imperative). Only in more complex cases do we find contents that are purely indicative or imperative. Suppose there is no specific outcome that a consumer is supposed to bring about in response to a given icon. The icon is instead supposed to enter into something like a process of inference, with the actions resulting being affected by other icons and other states of the system. In that case the sign will have a purely indicative content. The kind of dual content that Millikan recognizes in simple sender–receiver cases is different from Skyrms', and conforms better to the role that signs have in such systems. The sign does not say which actions have become more likely; it instructs the receiver what to do.

From now on I will mostly discuss the indicative side of the content of these simple signs. This content is not determined by which states of the world are made *likely* by the signal. Instead, the content depends on *how the world has to be* if the icon is to affect the activities of its consumers in a way that leads to them to perform their functions in a historically normal way – in the way that led to that pattern of icon-use being favored by some process of selection. When thinking about this idea, I find it helpful to approach it in the following order. Hold fixed the fact that the consumers will act in a particular manner as a result of the form of each icon they receive – that they will apply a particular "interpretation rule." Suppose that *given* the rule that the consumers are following, the icon will only affect the consumers in a way that leads to them successfully performing their functions in a historically normal way if the world is such-and-such a way. Then the icon says that the world is that way (see Figure 2.2). To put things very roughly, the condition under which an icon is *true* is the condition under which the behaviors that the consumers will be induced to perform will be *successful*.

Putting things less roughly and in Millikan's preferred language, icons are "supposed to map" the world, and the mapping is one that involves "transformations" of both icon and world. For example, a bee dance says there is nectar at location X at time t. The fact that this is the dance's content depends on the fact that there are *other* bee dances that could be produced at different times and with different "shapes," which have different success-conditions. If the dance had been performed at time $t+1$, it would have the content that there is nectar at X at $t+1$, and it would only be useful for the consumers to act on this dance by its usual rule if there was nectar at X at $t+1$. It is this rule mapping transformations of icons to transformations of "world affairs" that natural selection

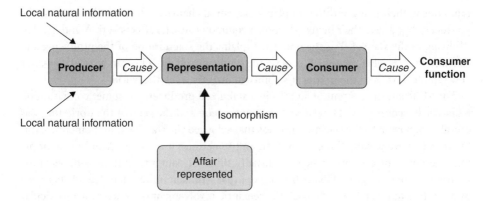

Figure 2.2 Millikan's schema for "descriptive" representations.
Source: Ruth Garrett Millikan, *Varieties of Meaning*. The 2002 Jean Nicod Lectures. Figure © Massachusetts Institute of Technology, by permission of the MIT Press.

(or some similar process) singles out.[11] What selection directly acts on is the way that producers produce icons and the way consumers interpret them. Most of the possible combinations of ways of producing and consuming icons are pointless or worse. A few ways, however, are adaptive. This stabilizes the pattern of icon production and consumption. When this stabilization has occurred (according to Millikan) there will be a systematic relation visible between different forms the icon can take and different ways the world can be. That relation between "transformations" is what determines the content of a particular icon.

As the danger-signal cases make clear, for an icon to have one of these historically normal success conditions it is not necessary that the icon's presence makes that condition probable. A low-reliability danger-signal can still be useful, and can still be a *danger*-signal rather than a signal whose content is determined by the most common cause of false alarms. Millikan's approach deals with the possibility of false content very easily. When an icon is produced and the world is not in the historically normal success condition described above, then the icon says something that is not true. This might happen 99 percent of the time.

The way Millikan's theory handles the possibility of false content seems an advantage over the information-theoretic approach. In some cases, though, the way Millikan's approach to content focuses on the states of the world that *explain* stabilization is at odds with reasonable-looking intuitions about content. Modifying an argument due to Pietroski (1992), suppose a producer observes the state of some variable X, and sends a sign that carries information about it. The consumer acts, and this action is coordinated with the state of X. The producer and consumer benefit – but not because of anything about the practical importance of X. They do well because of a correlation between X and another variable, Y, whose state cannot ever be observed by this producer and consumer but whose state has important practical consequences. And the correlation between X and Y

---

[11] The role of this talk of "mapping" and allied talk of "picturing" is discussed in detail in Shea (this volume).

is not due to their being at different places in a causal chain (Y affects X which affects the producer), but due to their being effects of a common cause (Z affects both X and Y). For Millikan, as the state of Y is the one that explains the stabilization of the producing and consuming, Y is what the signals sent by the sender are about. This can be true no matter how remote Y is from the producer's ability to directly track the world.

For Millikan, the content of an icon in a stabilized producer–consumer configuration is always determined by the relation between the icon and the state of the world that was causally important in the selection process that achieved the stabilization. Sometimes this fits with how we usually think of content, and sometimes it does not. Millikan can reply that she does not care about how we usually think of content – fitting with semantic intuitions is not her goal. That is fine, except that Millikan quite often defends her view by appealing to its ability to "focus" the semantic involvement of an icon on a particular part of the world, and to handle the problem of error in an intuitive way. In both those cases a defender of an information-theoretic view is also at liberty to say that he does not care about "familiar intuitions" about content.

Both views make use of real features of sender–receiver configurations in their explanation of content, and both schemes of interpretation are partly in accord with familiar intuitions about content and partly at odds with them. The features they draw on have different roles within a complete description of what sender–receiver configurations are and how they work. Millikan connects the content of a sign to the history of stabilization of the sender–receiver configuration; Skyrms links it only to the operation of the configuration at a given time. The information-theoretic approach links content to *how a sign manages to inform* a receiver; the teleofunctional view links it to what the sign is *supposed to do*, and how the world must be if the sign is to be able to do its job.

States of the Mind and Brain

The sender–receiver approach can be seen as a way of explaining just some semantic phenomena, or all of them. It might seem an unlikely way to approach the explanation of the content of thought. Senders and receivers are found in situations where there is *communication*, and the semantic phenomena involved in thought seem to be of a different kind. If we think about the "directedness" of thought on the world seen when someone believes that (say) New York is south of Boston, this does not seem to be a matter of one entity *telling* something to another.

One response to this is to say that surely the brain can be seen as a signaling device. Neurotransmitters transmit signals between neurons, for a start. But whether this kind of activity fits into the sender–receiver configuration discussed above is not so clear. If we look inside a brain and find a huge network of neurons, each affected by some and affecting others, it appears that any one neuron's firing might be described as either a signal, or the reception of a signal by a receiver, or the sending of another signal, depending on how one divides things up. That is not how things are in the clear cases of sender–receiver configurations.

There is a different way of making a connection between internal states and the sender–receiver model, however. The first move is one that many philosophers will accept: endorsement of a "representational theory of the mind." On this view, our minds contain internal representations that have both semantic content and a role in cognitive processing. In virtue of using these representations in a certain way, we have propositional attitudes like the belief that New York is south of Boston. The strong and commonly held version of this idea is not just that beliefs exist as a consequence of the use of internal representations, but these inner representations have the same kind of content as that ascribed to the whole agent. Very roughly, because a representation with the content that p is in the right place inside the agent, the agent believes that p.[12] The match between what an inner sign says and what the agent believes need not be exact, and a representationalist of this kind can allow that many ordinary belief ascriptions are of "implicit" beliefs, whose contents follow from a smaller set which are "explicitly" represented. But the aim is to explain the content of thought in terms of the cognitive role of inner signs with closely related contents.

Suppose a representationalist view of this kind is provisionally accepted. Such views can be developed in several ways. Recent work on this topic has often been influenced by ideas from computer science. But there is also an older tradition. A number of philosophers[13] have been attracted to an analogy between beliefs and maps; Frank Ramsey (1929/1965, 238) suggested thinking of beliefs as maps "by which we steer." Looking further back (as Millikan 1986 does), Plato used an analogy between knowledge and impressions on a wax tablet. So one way to have a representationalist view is to hold that inner representations are inner pictures of what they represent. Millikan thinks that this idea is still quite viable, provided that we have a suitably abstract sense of "picture," one in which spatial relations between elements of a picture need not be what is important, for example. Whether or not the idea of picturing is used, it seems possible to argue that thinking involves an interaction between parts or devices in the brain, with some producing or inscribing representations and others consuming or reading them. The language of "sender and receiver" is not especially natural here, but the model itself is applicable. Randy Gallistel has recently emphasized the importance of "read–write" memory systems in the explanation of intelligence (Gallistel 2006, Gallistel and King 2009). Gallistel's read–write (or write–read) architecture is another variant on the sender–receiver picture. Another link to the idea of signaling can be made by noting that *memory* might be thought of as the sending of messages through time. Many of the phenomena philosophers discuss in terms of "belief" can be understood as the operation of memory.

So a possibility has been sketched, but what would these inner senders and receivers (writers and readers, producers and consumers) have to be like? One way to put the point is to ask whether there are things we might discover that would show that the sender–receiver model was *not* applicable to thought, and if we discovered one of those things, what that would show about beliefs.

I will compare how these issues are handled in some different parts of Millikan's work. In her 1984 book Millikan discusses, with some caution, the application of her analysis of

---

[12] See Fodor (1981) and Field (1986).
[13] Armstrong (1973), Dretske (1988), Braddon-Mitchell and Jackson (1996).

intentional icons to beliefs (and similar states) in human beings. She outlines a way in which the analysis *could* apply, a way that depends on the existence of a language-like medium for thought and some specific mechanisms that interact with inner sentences. The "interpreter" of beliefs she hypothesizes is a device called a "consistency tester" (LTOBC, 144–146). A consistency tester looks for ways in which contradictions (which are to be avoided) may arise in a single mind, by comparing different ways in which inner sentences can be formed.

If "consistency testers" are real parts of us, they are presumably very contingent parts of the machinery. In Millikan's discussion, their usefulness derives from the fact that we use inner sentences that contain newly coined mental "terms." It seems that humans could get by fairly well without consistency testers, even if we think in a language-like medium, though we might produce more contradictory combinations of inner sentences without them. It seems unlikely that our having beliefs depends on their presence.

In the same discussion Millikan also describes some fictional analogues of humans, whose concepts are all innate and products of evolution by natural selection. These fictional agents are said to contain some components that produce inner sentences, and other components that use the sentences in guiding behavior (LTOBC, ch. 8). Millikan takes this to be unproblematic in principle, and so do I; it is a possible psychology. So in Millikan's 1984 book we have inner sentences that qualify as icons within a fiction, and empirical hypotheses about a "consistency tester" which, if real, would make human beliefs into intentional icons as well. What is missing is something in between the acknowledged fiction and the very strong hypothesis about consistency testers.

Millikan's follow-up work about the mind (1986, 1989a, VOM) gives a different treatment of interpreter or consumer mechanisms. Consistency testers are not made central, and neither is a language-like medium for thought. In some of these discussions, the notion of a consumer is apparently understood in a very broad way; all the machinery guiding action and inference that is affected by a representational state is a "consumer" of that state. In "Biosemantics" Millikan says: "Let us view the system, then, as divided into two parts or two aspects, one of which produces representations for the other to consume" (Millikan 1989a, 285). I am not sure whether this is meant as an idealization which will help us think about the problem, a commitment to something like a "belief-box" model of thought, or as a minimal commitment that must surely be accurate in virtue of uncontentious facts about the causal flow in the brain from senses to effectors.

The last of these options seems unlikely; a "consumer" should not just be everything downstream of an inner state. And, given that, we should think about the possibility that what evolution has done in the case of brains is move away from the producer–consumer configuration. Perhaps, for instance, it has built something in which the representation role "leaks into" that of producer or consumer. The configuration may also be present in a partial or attenuated form.

I will follow this idea up by looking at some work in psychology, useful here because it seems to be a promising case for both representationalism and Millikan's producer–consumer model. Earlier I mentioned philosophers' description of beliefs as "maps by which we steer." Psychologists also talk of inner maps, especially when studying navigation. This work has been discussed rarely by philosophers, but it provides a useful alternative empirical angle on the idea of internal representation, distinct from "language of thought" views influenced by linguistics and classical AI.[14]

Psychological work on inner maps originates mainly with E. C. Tolman, especially in his "Cognitive Maps in Rats and Men" (1948). This paper was part of an attempt to show that "stimulus-response" models are inadequate for both kinds of animal. In Tolman's discussions, evidence for an inner map is behavioral. Experiments show that rats can be smarter in dealing with space than a stimulus-response view allows. For example, in a "starburst maze" experiment, a rat first learns a highly indirect route to a food source, and is then presented with a range of novel paths that lead more or less directly to the food. Rats choose a nearly direct path much more often than chance would predict. This, for Tolman, shows that the rat stores information about its environment in a map-like representation that can later be used to work out how to behave.

Some years after Tolman, evidence was found that whatever the spatially smart rats were doing, the hippocampus was the likely part of the brain where it happened. The hippocampus contains "place cells," which fire when the animal is in a particular location, regardless of the animal's orientation (O'Keefe and Nadel 1978). Neurobiological work has continued. A feature of this work, as I understand it, is that a picture has *not* emerged in which there is a separation between an inner map and a "reader" of the map. Maybe such a picture will emerge, but it has not so far. One alternative is that the important neural structures have map-like and reader-like features mixed together. "Place cells," for example, are often described informally in this mixed sort of way – as representations of where the animal is, which may be read, and as entities which have done some reading and know where they are.

I will look closer at the issue with the aid of a computational model of how a system of mapping of space might work in the brain, a model due to Reid and Staddon (1997, 1998). Their aim is not to model how the map is formed, but how a map can be used by the rat, once it exists.

The model assumes a connected two-dimensional lattice of units. These units are a bit like neurons, or might better be seen as small assemblies of neurons. Each unit is connected to eight neighboring units, its N-S-E-W neighbors plus diagonals. Each unit is also connected to sensory input and motor output. At any time, just one unit in the lattice is *active*, and the others are *inactive*. For each unit there is a unique sensory stimulus that makes it active. This amounts to the animal's ability to track its present location in the environment. Adjacent units in the lattice are cued by their stimuli to adjacent locations in space.

Each unit in the lattice has at each time a value of a variable V. As time passes, the value of V for all the inactive units changes by a kind of "diffusion." Each unit is affected by its neighbors in a gradual ongoing averaging process. If a low-value unit is between a lot of high-value ones, it will approach their higher value and they will be slightly affected by its lower value. There is an exception to the diffusion rule. If two neighboring units correspond to locations that are separated by a barrier that the animal has found that it cannot move through, there is no diffusion between those units.

The *active* unit at a given time has its value of V set differently. The animal is either rewarded or not rewarded at each time step. If the animal is rewarded, the V value of its active unit is set to a high value, and it is set to a low value if the animal is not rewarded.

---

[14] Exceptions include Bermúdez (2003), Rescorla (2009), and a short discussion in Millikan's VOM.

Lastly, there is a rule of movement for the animal. It moves at each time step in a direction determined by the neighbor with the highest value of V at that time step (with the proviso that movement cannot occur across a barrier). This is an entirely local process; the active unit at a time step sends a motor output determined by the V values of all its neighbors, and by nothing else. No inactive units emit motor outputs. If the active unit has a higher V than its neighbors, the animal stays put. So, for example, if at time t the active unit has a neighbor in the lattice to its northeast which has a higher V value than all the other neighbors and itself, the animal moves in a direction which makes that northeast neighbor the new active unit.

Reid and Staddon show how this simple process suffices to generate some quite smart navigation behaviors, including shortcut and detour behaviors, and solution of some maze problems (including the "starburst maze"). They show this with simulations in some cases and verbal arguments in others. The model assumes that an animal goes through an initial exploration or training period, and in a "trial" phase is released and allowed to look for food. The feature of the mechanism that generates apparent spatial "insight" is as follows. If an animal learns the location of food, by being present at that place during the training period, the corresponding unit will acquire a high V, which will then diffuse out from the unit corresponding to that location after training has ceased. This generates a "landscape" of V values that is affected by encountered barriers as well as the distance and direction of food. In the trial phase, the animal moves by hill-climbing from lower to higher values of V, leading it from a starting location to food. "Thus, Tolman's contention that something more than *S-R* [Stimulus-Response] principles is required to account for maze behavior is only partly true. A map *is* required (in our scheme), but beyond that, no 'insight', no internal cyclopean eye with an overview of the situation as a whole, is necessary" (Reid and Staddon 1997, 227).

Suppose that an evolved system did deal with space in the way described by Reid and Staddon. How would this relate to a scheme of interpretation for inner states based on a producer–consumer model like Millikan's?

The state of the lattice and its relation to the rest of the animal embodies the animal's knowledge of the spatial structure of its environment. That seems true. It is also true that in one sense, the lattice is a map of the landscape. *You* could look at the lattice in the brain and use it as a map; you could look at the lattice and read off the pattern of barriers, because the channels through which diffusion occurs are presently blocked or broken there. But, I will argue, there is no reader of the map, separable from the map itself, inside the animal.

There may be *a* kind of reading going on, at a low level. At each time-step, one unit is active. The active unit compares the V values of neighboring cells, and responds to these values by choosing a direction of movement. With the sensory stimulus that ensues, a new unit then becomes active. The new active unit then reads the V values of its own neighbors, leading to a new motor output.

I said just then that the active unit "reads" the Vs of its neighboring cells. That way of putting things might be denied – certainly it is not a clear case of the producer–consumer configuration. It is hard to say what the "producer" is, as the entire lattice and its history of reinforcement and exploration is responsible for the V values found at a unit at a time. Perhaps it is better to say that the units form an interacting network that achieves a kind of computation, but no sending and receiving of messages. (Or maybe messages exist only at an even lower level, where a neurotransmitter is an icon-like intermediary telling unit i what the present V value in unit j is.) But rather than making this a sticking-point, let's instead accept that there is a low-level reading of signs going on when an active unit tracks and responds to the values of V around it. Then the V values that are read by the active unit could be interpreted as having content involving environmental conditions. We might initially say that each V value that is read maps *the closeness of reward along a viable path*, at that time. That would not be right, though, as a given V value has no significance on its own, and what matters is the comparisons between the Vs of the neighbors. The overall level of V across the landscape becomes more similar as the animal moves further in time from its most recent reward. So it would be more accurate to say that the total set of V values of an active unit's neighbors maps the direction of a viable path to reward. Employing Millikan's schema, the active unit is supposed to respond to these V values in a way that will lead to the system doing well only if such-and-such *is* the direction of a viable path to reward.

So at this unit-to-unit level, a description in terms of the consumption of signs might be given. But the reading event that occurs at each time-step is sensitive only to the V values (and any barriers) in the immediate neighborhood of the active unit. Whenever the animal moves, a new unit becomes active. What was *part of the map* at step t – an element important only in virtue of its readable V value and sensitivity to the diffusion process – becomes *the reading device*, the consumer of information, at step $t+1$.

Reid and Staddon call one of their papers "A Reader for the Cognitive Map" (1997). But what they have described is more like a self-reading map, or a map that works without being read at all. The total set of V values in the lattice at a time is a map of where rewards are found in the landscape according to the animal's experience. Again, we as outsiders could look at the lattice and see not just where the barriers are, but also how to jump to food (find the high point which V is diffusing away from). But the animal can't do this. Only the presently active unit treats the local V levels as informative. That unit instructs a small step in some direction, and that unit then falls back into being part of the map as a new unit becomes active. The new active unit, which will be one of the former active unit's neighbors, will read the former active unit in turn. So there is lots of low-level *reading of V-values*, by readers that are separable from the things being read, but there is not in the same sense a *reader of the map*.

Reid and Staddon's is a "how-possibly" model. It illustrates one way that things might turn out in the explanation of how brains work, and this includes interesting possibilities with respect to the relation between personal and subpersonal levels. Looking at the rat and its behavior, we initially think: there is a map of the landscape in there, which has a reader, and that is how the animal knows where food is. Then we imagine finding a Reid–Staddon mechanism inside the rat. We find there is a low level at which a producer–consumer structure can be seen, though of an unclear kind. The contents of the inner icons we recognize there *might* also be seen as corresponding to belief-like states of the whole animal – as things the animal thinks or knows. But the organism-level contents we thought we were analyzing – knowledge of the layout of the environment, knowing where the food is – are not being found. There are no icons in the system with contents like that. The Reid–Staddon model posits a map that *would* be readable, if a suitable reader was present. But the rat cannot read the map in the way we can. (In this respect the Reid–Staddon map is like the "somatosensory map" in the human cortex.)

On seeing such a mechanism, we might decide it was a mistake to say the animal knows the layout of the landscape, and knows where the food is. The animal does not know anything like that, we might decide, but only knows at each moment the direction of the best path (which may be different from the direction of food). We might instead decide that our initial attribution was not *wrong*, but move to a more instrumentalist or behaviorist treatment of attributions of these belief-like states. Or we might say that in virtue of all the facts about how various subpersonal signal systems work, along with much else, the rat really does have the whole-organism beliefs we initially attributed. Then the producer–consumer model will be telling us about subpersonal contents that provide a *basis* for the whole organism's intentional states, even though the contents of those states are not the contents of any inner signs.

The picture illustrated in a simple and mild form here might turn out to apply more generally. This is a picture of cognition in which there is much low-level natural signaling – perhaps with leakage of sign into consumer, perhaps with unclear producers – but with no straightforward relation between subpersonal sign contents and the agent's intentional states. Or maybe that is generally true, but with an important exception: cases where the inner sign is an internalized public language sentence which functions in cognitive processing.[15] In those cases, an inner sign does become a vehicle by which an agent can have thoughts with a particular sophisticated content, and where the sign has the same content as the thought.

An analogy might be drawn with another controversial realization of the sender–receiver configuration, the case of genes. Protein synthesis within a cell is a process in which a kind of "reading" of the genes does go on, along with a quasi-logical regulation of gene action by regulatory mechanisms. This, again, is a questionable case from the point of view of section 2, but let's accept it here. It is also true that a lot of highly coordinated gene-reading of this kind, taking place within many different cells, results in the growth and development of an entire organism. But there is no reader of a "genetic message" at the whole-organism level. If there is a kind of content in the messages being read, it is restricted to the specification of the primary structure of protein molecules. Nothing is reading a specification of, or instructions to make, a hand or a kidney. We could have had a Grand Central Reader of the genome but we do not.

The sender–receiver configuration is a "natural kind," one that Millikan has done a great deal to help us understand. This kind is found in a variety of forms, arising independently many times. We have some account of how it comes to exist and is maintained. Our investigation of these set-ups is also affected, perhaps both positively and negatively, by habits of description and interpretation that derive from the place that sign systems have within everyday life. Among the many further questions that might be considered are those about when the clear, paradigmatic forms of the configuration can be expected to arise, and when instead the attenuated and blurred ones will, as a consequence of the role being played by the set-up, its circumstances, and the raw materials being fashioned into its various components.[16]

---

[15] See Dennett (1991), Carruthers (2002), and Yegnashankaran (2010).
[16] I am grateful to Rosa Cao, Ruth Millikan, Nick Shea, and Brian Skyrms for discussions and correspondence.

Reply to Godfrey-Smith

Peter Godfrey-Smith raises fascinating questions about the impact that the discovery of certain kinds of inner representations might have on our ordinary ways of thinking about knowledge at the personal or whole-animal level. Surely he is right that these ordinary ways of thinking might be – indeed, I am morally certain they will be – severely challenged given enough information about how brains actually work. But I will fuss just a little about the route Godfrey-Smith has taken to his conclusions. I will offer a word about "familiar intuitions about content," a word about the difference between what one would normally think of as a "sender–receiver configuration" and what I had in mind as a "producer–consumer" system, and a word about "consistency testers."

I think it important for understanding how nature really works in those areas where we are inclined to talk about "information," "signs," or "representations" – in the area where philosophers are inclined to use the term of art "content" – to recognize a rather sharp distinction between natural informational content and intentional content. If we mix these two up, we will fall into all kinds of deep confusions, not about words but about the phenomena we want to understand. What makes it super easy to mix these two up is the fact that, in many cases, not only will a single sign token carry both kinds of content, the two kinds of content will coincide. What will make the sign have this content as natural informational content will not, however, be the same as what will make it have this content as intentional content.

Natural informational content is usually thought of as a function of current causal or statistical relations between signs and the things they signify. There are various ways one might try to carve out useful notions of natural informational content. I have suggested one in chapters 3–4 of VOM. Brian Skyrms works with another, offering a description of various ways in which the operations of natural selection can produce organisms that emit signs carrying natural information, both information about the way the organism's world is and about what a cooperating organism will do in response to the sign. I won't comment

Millikan and Her Critics, First Edition. Edited by Dan Ryder, Justine Kingsbury, and Kenneth Williford.
© 2013 John Wiley & Sons, Inc. Published 2013 by John Wiley & Sons, Inc.

on Skyrms' analysis here.[1] I want only to distinguish it from a different kind of analysis, an analysis of (what it seems sensible to call) "intentional content."

Intentional content might be thought of very crudely as purposeful content, blurring together two meanings of "intentional." It is not defined by reference to any current causal or statistical tendencies. It is defined by reference to the purposes of sign producers and sign consumers rather than to their actual achievements, these "purposes" being analyzed as what the producers and consumers were selected for doing by some means of selection (genic, cultural, learning). That these two kinds of analysis of "content" have completely different subject matters should be obvious from the fact that one exclusively concerns what is current, the other concerns only history. However (as explained in chapters 5–6 of VOM), there are good reasons why these two kinds of content often coincide for a given sign token. They will often coincide, for example, in the cases Skyrms discusses. Thus Godfrey-Smith notes that what Skyrms treats as natural information carried by a sign about what the signal receiver will do corresponds to the content of a directive – on my view, the intentional content of a directive-intentional icon.

So there is more than one way that a sign can be "about" something signified, and some signs are "about" something signified in more than one way. But Godfrey-Smith suggests that my description of intentional content covers cases in which one would not normally suppose there was any aboutness at all. These are cases in which the supposed sign, rather than being an effect of its supposed signified, is a second effect of a common cause. I think this can't, however, be a general problem. Lightning may be a sign of thunder and of rain; my minute minder's ding is a sign that the spaghetti is *al dente*; that the tomato is green is a sign that it is also bitter; and I see, that is, I perceptually represent, that the ball will land in left field. All of these signs concern effects of common causes. We could not have beliefs about the future if there were no intentional signs (Normally) produced by means of common causes. There are also signs that have no causal relation at all to their signifieds. The way the compass points is a good sign of which way is North, but there is no causal connection. (In connection with Godfrey-Smith's discussion of this, we might note that Pietroski's tale of the snorfs and the kimus is an interesting example of inappropriate ascription of subpersonal content to the whole animal, to which Godfrey-Smith calls our attention in his last paragraphs. Pietroski's tale also misleads in several other ways, as described in OCCI, Appendix B; cf. Millikan 2001.)

The sender–receiver image, maybe also the producer–consumer image, has a tendency to distort the nature of the phenomena central to intentionality in suggesting two separate entities passing a third thing between them. I intended the producer–consumer image to be read much more abstractly. I thought of producers and consumers as mechanisms or systems. Systems can overlap in their components and may operate in several capacities even at the same time. Producers (of descriptive intentional icons) are mechanisms or

---

[1] As Godfrey-Smith has mentioned, I have doubts about the use of probability notions in this connection (VOM, ch. 3; Millikan 2007). First attempts at an entirely new view of natural information can be found in Millikan (2012; forthcoming a).

systems having as proper functions to produce specified kinds of states of affairs that map by rule onto certain other states of affairs. (Intentional icons are not properties but little states of affairs, VOM, chs. 3–4). Consumers are mechanisms or systems for which these produced states of affairs supply initial conditions for operation. If the produced states of affairs (the icons) don't map as they should, the consuming mechanisms will not be able to serve their proper functions in the historically Normal way. Nothing rules that the consumer of some icons is not also the producer of others (consider inference mechanisms, "information processing"), that the producer of an icon is not, in another capacity, also its consumer, or that the state of affairs that is the icon is not a state of the system itself. Certainly there is no suggestion of a "grand central reader." Reid and Staddon's model that "reads" its maps by causing the animal nearsightedly to clamber up just one more small gradient at a time (this is how I follow driving directions) is as well-formed a consumer as any other. But Godfrey-Smith's warnings about the bad fit between the contents of many kinds of inner intentional icons and ordinary talk about what an animal knows is dead on and important. The animal is not a homunculus that reads its inner representations.

Last, I must defend my poor misunderstood consistency testers! Consistency testers form part of a theory of the nature of empirical concepts and of empirical concept formation that rests on the general biosemantic position but is not implied by it. Their home is very far from the kinds of pushmi-pullyu signals and icons that Godfrey-Smith mostly examines here. They are needed for the development of "theoretical concepts," ones that may never have been put to any practical uses (see VOM, chs. 18–19). An important difference between natural informational content and intentional content is that signs bearing intentional content must have consumers with proper functions, for which functions there must be Normal explanations, which Normal explanations must have been determined by a history of usage. Theoretical concepts may never have been used in any practical ways, so how could the beliefs they participate in have intentional contents?

Empirical concepts, I have claimed, paradigmatically consist, in part, of capacities to recognize the same distal object, property, kind, and so forth, through a variety of different sorts of proximal manifestations presented to one's various senses at different times. The capacity to keep track of the same thing despite its diverse proximal effects on one's senses, not confusing it with other things and so forth, is a Herculean task, an accomplishment that must somehow be achieved by the cognitive mechanisms in their role of representation producers. Since we humans, at least, are not born possessing most if any of our empirical concepts, the question arises how these conceptual skills are learned.

One way empirical concepts can be tested (or evolved), I have suggested, is through practical experience. If I am right that it is the same thing I am encountering again, then behaving toward it in the same way should achieve the same result, for example, the same reward. But it is also true that the more mistakes one makes in identification, the more one will be liable to contradict oneself in judgment, making consistency in thought very difficult to achieve. Turning the coin over, consistency in judgment serves as another strong test of conceptual adequacy, and it can apply also to concepts that have not yet found practical uses. Reliably consistent judgment is evidence that our empirical concepts are succeeding in their reidentifying tasks.[2]

Now Godfrey-Smith's idea that the consistency testers are "interpreters" for beliefs has something right about it, although these certainly are not the only or final interpreters of beliefs. The most basic function of beliefs requires them to be taken up in processes of inference eventually leading to the guidance of practical activity. But it may be that most human beliefs never achieve this. (Recall that proper functions are not, as such, functions usually served.) How do beliefs containing concepts that have never been used in practical activity acquire Normal explanations for proper function, hence semantic values? Granted that the functions of beliefs are compositional (LTOBC, ch. 4, Appendix A), as are the beliefs themselves, beliefs containing only concepts that have been acquired through practical activity do not pose a problem. Their proper functions and Normal explanations for proper function are derived from past uses of various whole beliefs or other intentional attitudes that contained them as functional parts in helping to govern successful behaviors. But purely theoretical concepts have been selected only because, so far, they have helped in making stable judgments, in fixing beliefs that were not contradicted. The consistency testers have been their only actual consumers. Since helping to pass consistency tests is what they were selected for, however, passing consistency tests, helping to make stable beliefs, is for them a proper function. What judgments containing these concepts mapped onto consistently, explaining why they were able to pass consistency tests, is what determines the semantic contents of these concepts.

An analogy will make this clearer. Sweet tastes are natural reinforcers for human behaviors. The reason, we assume, is that sweet tastes usually indicate calories, which are important for human nutrition, so a taste for sweets was selected for during our evolutionary history. The basic biological function of liking hence reaching for sweets is intake of calories. On the other hand, through operant conditioning, natural reinforcers tend to increase the frequency of the behaviors that have produced them. They constitute part of a secondary selection mechanism that rides piggyback on genic selection. Focusing on this level of selection, the function of reaching for sweets – that which this behavior has been selected for – is getting a sweet taste into one's mouth. The function of having a mechanism with one function – procuring sweet tastes – is that it should ultimately serve another function – enhancing nutrition. But, of course, the first function can be served without the second (say, in reaching for the saccharine). Similarly, the function of the secondary selection mechanism that is the consistency testers is to hone conceptual tools for use in making focused maps of the world, the function of which maps is the guidance of action in the world. Lots of these maps are like having saccharine in your mouth, however, in failing to serve their root biological function (though unlike with saccharine, there may be no reason in principle why they could not.)

---

[2] One thing more is required. These concepts must sometimes be employed in intentional representations that are subject to a negation transformation, which means that they must have subject–predicate structure (LTOBC, ch. 14). All this is properly spelled out most clearly, perhaps, in VOM, chs. 18–19.

3

Millikan's Isomorphism Requirement

NICHOLAS SHEA

Introduction

Millikan's theory of content purports to rely heavily on the existence of isomorphisms between a system of representations and the things in the world they represent – "the mapping requirement for being intentional signs" (VOM, 106). This chapter asks whether those isomorphisms are doing any substantive explanatory work. Millikan's isomorphism requirement is deployed for two main purposes. First, she claims that the existence of an isomorphism is the basic representing relation, with teleology playing a subsidiary role – to account for misrepresentation (the possibility of error). Second, Millikan relies on an isomorphism requirement in order to guarantee that a system of representations displays a kind of productivity. This seemingly strong reliance on isomorphism has prompted the objection that isomorphism is too liberal to be the basic representing relation: there are isomorphisms between any system of putative representations and *any* set, of the same cardinality, of items putatively represented. This chapter argues that all the work in fixing content is in fact done by the teleology. Deploying Millikan's teleology based conditions to ascribe contents will ensure that there is an isomorphism between representations and the things they represent, but the isomorphism "requirement" is playing no substantive explanatory role in Millikan's account of content determination. So an objection to her theory based on the liberality of isomorphism is misplaced.

The second role for isomorphism is to account for productivity. If some kind of productivity is indeed necessary for representation, then functional isomorphism will again be too liberal a constraint to account for that feature. The chapter suggests an alternative way of specifying the relation between a system of representations and that which they represent which is capable of playing an explanatory role in accounting for Millikan's type of productivity. In short, the liberality of isomorphism is no objection to Millikan's

---

Millikan and Her Critics, First Edition. Edited by Dan Ryder, Justine Kingsbury, and Kenneth Williford.
© 2013 John Wiley & Sons, Inc. Published 2013 by John Wiley & Sons, Inc.

teleosemantics, since the isomorphism "requirement" need play no independent substantive role in Millikan's account of representation.

Isomorphism and Functional Isomorphism

An isomorphism is a mathematical function or mapping between two sets of items. The items in these sets can be concrete or abstract. A function takes each element in the *domain* to an item in the *range*. So the mapping from people to their mothers is a function. Mother(x) takes each person to the one person who is their biological mother. If every item in the domain gets mapped to a different item in the range, then the converse map, taking each element in the range back to its corresponding element in the domain, will also count as a function. The map from adults to their social security numbers is like that. SocSec(x) takes adults as input and delivers for each a unique social security number. The inverse map takes as input social security numbers that are in use, and for each delivers a unique person.

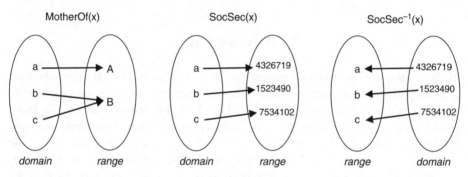

Figure 3.1

We can use f^{-1} for the inverse of f. So SocSec^{-1}(x), taking as its domain the social security numbers that are actually in use, is a well-defined function. Contrast Mother(x). Its inverse is not a function, since it maps some people to more than one person (it maps my mother to all my brothers and sisters and myself).

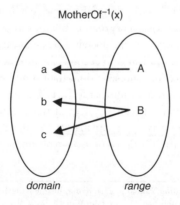

Figure 3.2

When a function has an inverse that is also a well-defined function, notice that we have a correspondence relation. Every element in the domain corresponds to a unique element in the range, and vice versa:

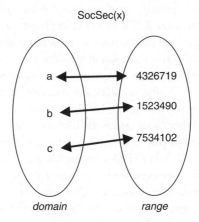

Figure 3.3

At its simplest, an *isomorphism* is such a 1–1 correspondence between two sets of items.

Whether there is such an isomorphism will depend on what we are taking the domain and the range to be. Consider my son's parent and child music class in which each parent brings only one child and vice versa. Defined on the domain of children in the class, the parent of relation picks out a function onto a range consisting of adults enrolled in the class. The inverse is also a function, from adults enrolled in the class to children in the class (but would not be if a parent happened to bring more than one child). That sets up a 1–1 correspondence or isomorphism between the parents and the children.

These are the standard kinds of examples used to introduce the concept of an isomorphism. They can be misleading in an important way. In each, the isomorphism described corresponds to some real relation: the natural relation of being a parent, the society based relation between people and unique social security numbers. But an isomorphism need not correspond to any natural or systematic relationship at all. Any way of lining up each element in the domain with a corresponding unique element in the range defines an isomorphism. So when there is one isomorphism between multiple items, there will always be many others too. If we were to pair up each child in my son's music class with a different adult chosen at random, that too would be an isomorphism:

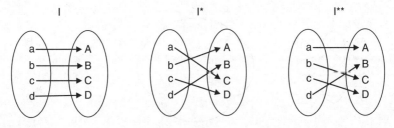

Figure 3.4

The crucial thing to notice is that there *are* all these isomorphisms – all at once, we might say. If we allow ourselves to talk about the *existence* of an isomorphism (some kind of abstract mathematical object), I, I*, and I** all exist (without the adults and children in the class having to change in any way). There is, indeed, something special about isomorphism I. It corresponds to the natural relation of parenthood. But as an isomorphism – a mathematical object – I is absolutely on a par with I*, I**, and all the many other isomorphisms between the two sets.

The term 'mathematical function' can mislead in a similar way. Familiar examples of mathematical functions do something regular and systematic to the elements they are defined on. They are functions like $x + 10$, $7x$, x^2, and $\cos(x)$. The map from 1, 2, 3, 4 to 1, 4, 9, 16, respectively, is a mathematical function (squaring), but the map from 1, 2, 3, 4 to 76, $\sqrt{2}$, $-\frac{5}{8}$, Π^e, respectively (chosen so that there is no systematic relation – that I can see) is just as much a mathematical function as is squaring. Indeed, the mathematical concept of function equally includes the mapping from 1, 2, 3, 4 to Tibbles the cat, this orange on my desk, the sun and the Queen. That mapping, which also happens to be an isomorphism, exists alongside squaring in the abstract realm of functions. Something more than function talk is needed if we are to say why squaring is more interesting in some way. So the question of whether *there is* an isomorphism between two sets of items is rather straightforward. If the two sets have the same number of items (or the same cardinality, to include transfinite sets), then there will be an isomorphism between them. And if there's one, there will be many isomorphisms between the two sets.

Before turning to Millikan's theory, we need one final, crucial piece of mathematical machinery. So far, we have been considering just unstructured sets of elements in the domain and range, in which case any 1–1 map is an isomorphism. But the more useful and important concept of isomorphism is that of a structure-preserving map. Once we consider the sets of elements in the domain and range as being endowed with some structure, we have to ask whether those structures also correspond under I. Unlike the bare existence of a 1–1 map, if we have two structures in mind, it is a very demanding constraint that there should be a structure-preserving map between them. The structures in question may be relations between or operations on the elements, for example. Philosophy has paid particular attention to "functional isomorphisms" – where the "function" is for mappings between elements *within* the domain of the isomorphism, and for corresponding functions between elements within the range of the isomorphism. Consider again my son's music class, with parent–child pairs sitting around in a circle. There is a function f within the children that takes each to their left-hand neighbor (the range and domain of f are the same set – the children in the class). And there is a corresponding function g within the adults, taking each parent to the parent on their left:

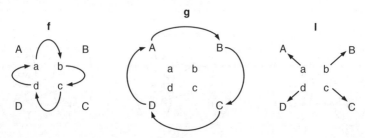

Figure 3.5

So here we have not only a 1–1 correspondence between parents and children, I, but also a correspondence between the functions *f* and *g*: under I, a maps to A, and the child to the left of a, *f*(a) (i.e., b) maps to the parent to the left of A, *g*(A) (i.e., B). Or to put it more formally:

$$I(f(x)) = g(I(x))$$

In that case, we say there is a *functional isomorphism* between function *f* on the children and function *g* on the parents.

Recalling the liberality of the mathematical concept of a function, it will be obvious that for every function defined on the elements of the domain of some isomorphism I, there is a corresponding function on the elements of the range of I. We can use *f* and I to generate such a function *g* as follows. For each y in the range of I we use I⁻¹ (the inverse of I) to see which element is mapped to y (e.g., to see which child belongs to that parent), we perform the given function *f* on that element (e.g., *f* takes us to the next child to the left), and then we use I to map that element back to the corresponding element in the range of I (the parent of the child to the left of [the child of the parent y we started with]). To give a graphical example, we can see that *g* will take element A to element B:

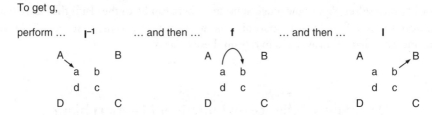

Figure 3.6

In short, we can use *f* and I to define the corresponding *g* as $g(y) = I(f(I^{-1}(y)))$.

Given the liberality of functions, whenever there is an isomorphism between two sets, there will be very many functional isomorphisms between them. And for any given function on the domain of I, there is some function on the range of I to which it is functionally isomorphic (also if we start with a function on the range). So it is a rather trivial question to ask whether there is *some* function on a range of elements D (for example, on a set of putative representations) which is functionally isomorphic to some function of interest defined on a domain of elements R (potential representeds), because there will be such a function iff there is an isomorphism between D and R, which there will be iff they have the same cardinality. We can, however, ask a more substantive question with respect to a function *f* on D *and* a function *g* on R, namely whether there is a functional isomorphism between those particular functions *f* and *g*. So we might ask whether a particular function of interest on D (e.g., the one given by the left-of relation) is functionally isomorphic to a particular function of interest on R (e.g., the one given by the left-of relation) under some 1–1 map between D and R (as there is: the parent-of function is a 1–1 map that preserves the rotational structure – it takes *f* to *g*). It is by no means guaranteed that the left-of function on the children corresponds to the left-of function on the parents. (Notice that if there were such a correspondence for some natural reason, then it would be something we could make use of, e.g., to find out which adult is to the right of A by checking to see which child is to the right of a.)

Whereas the existence of an isomorphism between sets of elements bare of any structure is a very liberal requirement, the existence of an isomorphism between two sets of elements endowed with structure is a very demanding requirement. The entities in mathematics are exhaustively characterized by their roles in structures. Accordingly, proving that two sets of such elements are isomorphic is a very strong result, because it means that everything true in one system is true in the other – there are no mathematically relevant differences between the systems. The liberality that arises when deploying isomorphism talk in the field of theories of content arises, to a first approximation, because the relevant structure on the set of elements represented is usually unspecified.

This careful walk through the mathematical concepts should give us the tools to be able properly to assess Millikan's claims about the explanatory work that can be done by isomorphisms and functional isomorphisms. The main lesson so far is that they are exceedingly liberal relations. Paradigm cases in which isomorphisms correspond to natural relations and functional isomorphisms subsist between naturally defined functions are prone to mislead us about how much substance there can be to the claim that there *is* an isomorphism or a functional isomorphism between two sets. In the next section we will examine ways that Millikan seeks to rely on these concepts.

Millikan's Reliance on Functional Isomorphism

Millikan requires there to be an isomorphism between a system of representations and that which they represent. This section sets out the explanatory work that Millikan claims her isomorphism requirement can perform. The central question for the chapter is whether the isomorphism requirement acts as a substantive constraint within Millikan's theory. Millikan relies on it in two main ways. First, it supports the claim that representational systems display productivity. Second, Millikan claims that a particular isomorphism between representations and representeds plays a role in explaining why the representations in a given system have the contents they do. This section sets out those claims.

Millikan most frequently turns to functional isomorphisms to account for productivity. She argues that representations must display productivity to be genuinely intentional. The kind of productivity she has in mind arises when representations come in systems, with transformations between elements in the system corresponding to systematic transformations of that which is represented. For example, in the honeybee's nectar dance, the angle of the dance to the vertical corresponds to the direction of nectar in relation to the sun. So a transformation of the dance, 10° further away from the vertical, say, corresponds to a transformation of the represented location, 10° further away from the direction of the sun. The argument is (i) that there are such functional isomorphisms; and then (ii) that they give rise to a kind of productivity. The first statement is in *Language, Thought and Other Biological Categories*:

> When an indicative intentional icon has a real value, it is related to that real value as follows: (1) The real value is a Normal condition for performance of the icon's direct proper functions. (2) There are operations upon or transformations (in the mathematical sense) of the icon that correspond one-to-one to operations upon or transformations of the real value such that (3) any transform of the icon resulting from one of these operations has as a Normal condition for proper performance the corresponding transform of the real value. (LTOBC, 107)

The claim is that there will always be a 1–1 correspondence (isomorphism) between a function on the system of representations (the domain) and a function on the worldly affairs that are represented (the range). That is the basis for a kind of productivity. Millikan treats compositionality as a species of productivity. Compositionality arises from replacing parts of a representation: if I can think < John is tall > and I have the concept MARY, then I can think < Mary is tall>. Millikan sees variant aspects of the representation playing a similar role: the angle of the bee dance changes, allowing different dances to represent different directions. The representation "... is not articulated into parts but into invariant and variant aspects" (LTOBC, 107). The bee dance always says *nectar at ...* – that much is invariant – but different dances say the nectar is at different distances, and in different directions. Millikan is strongly committed to the idea that this is genuine productivity, on an equal footing with the more obvious productivity of natural language sentences (LTOBC, 108). Whether she is right about that is a matter for another day. For now, the point is that an isomorphism is claimed to underpin a particular feature of systems of representations: that relations between the representations correspond to relations between the items represented (a relation-preserving isomorphism). That claim recurs throughout Millikan's work (LTOBC, ch. 5; 1989a, 287; VOM, ch. 6; LBM, ch. 5).

Between which items is this isomorphism supposed to subsist? In the domain, are we dealing with actual bee dances (for example) or merely potential bee dances? And in the range, do we find actual locations of nectar, or rather potential or represented locations? It is clear that we are not dealing with a correspondence between actual dances and actual locations, since Millikan has designed her theory to allow for misrepresentation, so there will be instances when there is no nectar at the location that an actual dance is supposed to correspond to. We could restrict our attention to the types actually found in the history of the species: dances danced and nectar locations on occasions that led systematically to survival and reproduction. These are the cases that serve to fix the correctness condition of the various kinds of dance. But Millikan wants dances that happen not to have occurred at all in the evolutionary history to count as instances, provided they fall within the general overall pattern. Perhaps, by chance, no dance was ever danced at 84° to the vertical. Still, that potential dance has an associated truth condition, given by the general rule that angle of dance to the vertical maps to direction of nectar with respect to the sun. For these reasons, we should think of the relevant isomorphism as subsisting between *potential* dances and *potential* locations of nectar. And the structure preserved is rotation: the operation of rotation on the angle of potential dances corresponds to the operation of rotation on the potential direction of nectar. That defines a functional isomorphism from {potential dances with rotational structure} to {potential directions of nectar with rotational structure}.

This way of understanding the domain and range of the isomorphism underpins the reason why Millikan sees variant structure in the domain of representations as a kind of productivity. The isomorphism specifies a way of giving the content of a new dance, not previously danced in the history of the species. That is, once the relevant correspondence relation between potential dances and potential locations has been established, that correspondence relation may assign contents to new representational items. A question remains about how such a correspondence is established. It looks as if a theory of content is needed to determine what is the represented location that corresponds to each of the putative representations (dances). We will see below that relational proper functions may establish correspondence relations that extend to genuinely new cases. What Millikan's isomorphism requirement adds, then, to the mere existence of a correspondence relation established by the theory of content, is that there should be some structure over the representations that recapitulates some structure over the items that are represented. In examples like the bee dance, a natural relation on the system of representations (rotations of the dance) is mapped to a natural relation on the items represented (rotations of the direction of nectar). But, as we saw above, the mathematical concept of isomorphism does not bring with it a restriction to natural functions or relations. Once a theory of content delivers a correctness or satisfaction condition for each representation, there will be lots of functions (in the mathematical sense) on the representations that correspond to functions on the items represented, and vice versa.[1] So it doesn't look like the requirement that there be a functional isomorphism can be doing any explanatory work. Any adequate theory of content delivers a 1–1 map between representations and the potential states of affairs they represent, and that map will preserve *some* structure on the elements, and so count as a functional isomorphism.

The existence of such isomorphisms may have more bite if we restrict our focus to certain kinds of structure on the potential representations/representeds. Many of Millikan's examples are ones where natural relations on the representations correspond to natural relations among the individuals or properties represented. I take up below the question of the explanatory work that may be performed by such a restriction (section 4).

So far, we have raised the suspicion that the isomorphisms that Millikan says underpin a kind of productivity are no more than a necessary consequence of the fact that a theory of content sets up a correspondence between potential representations and truth conditions. This is where Millikan's second role for isomorphism comes in, because she claims that mapping functions enter into an explanation of why representations have the content

---

[1] Strictly, the relation picked out by the theory of content will not be an isomorphism unless the theory assigns a unique content to each different representation (otherwise, the inverse would not be a well-defined function). To deal with cases of redundancy – different representations with the same correctness or satisfaction condition – we need to generalize to include *homomorphisms*.

they do (LTOBC, 99–100, 246; 2000b, 6;[2] VOM, ch. 8; LBM, 102). That is, they are, so to say, prior to the theory of content, rather than the result of it.

Recall the structure of Millikan's teleosemantics (simplifying considerably). A behavioral system separates into cooperating producer and consumer subsystems, with the producer giving rise to a range of intermediates, which cause the consumer to behave in a variety of ways. Among all the ways in which a consumer might behave in response to a given intermediate R we focus on the types of behavioral response B that, in the evolutionary past, led systematically to the survival and reproduction of the system (i.e., that are part of a Normal explanation). Doing B is then a candidate for the imperative content of R, and worldly conditions that enter into a Normal explanation of why B leads to survival and reproduction are candidates for the indicative content of R. Millikan claims that isomorphisms (mapping rules) enter into that Normal explanation. One of the most concise statements in LTOBC concerns beliefs (although the point applies to representations in general):

> In order to ... [be] an intentional icon, the belief token ... must have as a proper function to adapt an interpreter device to conditions in the world so that this device can perform proper functions, and it must be part of the most proximate Normal explanation for the interpreter's proper performance that the belief – the inner sentence – maps conditions in the world in accordance with some definite mapping rule. (LTOBC, 146)

Here we have the idea of a mapping rule entering into a Normal explanation of the operation of the system. That is supposed to be prior to content determination. What makes it the case that representations have the content they do is that there is a Normal explanation that mentions a particular mapping rule. The consumer system acted on intermediates in a way that led historically to successful behavior, a mapping rule is part of an explanation of that success (LTOBC, 99–100), and in virtue of all of that, that mapping rule becomes the rule for delivering truth conditions or satisfaction conditions. There is an important distinction here between a representation actually *mapping* onto a condition and a *mapping rule* (a distinction that sometimes turns into an ambiguity in the use of the term). A *mapping rule* is an isomorphism between two sets of items. There being an isomorphism between two sets of items does not depend upon them being concurrent or co-present, or upon their standing in any other natural relation. (Indeed, we remarked above that the relevant isomorphism was between types of entities rather than tokens: between potential dances and potentially represented locations, for example.) By contrast, a representation *maps* some condition, in Millikan's terminology, when the representation is actually tokened and the condition actually obtains. For example:

> Intuitively it is clear that in some sense of "mapping," *the bee dance that causes* watching bees to find nectar in accordance with a historically Normal explanation *is one that maps* in accordance with certain rules *onto a real configuration* involving nectar, sun, and hive. As such it is an indicative intentional icon. The bee dance also maps onto a configuration that it is supposed to produce, namely, bees being (later) in a certain relation to hive and sun – that is, where the nectar is. So the bee dance is also an imperative intentional icon. (LTOBC, 99; emphasis added)

---

[2] Page references to Millikan (2000b) refer to the readily available version available in pdf format from Ruth Millikan's website at uconn.edu.

Millikan here uses "maps" for a relation obtaining between a particular dance and real nectar at an actual location. The properties that are connected by a mapping rule can be instantiated: a dance is performed in a particular way and there is nectar at a particular direction and distance from the hive where it is performed. When the types so instantiated are indeed connected by a mapping rule, Millikan says the dance "maps" the location. Whether a particular dance maps a particular location depends upon which mapping rule is in question. Is Millikan right to claim that the existence of a mapping rule is a substantive constraint on the theory of content?

That there should be such an isomorphism does not act, in Millikan's theory, as a substantive constraint on content. On occasions that led to survival and reproduction, particular dances were performed and nectar was found at particular locations. We can indeed generalize across those instances: for a dance at $\theta°$ to the vertical, there was nectar at $\theta°$ from the direction of the sun. That generalization does enter into a Normal explanation of the success of the behavior: operation of the producer–consumer system led to survival and reproduction when producer bees produced dances at $\theta°$ to the vertical and there was nectar at $\theta°$ to the hive. It is this natural relation between dance-types and nectar-location-types, not the isomorphism as such, that explains success (and hence gives the content).

There are two points here. The first is that it is not the isomorphism – the 1–1 mathematical function – that enters into the Normal explanation, but instead some natural relation between dance types and locations. The second is that it is the Normal explanation that picks out a particular isomorphism. It would get things precisely the wrong way round to claim that there being an isomorphism is a substantive constraint on the theory of content. Instead, Millikan's account of content delivers an isomorphism. It may be relatively benign to talk as if it is the isomorphism as such, rather than a corresponding natural relation, which explains successful behavior. But that leads to the second, more serious mistake, which is to claim that the existence of isomorphisms is part of what makes it the case that representations represent as they do – that they are prior to the theory of content.

Millikan sometimes writes as if isomorphism (or picturing) is the basic representing relation, and that teleology only comes in to account for error:

> … [N]aturalist theories of the content of mental representation are often divided into, say, picture theories, causal or covariation theories, information theories, functionalist or causal role theories, and teleological theories, as though these divisions all fell on the same plane. That is a fairly serious mistake, for what teleological theories have in common is not any view about the nature of representational content. "Teleosemantics," as it is sometimes called, is a theory only of how representations can be false or mistaken, which is a different thing entirely. Intentionality, if understood as the property of "ofness" or "aboutness," is not explained by a teleological theory. …

> … What teleological theories do not have in common is an agreed on description of what representing – what "ofness" or "aboutness" – is. They are not agreed on what an organism that is representing things correctly, actually representing things, is doing, hence on what it is that an organism that is misrepresenting is failing to do. To the shell that is "teleosemantics" one must add a description of what actual representing is like. When the bare

teleosemantic theory has been spent, the central task for a theory of intentional representation has not yet begun. Teleosemantic theories are piggyback theories. They must ride on more basic theories of representation, perhaps causal theories, or picture theories, or informational theories, or some combination of these. (VOM, 63, 66)

[Teleological] theories generally begin with some more basic theory of the relation between a true thought, taken as embodied in some kind of brain state, and what it represents, for example, with the theory that true mental representations covary with or are lawfully caused by what they represent, or that they are reliable indicators of what they represent, or that they "picture" or are abstractly isomorphic, in accordance with semantic rules of a certain kind, with what they represent. The teleological part of the theory then adds that the favored relation holds between the mental representation and its represented when the biological system harboring the mental representation is functioning properly.... (Millikan 2003b, 1)

Millikan relies on isomorphism or picturing as the basic representing relation on which her teleosemantic theory is based (LTOBC, 9, 11; 2003b, 3; VOM, 79). That has led to the criticism that the concept of isomorphism is far too liberal for it to be able to do any substantive explanatory work. We saw the reasons for that liberality in section 1. Godfrey-Smith provides the clearest statement of the objection (Godfrey-Smith 1996, 184–187).

However, a better interpretation is that Millikan is not taking the existence of a preexisting isomorphism to be a substantive constraint on the theory of content. Instead, the facts about Normal explanations pick out an explanatorily significant natural relation (e.g., between dances at $\theta°$ to the vertical and nectar at $\theta°$ to the sun), and this relation specifies a particular isomorphism I between representation types and properties represented. The fact that there is such an isomorphism does no explanatory work. (After all, there is an isomorphism between the system of representations and very many other ways of assigning content to them.) What enters into the Normal explanation is that there was nectar at $\theta°$ to the sun when a dance was produced at $\theta°$ to the vertical. That reading of Millikan is clearest from the following:

> Papineau and Millikan claim that it is only the uses to which mental representations are put that is relevant to their content. Millikan claims that a true representation maps onto its represented in accordance with semantic rules *determined by* the way the systems using the representation are designed to react to it in guiding, perhaps first inference processes, but ultimately behavior. (Millikan 2003b, 8; italics added)

That passage makes it clear that the relevant isomorphism ('semantic rule') is determined by the way consumer systems are designed to react to the representation. So the *existence* of isomorphism I plays no explanatory role. The existence of I is not part of what makes it the case that the representations have the content they do. But the theory of content will deliver a correspondence, giving a correctness or satisfaction condition for each representation in the system to which the theory applies. So this content assignment will indeed pick out an isomorphism, which Millikan calls the "semantic rules" or "mapping rules" for the system of representations. There is indeed an isomorphism or correspondence relation that gives the content of a system of representations. But it is wrong to suggest that the isomorphism is basic and the teleology is needed only to account for error. The bare existence of an isomorphism between a system of representations and potential contents they could represent plays no explanatory role.

Once we have dismissed the idea that the existence of an isomorphism is prior to or explains content determination, the only substantive role left for an isomorphism requirement is the first one identified above – to account for productivity. In the next section we examine that role.

Isomorphisms and Productivity

So far, we have seen that Millikan's theory of content will serve to identify a particular isomorphism between representation types and represented properties (/objects/states of affairs) as privileged – as giving the content of the representations (as will any theory of content, subject to the point about one–many mappings/homomorphisms). We have also seen that it is an automatic consequence that there will be operations on the domain of representations that correspond to operations on the range of representeds (indeed, there will be many such operations). We noted that the bee dance has an additional feature. There, the operations which are functionally isomorphic as between representations and representeds correspond to natural relations (rotation of dances and rotation of direction). Is it a further requirement on intentionality that the functions that correspond under the relevant functional isomorphism be ones that correspond to natural relations? Millikan is clear that it is not: "Isomorphisms can be defined by functions that are as bizarre, as gruelike, as you please" (VOM, 84).

That is what she should say. It follows from the fact that the relevant isomorphism ("semantic rule") is determined by the way consumer systems are designed to react to intermediate representations in guiding behavior. Suppose bee dances represent nectar at the distances given by I (on the left below):

I		I*	
1 waggle	300 m	1 waggle	75 m
2 waggles	150 m	2 waggles	300 m
3 waggles	100 m	3 waggles	60 m
4 waggles	75 m	4 waggles	150 m
5 waggles	60 m	5 waggles	100 m

An explanation of how the bee dance system led to survival and reproduction appeals to the fact that, in the historical past, consumers of dances of 3 waggles were disposed to fly off 100 m before searching for nectar, and such dances contributed systematically to survival and reproduction when there was nectar 100 m from the hive; similarly for each of the other variants. But had the consumer bees had the disposition to fly the distances given under I* above, then the bee dance system would have contributed systematically to survival and reproduction when there was nectar at the distances given by I* (provided producer bees ensured the number of waggles correlated with nectar at those locations often enough to be worth acting on). So I* would give the content. That is just to reiterate Millikan's point that the relevant isomorphism is determined by the way consuming systems are designed to react.

Of course, the set of dispositions given by I can be more compactly described: consumer bees were disposed to fly off a distance given by 300 m/# waggles. A natural relation among the dances (# waggles) corresponds to a natural relation among the locations (distance^{-1}). But I* is just as much a functional isomorphism as I is. The function f on the number of waggles corresponds to the function g on the distances, where f and g are given by:

f:
1 waggle → 2 waggles
2 waggles → 3 waggles
3 waggles → 4 waggles
4 waggles → 5 waggles

g:
75 m → 300 m
300 m → 60 m
60 m → 150 m
150 m → 100 m

Of course, g is not a very natural function; but it is just as much a function for all that. The example demonstrates why Millikan has to accept bizarre or gruelike isomorphisms. So the functional isomorphism "requirement" – that there must be operations on the representations that correspond to operations on the representeds – is not a substantive constraint on the class of isomorphisms that are admissible as giving the content of a system of representations.

When it comes to relying on an isomorphism to account for productivity, it seems that isomorphisms that have a natural structure like I have an advantage over more arbitrary mappings like I*. Cases like I arise because there is some mechanism that acts in a systematic way on the intermediate representations. So the function that describes I compactly – distance = 300 m/# waggles – may correspond to a single mechanism by which the system operates.[3] If so, the facts about selectional history will establish that the dance-producing mechanism has a *relational proper function* – to direct the consuming bees to a place at the distance given by the rule. Similarly for direction, dance producers have the relational proper function of causing consumer bees to fly in a direction corresponding to the angle of the dance. A particular dance has its own proper function, deriving from this relation – a *derived proper function*.[4] For example, a dance at 84° to the vertical has the derived proper function of sending consumer bees in the direction 84° from the direction of the sun. The particular dance has that derived proper function, whether or not a dance of 84° ever occurred in the evolutionary history. The upshot is that when a system of representations has a proper function arising from natural properties of the system during episodes of selection, then the contents of those representations may be expressed in terms of that

---

[3] It has been suggested that there is a straightforward explanation of why *incoming* bees first behaved in the corresponding way. The further they have come from the nectar, the less energy they have left when they reach the hive, so the fewer waggles they perform.

[4] It also has an *adapted proper function*, but we do not need that part of Millikan's terminology in what follows.

relation. And that does indeed give rise to a kind of productivity. Current instances which have not occurred in the history of selection may nevertheless fall under that relation, and so have derived proper functions, hence content. Notice that the list-type cases will give rise to relational proper functions, too. In the case of I* above, the relevant relational proper function is given by I*. But such list-type isomorphisms will not apply to any new cases. The application to new cases arises from there being something systematic in the natural operation of the system of producer–representation–consumer during the history of selection.

The bee-dance producer has two relational proper functions (the ones for direction and for distance). That results in further opportunities for productivity. Even if nectar at 84° is part of the selectional history, and nectar at 300 m is part of the selectional history, it may be that nectar 300 m away and at 84° happens not to be part of the selectional history. If so, such a dance will have a derived proper function, and a content, even though no instance of that dance was ever the basis of selection. That is a kind of productivity.

It also looks like a kind of compositionality. But it is important to notice that it is not compositionality in the regular sense. In conceptual thoughts and natural language sentences, the constituents make no claims on their own. They are unsaturated. For example, the predicate 'is tall' is perfectly meaningful, but it has no truth condition or satisfaction condition. Contrast the bee dance case. There, each relevant dimension of variation has its own truth condition. A dance of 3 waggles says that there is nectar 100 m away, irrespective of the direction in which it is performed. There are no separate unsaturated terms, like a subject and a predicate. For some purposes it will be important to distinguish between these cases.[5] Sometimes a system of representations contains two or more dimensions of natural variation, but each dimension has its own fully formed representational significance (truth, satisfaction, etc.). Multiple variant aspects build up a list of truth conditions, roughly as if they were terms in the propositional calculus connected by conjunction. In other cases like the conceptual constituents of a belief, the variant aspects Millikan describes are each unsaturated, and build up only together into a truth-evaluable thought, like terms in the predicate calculus.

To recap, we saw in section 2 that it would be a mistake to think that some isomorphism between representations and represented is part of an explanation of why representations have the content they do. Rather, a Normal explanation will advert to physical properties of the system of producer and consumer that make it the case that representations have the content they do – the theory picks out a special isomorphism, rather than being based on the existence of an isomorphism. In this section we saw that, where the selectional story supports a mapping rule that extends to new cases, there is a kind of productivity. But that works via relational proper functions arising from natural relations, rather than being based on the existence of some preexisting isomorphism. Finally, we saw that it is not built into the theory that bizarre or gruelike isomorphisms are excluded. The concluding section suggests that more natural isomorphisms may have a further significance.

---

[5] For example, the idea of separate relational proper functions applies unproblematically to the first case. It is less clear how it applies to unsaturated representational constituents.

Exploiting 1–1 Maps Which Preserve Natural Relations

Recall the contrast between isomorphisms I and I* above. Under I, a natural relation on the system of representations corresponds to a natural relation on the properties represented. Rotation of the dance away from the vertical corresponds to rotation of the direction of nectar away from the sun. Under I*, there is no such correspondence between natural operations. Given an appropriate consumer, I* would qualify as giving the content of the range of representations. But surely there is some natural advantage of I over I*? It looks less likely that there would be a consumer that reacts in accordance with I*. This section sketches a reason why that reaction may be justified.

A first reason is just that smooth or natural correspondences are more likely in nature. The consumer system has to react to a range of intermediate representations. Natural relations between representation-type and appropriate behavior may be easier to implement, just from an engineering standpoint. So although a consumer-based theory of content tells us what the contents would be if there were some gruelike connection between representations and behaviors of the consumer that figure in a Normal explanation, in natural cases we are more likely to come across natural relations. So, at the very least, there may be natural reasons to expect many cases to be like the bee dance.

Does the point run deeper? Isn't it some kind of achievement to produce a structure (a system of representations) that is isomorphic to things in the world? It took years of work involving many pieces of technology for the Survey of India to produce systems of representations (maps) that bear rich isomorphisms to portions of the Indian countryside. And having these artifacts was immensely useful in controlling and defending the territory. If isomorphisms are so cheap, why is it such an achievement to produce items that stand in such relations?

Although Millikan has emphasized the role of consumers in her theory of content, she also writes in places as if it is an achievement of representation producers that they give rise to a system of representations that are actually isomorphic to some set of things in the world (LTOBC 146; 1995, 288). We saw in section 2 that such isomorphisms are found in specifying the normal conditions for the producer to perform its function. That does not imply that producers actually do produce a system of representations that bear any interesting isomorphisms to properties of interest. From the discussion in section 1 it should be obvious that to produce a system of representations for which there was some isomorphism to relevant objects and properties in the world is utterly trivial. What more substantial thing might producers be doing? Godfrey-Smith talks about the producer–consumer framework as being vindicated when the producer system produces a range of states that bear some *exploitable relation* to features of the world that are relevant to the system (Godfrey-Smith 2006a). The mere fact that there is *some* isomorphism between the representations produced by the producer system and relevant properties in the world is not by itself enough. Isomorphisms are cheap, but most are not exploitable. Is there some special class of isomorphisms such that, to produce a system of representations bearing such an isomorphism to relevant objects and properties really would be an achievement, and the structure that resulted would thereby bear an exploitable relation to the world? The importance and value of cartography suggests that there is.

In the bee dance case, the consumer system is exploiting the fact that producer bees produce dances whose angle carries information about the direction of nectar (and about distance). Millikan sometimes claims that consumers are making use of the fact that representations are produced to carry information in something like Dretske's sense (LTOBC, 146; VOM, 79).[6] Subsequently, Millikan has developed the notion of "soft natural information" (OCCI, Appendix B; cf. Millikan 2001) or "local natural information" (VOM, ch. 3), generalizing Dretske's notion; and there are closely related notions in the mathematical theory of information, like mutual information based on Kullback–Leibler distance (Cover and Thomas 2006, ch. 2). I use the general term "correlational information" for all such relations, which trace back to Shannon's seminal work (Shannon 1948). Carrying correlational information is an exploitable relation in Godfrey-Smith's sense.

In the bee dance case the consumer is exploiting the correlational information carried by the dances it observes. The functional isomorphism between {rotation on dances} and {rotation on directions of nectar} is not being exploited as such. The gruelike isomorphism I* above would be just as good for consumers, if the correlational information carried by each dance was as reliable. So this is a case where, although there is an interesting isomorphism between natural relations on the representations and some relevant natural properties, that natural relation is not being exploited as part of what allows the consumer system to perform its functions.

In other cases, it may be important to see that an isomorphism really is being exploited, rather than just being a side-effect of the natural operation of the system. There is not space here to do justice to this idea, so a sketch will have to suffice. There are two elements to the picture. (1) What distinguishes the exploitable isomorphisms from those that are ubiquitous and cheap? (2) What is it for a consumer mechanism to exploit an isomorphism from this limited class? I have already hinted at an answer to the first question. When natural relations on a set of putative representations correspond to natural relations on a set of things in the world, then that is indeed something that a natural system can make use of. Any old isomorphism would be useful if the consumer knew the code. But correspondences between natural relations are easier codes to crack. Such a correspondence will be useful to the consumer if a natural relation on the representations is something that it can detect and respond to differentially with different output behaviors. And the natural items in the world to which the representations are isomorphic must be *relevant* to the consumer, in the sense of potentially making a difference to whether its behaviors are successful (ultimately a matter of survival and reproduction, in Millikan's teleosemantic framework).

This is the sense in which my map of Oxford bears an exploitable relation to the city of Oxford. The natural relation of 2D spatial position on the map is one that I can easily read and make use of. And this readable relation corresponds to properties of the

---

[6] But it is no part of the theory that representations must carry correlational information (1995; VOM, 67–68; 2007).

world that are relevant to the success of my behavior, namely the spatial relations between places in the city. I exploit those relations whenever I use the map to calculate a route. Suppose I am inclined to trace routes on the map, measure them, and take the shortest. Then I am making use of the fact that the shortest route in centimeters on the paper will correspond to the shortest route in meters on the ground. A map can also be rigged up to carry correlational information, of course. That happens when a car satellite navigation system uses GPS to plot your current position on a map. But online correlational information cannot be the whole story of how a map is used to calculate a new route.

In short, an isomorphism between readable natural relations on a system of representations and real world items that are relevant to the consumer of those representations (a "natural isomorphism") may constitute an exploitable relation, and may actually be exploited by consumer systems in some cases. If that is right, then there are indeed some representing systems for which the existence of an isomorphism – of this much more tightly constrained sort – plays a substantive role in the theory of content.

A final caveat is that these natural isomorphisms need not be exact. It does not make a map useless that one of the points on it is slightly out of place. And any map will be imperfect at some level of detail. Some of these inaccuracies may be accounted for by individuating the isomorphisms at an appropriately coarse level of grain. But there may be other cases where we need a notion of the isomorphism itself only holding approximately. That suggests that the mathematical concept of isomorphism may not be the most useful explanatory tool here. It may be more useful to treat the system of representations as a model of its target (Godfrey-Smith 2006b, Weisberg 2007). Notice that this point about replacing isomorphisms with models does not arise in the discussion above, where the consumer fixes the isomorphism, since it is facts of the matter about the evolutionary story that deliver the appropriate isomorphism. There, the theory of content works so as to deliver a precise isomorphism, to which operation of the system will have been at least approximately true. We are not thinking of the isomorphism as some preexisting relation that the consumer makes use of. Once we do look for preexisting exploitable relations, as suggested here, it is clear that the correspondence need not be exact to be exploitable.

Conclusion

The mathematical concept of isomorphism or functional isomorphism is unsuited to playing a substantive role in Millikan's theory of content. Instead, various natural relations adverted to by the theory (in a Normal explanation of the operation of the system) make it the case that each representation has a particular content. That picks out a particular isomorphism as suitable for giving the content of the representations. But we saw two ways in which more constrained notions do have a role to play. Firstly, where natural relations between a system of representations and things in the world enter into an explanation of how the producer–consumer system managed to survive and reproduce, then that evolutionary story may give rise to relational proper functions. In such cases, the theory will assign contents to new representations that fall within the relational type, which is a kind of productivity. Secondly, natural isomorphisms are exploitable relations in their own right, and there are probably cases where consumer systems make use of such relations, in a way that cannot be fully explained by appeal to correlational information. These points do nothing to undermine Millikan's theory of content. Rather, they clarify the theoretical machinery needed to deliver her results.[7]

---

[7] Many thanks to Peter Godfrey-Smith and Ruth Millikan for very helpful comments.

Reply to Shea

Nick Shea's essay takes us into a critical area in the theory of representation, asking whether or in what sense a representational system, understood the biosemantic way, must run isomorphic to the domain of its representeds. I am enormously grateful to him for laying out the tools and the questions in this area so carefully. His essay caused me some strenuous thinking. Using some of his tools, I will explain how I have now come to see the main features of the territory he has been prospecting.

Shea says, "It would get things precisely the wrong way round to claim that there being an isomorphism is a substantive constraint on the theory of content. Instead, Millikan's account of content delivers an isomorphism.... [T]he theory picks out a special isomorphism, rather than being based on the existence of an isomorphism." If the account delivers an isomorphism, picks out a special isomorphism, this is certainly different from *that-there-be-some-isomorphism-or-other* being *all* that the theory requires, or an additional constraint added to what else the theory requires. On the other hand, that there must be just this special kind of isomorphism, doing just the work described by the theory, is certainly a substantive constraint. I think this is what Shea intends and I agree.

Shea also says "… it is wrong to suggest [as Millikan does] that the isomorphism is basic and the teleology is needed only to account for error." What I had in mind in saying teleosemantic theories are piggyback theories that must ride on more basic theories was that the teleology only explains the *intentionality* of intentional signs, that is, it explains only how there can be said to be representations *even when there is nothing (real) that these representations represent* (the earmark of the "intentional" according to Brentano). I was thinking that the more basic kind of representation, which does require the reality, the "truth," of what it represents is a natural sign. In *Varieties of Meaning* I argued that systems of "locally recurrent natural signs" run isomorphic to what they represent, and that intentional signs, when true in accordance with a Normal explanation, are locally recurrent natural signs. But given that I allowed the definition of "locally recurrent natural sign" to cover signs whose recurrence was owing to the work of systems having as proper functions to make them recur, Shea is right (though perhaps for the wrong reasons?) that there is teleology in the analysis prior to the account of error.

However that may be, Shea is surely right that the central problem is to define the kind of isomorphism intended. In previous work, my understanding of what an isomorphism (I often just said "a mapping") between domains consisted in was the naive one that requires the isomorphism to correlate structures over domains, where these structures are defined by *real*, as opposed merely to nominal, relations. It seems to me that the "relations" of set theory are nominal, and probably best understood as set theoretic *models* of real relations rather than as relations themselves. No mathematician myself, and certainly no nominalist, I was gratified to find very quick, though informal, support for this in the article called "Relation (mathematical)" in *Wikipedia*.[1] Certainly the relations that define the "semantic rules" or "mapping rules" that correlate the representations in an actual representational system with their representeds have to be real relations, since they figure in Normal explanations for proper functions of their consumers, and Normal explanations are causal explanations. These relations figure as causes or initial conditions in causal processes. But Shea has illustrated deeper problems with the way I attempted to describe the isomorphisms I had in mind.

The original description of "indicative intentional icons" (= descriptive representations – close enough) in LTOBC ran as follows:

> The real value [= the state of affairs represented by a true icon] is a Normal condition for performance of the icon's direct proper functions. (2) There are operations upon or transformations (in the mathematical sense) [bad parenthetical: I just meant that the transformations were not causal processes] of the icon that correspond one-to-one to operations upon or transformations of the real value such that (3) any transform of the icon resulting from one of these operations has as a Normal condition for proper performance the corresponding transform of the real value. (LTOBC, 107)

The "operations upon or transformations of" were to correspond to real relations between pairs of icons, and between pairs of real values (represented states of affairs). What I also had in mind but did not say was that each mapping of a relation from icon to icon onto a relation from world affair to world affair was to be an instance of a *general rule* of correspondence, applicable also to other pairs of icons interpretable by the same consumer (interpreter) systems. And what I also *should* have had in mind but didn't have, or not clearly, was that although these correspondence rules should be multiply applicable – that was where the "productivity" came in – they needn't be universally applicable. There

---

[1] The article "Relation (mathematics)" from *Wikipedia*, November 2008, "sets out the set-theoretic notion of a relation." "Because relations arise in many scientific disciplines as well as in many branches of mathematics and logic," it says, "there is considerable variation in terminology. This article treats a relation as the set-theoretic extension of a relational concept or term. A variant usage reserves the term 'relation' to the corresponding logical entity, either the logical comprehension, which is the totality of intensions or abstract properties that all of the elements of the relation in extension have in common, or else the symbols that are taken to denote these elements and intensions. Further, some writers of the latter persuasion introduce terms with more concrete connotations, like 'relational structure', for the set-theoretic extension of a given relational concept."

might be sub-domains in which one correspondence rule applied between pairs of transformations but other sub-domains in which other rules applied, or in which that particular transformation would not yield a (significant) icon.

But that may be just the beginning. It seems, after Shea's reflections, that we also have to say that the relations that are correlated must be ones that are, as I shall put it, "operative" under the descriptions we give them. This means that the universal correspondence (perhaps within a sub-domain) between these relations follows directly from a Normal explanation (Normal explanations are causal) of how the various parts and aspects of the interpreter system work when they perform properly and – just to be sure – follows from this Normal explanation as a general rule, not case by case. That is, this correspondence between real relations is operatively involved, or directly implied by what is operatively involved, in the causal mechanism of interpretation, that is, of use.[2] (The causal mechanisms of interpretation, it should be remembered, engage in activities that require the existence of the condition represented if they are to produce their proper result.) The relevant differences between icons that correspond to differences between world affairs are differences that the interpreter mechanisms are causally sensitive to, changing their behaviors with these differences. In the past I have sometimes called these operative relations "significant relations" or spoken of "significant transformations" of an icon. Godfrey-Smith and Shea say "exploited" relations or speak of "exploited isomorphisms."

(*Contra* Shea's discussion of "exploitation," however, the carrying of natural information plays no role in determining which relations are "operative." Accidentally true intentional icons guide inferences and behaviors such that they synchronize with actual world affairs/structures just as effectively as icons that carry natural information. That the icons Normally carry natural information does not explain how they work, but how the producing systems that make them Normally work. True icons can do their work because they correspond to the world as they do, not because of how they came to correspond.)

Thus it is that the numberless cheap isomorphisms that are definable given a set-theoretic model of relations are not an embarrassment for teleosemantic theory as I formulate it, nor is the ubiquity of real relations an embarrassment. Relations defining the structures of causally operative initial conditions that have figured in univocal explanations across many historical cases are by no means ubiquitous.

Shea makes some interesting observations and claims about different types of correspondence between icon–icon relations and world–world relations that may be involved in representational systems. I would like to suggest that there are two broad types of such correspondence. These may pair off with Shea's distinction between "isomorphisms that have a natural structure" and isomorphisms involving "more arbitrary mappings." However, I should like to keep more in the foreground two points. First is that not only are both of these types often instanced along different dimensions of a single representational

---

[2] It's possible that this constraint is redundant, that it follows from the restriction that the general semantic or mapping relations between icons and world affairs (these semantic relations being settled given the correspondences between icon–icon and world–world relations) should serve as Normal conditions for correct operation of the interpreting devices. Normal conditions figure in Normal explanations, which though often relational, are exactly the same over all cases.

system, both may be instanced along a single dimension. (I will give examples.) Second is that the style of mapping that is involved in a system of intentional representation is a function of how its interpreters read the representations. It is not determined merely by the structure of the natural information carried when these representations are true for Normal reasons. For a fanciful example, a creature might use representations that correspond to the roads in its territory in exactly the same way Shea's map of Oxford corresponds to the Oxford roads, but it might be able only to find possible routes that way, and not to judge distances. It might use a map showing metrics merely as a topological map. What is represented naturally may not be represented intentionally.

The first broad type of transformation-to-transformation correspondences I will call "projected," the second "substitutional." To anticipate, almost all of the correspondences found in natural language are substitutional. But let us look at projected correspondences first.

Projected correspondences turn on transformations that move us from an icon having one property in a certain determinable range to an otherwise (significantly) identical icon having a different property in that same range, where the second property is determined by its bearing a certain (real) relation to the first. For example: *add one waggle to the dance = add 1,000 yards to the distance of nectar*. If this equation applies to every possible bee dance, it describes a projected correspondence between dances and distances. If it applies, there are, of course, also other equations that apply such as: *double the number of waggles = double the distance of nectar* and *x number of waggles = nectar at x times 1,000 feet*. However the watching bees actually figure it all out, unless they do it by accident these equations all capture the same set of operative/significant transformations that figure in a "projected" correspondence. What's important to hang onto here is that for projected correspondences you have to look at the determinate property of the icon that is going to be (logically) transformed and determine the result of the transformation *from it* by a rule (a real rule, not a set-theoretic function), and look at the determinate property of the world affair that is to be (logically) transformed and project the new world-affair property *from it*. Such a correspondence may, as Shea puts it, "correspond to a single mechanism by which the system operates." Unpacking this, we can imagine that the transformations between numbers of waggles might move from one value of a variable to another value of the same variable, both values always to be taken as input to exactly the same kind of physical system, always operating in accordance with exactly the same principles, calling on exactly the same causal laws governing exactly the same parts of the system, at the same points in the process – always exactly the same Normal explanation for successful output.

When I wrote LTOBC I borrowed the term "substitution transformation" from a topologist,[3] who certified my use as, roughly, hers in mathematics, so I'll stick with it. Substitutional transformation-to-transformation correspondences are correspondences between substitution transformations. Shea suggests for bee dances a list giving arbitrary correspondences between numbers of waggles and distances. This would be a list of correlated substitution transformations: *substitute one waggle for whatever waggles were there before = substitute 500 yards into the distance-away place; substitute two waggles for whatever*

---

[3] From Stephanie Troyer, whom we have since tragically lost from cancer.

waggles were there before = *substitute 150 yards into the distance-away place*; and so forth. Note that the substitution transformations ignore, rather than projecting from, the replaced values of the original icons or world affairs.

That doesn't mean, however, that they ignore the original icons. In this context, it becomes important that we are not beginning with parts or aspects of icons that correspond to parts or aspects of the world and then building these up into complete representations. (That way lies Bradley's paradox, as Wittgenstein recognized in the *Tractatus*.) Rather, we begin with whole representations and consider correspondences between transformations on them and transformation on the world. A substitution transformation operates on a whole icon, changing one part or aspect for another but leaving the rest intact. The substitution transformations that plug in one number of waggles in place of whatever other number leave the rest of the dance intact, the part that tells direction and, it is important never to forget, the part that tells *when* the nectar is where the rest of the dance says it is. Substitutional correspondence rules are productive rules for a reason to which Shea alludes in another connection. He writes:

> Even if nectar at 84° degrees is part of the selectional history, and nectar at 300m is part of the selectional history, it may be that nectar 300m away and at 84° happens not to be part of the selectional history.... That is a kind of productivity.

Arbitrary waggle-number to distance substitutions would be productive for the same reason. They apply by rule, over and over, to any number of possible dances occurring at different times and/or pointing in different directions.

Shea's paragraph continues, however:

> It also looks like a kind of compositionality. But it is important to notice that it is not compositionality in the regular sense. In conceptual thoughts and natural language sentences, the constituents make no claims on their own.... Contrast the bee dance case. There, each relevant dimension of variation has its own truth condition. A dance of 3 waggles says that there is nectar 100m away, irrespective of the direction in which it is performed.... [In some cases m]ultiple variant aspects build up a list of truth conditions, roughly as if they were terms in the propositional calculus connected by conjunction. In other cases like the conceptual constituents of a belief, the variant aspects Millikan describes are each unsaturated, and build up only together into a truth-evaluable thought, like terms in the predicate calculus.

But the various things that the bee dance says are not just additive. Suppose the angle of the dance says there is some nectar SSW, the waggles say there is some nectar 2,000 feet away, the time of the dance says there is some nectar now. These separate facts do not *add up* to there *now being nectar 2,000 feet SSW* – the same nectar! Further, there is no such thing as a (well-formed) bee dance that takes place at no time, or that has no orientation, or that has no number of waggles, any more than there can be a subject of a (well-formed) sentence without a predicate. The significant aspects of bee dances also "build up only together into a truth-evaluable" representation.

The "compositionality" that characterizes natural languages is just another example of the more general principle that characterizes all representational systems, from beaver-tail slaps (which are significantly transformed by changing their time or place of occurrence) through bee dances, to maps, graphs, musical notations, and so forth. What's unique to language is the ubiquity and variety of substitutional correspondences that govern it.

Let me now give an easier example of my two kinds of transformation correspondences, in the process making a couple of final points about them.

Consider a road map of an area the size of a county, the shapes and placement of lines on the map representing the shapes and placement of roads in the usual way. This representational system is governed by projected correspondences that involve a great variety of different significant projected transformations among positions, shapes, and distances. Suppose now that the lines on the map are also different colors and that the colors represent different road surfaces, such as dirt, gravel, paved, and so forth. Changing the color of a line to another color is a significant substitutional transformation. It is not a projected transformation. To know what a certain color means, it does not help to know what corresponds to some prior color, from which the new correspondence can be derived by a rule. Instead one needs to look at the key to the map for each new color introduced.

Now consider a map of the whole world. Suppose that it is Mercator projection. Now some of the kinds of facts that could be read off the county map mentioned above are no longer readable in the same way, say, by all the same consumer/interpreter systems. For example, *move one centimeter right* no longer corresponds consistently to *move one kilometer east*. The distance it corresponds to now depends on its vertical position on the map. Each transformation correspondence of the form *move right x far on the map* = *move east y far on the earth* applies only in two small sub-domains of the map, two narrow horizontal strips. Now it could be that some readers of such a map are able quickly to calculate, given the vertical distance from the equator on the map, what the transformation correspondence will be at that latitude. For such a consumer, the map tells east–west distances by projected correspondences alone, though the rules of use are not simple. What I myself would rather have is a handy key of horizontal to east–west distance correspondences, ideally one key for each latitude band and placed right beside it on the map. Or maybe I could just memorize the keys. That is, I would rather read the same map as governed by substitution transformation correspondences so far as east–west distance correspondences are concerned. Sometimes memorizing is easier than calculating. This also illustrates both the point that significant transformations on representations may be different in different sub-domains and the point that what constitutes the significant structure of an intentional representation depends on its consumers.

In sum, the main difference between simpler representations and the representations of human language is in the use of substitutional transformations. There is no basic difference in the kind of conpositionality involved.

4

MILLIKAN ON HONEYBEE NAVIGATION AND COMMUNICATION

MICHAEL RESCORLA

Insect Cognition

A central task for philosophy and psychology is to identify mental continuities and discontinuities across species. Current science attributes sophisticated mental activity to simple insects, such as bees and ants. Clearly, though, such creatures differ profoundly from higher-level animals, including humans. To what degree does insect cognition resemble our own? In what sense, if any, do insects "represent" the world? How do higher-level animals, including humans, differ in their representational capacities from lower-level creatures, including insects?

Ruth Millikan has pursued these questions throughout her career, delineating a view of mental representation so powerful and systematic that it commands attention from all who study animal cognition. She frequently illustrates her view with empirical case studies, ranging from bacteria to insects to lower mammals to humans. She deploys the case studies to propose various mental continuities and discontinuities across species.

I will explore one of Millikan's favorite examples: the remarkable honeybee waggle dance. Her discussions of this example are quite illuminating. However, I will question certain aspects of her analysis. I will suggest that Millikan blurs important distinctions and elides important commonalities, generating a misleading picture of the overlap between human and insect mental capacities.

The Science of Honeybee Navigation and Communication

In Nobel Prize-winning work, von Frisch (1967) discovered that the honeybee (*Apis mellifera*) performs a "waggle dance" whose properties reliably correlate with direction of

---

Millikan and Her Critics, First Edition. Edited by Dan Ryder, Justine Kingsbury, and Kenneth Williford.
© 2013 John Wiley & Sons, Inc. Published 2013 by John Wiley & Sons, Inc.

and distance from the hive to the location of some resource, such as nectar. Does the waggle dance have representational content? Do production and reception of the dance involve representational mental states? Answering such questions requires scrutiny of the empirical science.

Honeybee cognitive maps?

Honeybees explore large open terrain, reliably returning to the hive. During exploratory flights, they locate profitable food sources, which they can revisit from the hive or elsewhere. What cognitive mechanisms underlie these navigational feats? I will review three widely discussed navigational paradigms: *dead reckoning*, *route following*, and *map-based navigation*.

Dead reckoning

Also called *path integration*, dead reckoning maintains a running record of the creature's position, regularly updated by monitoring the creature's current motion. Using dead reckoning, animals can travel long, circuitous routes and then return directly home along a straight path. Dead reckoning is ubiquitous among vertebrates and invertebrates (Gallistel 1990, 57–102). To employ dead reckoning, an organism must detect its speed and its direction. Honeybees detect speed mainly by monitoring optic flow (Srinivasan et al. 1997). They detect direction through a "sky-compass" attuned to the sun's position and to patterns of light polarization (Wehner 1994). Converting speed and direction into displacement requires elementary integration or the discrete equivalent: iterated vector summation.

Dead reckoning is noisy and fallible. Errors are cumulative, rendering it unreliable over long distances. Moreover, in the special case of honeybee navigation, an overcast sky renders the sky-compass inoperative. Accordingly, the honeybee must supplement dead reckoning with additional navigational mechanisms.

Route following

During route following, the organism implements *sensorimotor vectors* that correlate sensory stimulations with motor instructions. A given stimulation triggers a given motor behavior. A simple example, displayed even by bacteria, is *beaconing*: the organism navigates toward a target by exploiting sensory input emanating from the target. A more sophisticated strategy, implemented by honeybees, is retinal image matching to a stored "snapshot" of the environment as seen from some location (Collett and Collett 2002). Another example: honeybees can learn to fly in some direction when confronted with a certain stimulation.

Honeybees can chain together sensorimotor vectors: an initial stimulus induces some motor behavior until a new stimulus induces a new motor behavior, and so on. In this manner, the bee divides its route into segments, each segment demarcated by a landmark that triggers some associated motor instruction (Collett and Collett 2002). Chained sensorimotor vectors can generate sophisticated route-following behavior.

Map-based navigation

Ethologists agree that honeybees deploy dead reckoning and route following. Collett and Collett (2002) argue that we can explain honeybee navigation solely through such navigational capacities. Other researchers disagree, insisting that we must also posit a "cognitive map" of the environment. More specifically, they argue that the honeybee exploits an *allocentric* cognitive map, anchored to the external environment rather than to the bee's own body (as an *egocentric* map would be). One reason this controversy has proved so recalcitrant is that scientists employ conflicting, and sometimes obscure, notions of "cognitive map." To a first approximation, a cognitive map is a unified mental item whose properties reliably correlate with the spatial distribution of objects and properties.

A related issue is whether honeybees can fly novel shortcuts in a familiar environment. Historically, many scientists have held that such an ability would indicate a "map-like" spatial memory. Yet certain shortcuts are explicable through a sophisticated route-following model that includes appropriate operations on sensorimotor vectors. For instance, a honeybee that learns routes from two distinct feeders back to the hive can, in certain circumstances, home directly to the hive when released from a novel site intermediate between the two feeders (Menzel et al. 1998). We can explain this behavior as resulting from a weighted average over sensorimotor vectors correlated with the two learned routes (Giurfa and Capaldi 1999). On the proposed model, the honeybee "interpolates" between its two learned routes, without deploying an integrated cognitive map.

For a period, most researchers concluded that the evidence did not warrant attributing cognitive maps to honeybees. Recently, however, improvements in technology and experimental design have transformed the debate, lending renewed support to the cognitive map hypothesis (De Marco and Menzel 2008).

Through modern radar technology, we can track flight paths of individual bees over large distances. Menzel and colleagues allowed bees to perform orientation flights in a new environment, so that bees could familiarize themselves with their surroundings. Researchers then divided bees into three test groups: VF bees, trained to a feeder placed at varying locations within 10 meters of the hive; SF bees, trained to a stationary feeder 200 meters from the hive; and R bees, recruited through the waggle dance by foragers trained to the stationary feeder. Researchers trapped bees en route to the feeder or en route to the hive, displacing them to various locations. Flight paths of displaced SF and R bees divided into three stages: (i) an initial straight vector that would have carried the bee back to the hive (or feeder) had the bee not been displaced; (ii) a circuitous search of the local environment; (iii) a relatively straight path toward the hive (or sometimes first to the feeder and *then* to the hive). VF bees did not exhibit stage (i), since they had not learned any routes in their training. SF, R, and VF bees performed equally well during stage (iii), even when displaced to locations too far from the hive for beaconing.

These results show that honeybees can navigate from arbitrary locations within the range of initial orientation flights. Since VF and SF bees performed equally well, this navigational ability must reflect persisting mental changes induced during orientation flights. Menzel and Giurfa (2006) conclude that orientation flights provide bees with a flexible "landscape memory," which correlates hive-centric vectors either with distal landmarks or else with patterns of proximal stimulation. Dead reckoning must play a dominant role in forming these correlations, since dead reckoning provides the bee's only initial source of spatial coordinates. Once the bee has formed its map, it can localize itself with respect to the map's hive-centric coordinate system, in the sense that it can compute a hive-centric vector corresponding to its own position in space. It can then perform vector addition to compute a course to the feeder. In this sense, the honeybee has a "cognitive map."

Navigation with respect to the cognitive map differs in several respects from route following. First, it does not consist in a correlation between stimuli and motor commands. VF bees have not mastered the sensorimotor vectors emphasized by Collett and Collett (2002), because they are not trained to specific routes. Their orientation flights do not include sensorimotor routines corresponding to elements in the cognitive map. Second, the map supports computations, such as addition of arbitrary vectors, that outstrip mere motor response to stimuli. In these two respects, landscape memory is more "cognitive" than route following. Its links to sensory stimuli and motor output are complex, indirect, and flexible.[1]

When route following is available, it dominates the cognitive map, as manifested by stage (i) flights of displaced SF and R bees. The cognitive map apparently serves as a "backup system" that guides behavior only when route following becomes inapplicable.

How does the honeybee form and update the cognitive map based on experience? How does the cognitive map interact with other navigational capacities, such as dead reckoning and route following? How does the cognitive map figure in the honeybee's path planning? So far, these questions remain unanswered.

The waggle dance

Under certain circumstances, a foraging bee returning to the hive from a desirable resource performs a waggle dance inside the hive. Paradigmatic resources include food, water, and potential new hive sites. I focus on food as illustrative.

To a first approximation, the dancing honeybee repeatedly traverses a figure-8 pattern whose two circles meet in a straight line. While traversing the straight line, the bee repeatedly waggles its abdomen. The average orientation of the bee's waggle run with respect to gravity reliably correlates with the food source's solar bearing (i.e., the angle one must fly relative to the sun to reach the food source from the hive). The average duration of the waggle run reliably correlates with distance from the hive to the food source. The dance

---

[1] Cheng objects to Menzel by correctly noting that the bee need not possess "... an overall map plotting the geometrical relations of all significant locations" (Cheng 2006, 204). Yet Cheng acknowledges that the honeybee must master a large collection of hive-centered vectors, each vector correlated either with distal landmarks or else with patterns of proximal stimulation. He also acknowledges that the honeybee must subsume the vectors under operations such as averaging and addition. He misleadingly assimilates these vectors to the sensorimotor vectors emphasized by Collett and Collett. He thereby elides the fact that VF bees have not mastered sensorimotor routines. The "landscape memory" acquired during orientation flights is not "route learning," because it does not consist in a correlation between sensory stimulations and motor routines.

recruits various bees, who promptly fly toward the vicinity of the food source. Upon arriving in the vicinity, recruits search for the food source. This search exploits various cues, including odor. However, the initial flight toward the food's vicinity does not exploit odor, as shown by carefully controlled radar tracking experiments (Riley et al. 2005). In some sense, the bee dance must "encode information" that correlates with the location of the food source, information which recruits can "decode."

Despite intensive research, this "encoding" and "decoding" remain mysterious. For instance, we do not know precisely which sensory stimulations caused by the waggle dance allow recruits to decode the dance. Sound, touch, and comb vibration apparently all play a role (Dyer 2002). Nor do we know exactly how dead reckoning, route following, and the cognitive map inform dance production and reception.

Despite these gaps in our knowledge, we know that considerable cognitive complexity underlies the waggle dance. A forager's evaluation of whether to perform a waggle dance and how vigorously to perform it depends on various factors, including quality of the food source, uncertainty of reward, distance from the hive, risk of predator attack at the food source, and the colony's current nutritional need (Abbott and Dukas 2009, De Marco 2006, Seefeldt and De Marco 2008). Specific properties of the waggle dance depend on the forager's past experience. In particular, the dancing bee relies in complex and poorly understood ways upon dead reckoning and past exposure to landmarks (De Marco and Menzel 2008). How recruits react to the waggle dance depends on their own history. The dance induces novice foragers to locate new food sources, whereas its main effect on experienced foragers is to reactivate interest in a previously visited site (Biesmeijer and Seeley 2005). A striking illustration occurs when a dancer carries the scent of a flower species previously encountered by a prospective recruit with extensive foraging experience. If the dancer indicates an unfamiliar location, then the prospective recruit typically ignores the dance, instead relying on its own past foraging experience (Grüter et al. 2008).

In sum, the waggle dance is embedded in a rich context of memories and motivational states. The surrounding cognitive context heavily informs dance production and reception.[2]

For a vivid example, consider the *solar ephemeris function*. As the sun moves through the sky, its compass direction changes. Dead reckoning with respect to a sun-compass requires mastery of the solar ephemeris function, which yields the sun's compass direction as a function of time. The solar ephemeris function varies according to season and latitude, but honeybees learn it quickly from a few environmental observations (Gallistel and King 2009, 220–226). They exploit it during the waggle dance, as illustrated by the following fact: if recruited foragers are trapped for several hours when leaving the hive, then upon release they fly in the approximate compass direction of the food source, not its

---

[2] The controversial "Lake Experiment" purported to demonstrate a particularly sophisticated cognitive structure underlying dance reception. Gould and Gould (1982) trained foraging bees to visit a feeder gradually moved further into a lake. Foragers successfully recruited other bees to the feeder when it was on land but not once it was located well into the lake. A natural interpretation, endorsed by Gallistel (2009), is that potential recruits evaluate a dance for "plausibility" by comparing it with a preexisting cognitive map. However, a recent experiment by Wray et al. (2008) using modern radar tracking casts the Lake Experiment into doubt.

current direction relative to the sun. In other words, they compensate for the delay, rather than choosing the solar bearing they would have chosen had they left the hive immediately. Apparently, the waggle dance is integrated into mental computations that draw crucially on the solar ephemeris function.

Representation and Truth-Conditions

I now want to examine the foregoing results for their philosophical import, with an emphasis on Millikan's work. I first offer some background remarks concerning the explanatory status of intentional content.

Folk psychology assigns a central role to *intentional explanations*, which individuate mental states and linguistic performances through their *truth-conditions*, i.e., conditions for correct or accurate representation of the world. Should scientific psychology likewise individuate mental states and linguistic performances truth-conditionally? Fodor (1987) and many others hold that it should, while Stich (1983), Field (2001), and others hold that it should not. I know of no convincing argument for the latter position. I also think that many impressive scientific theories already assign truth-conditions a central explanatory role. Two examples:

a) *Vision science* studies how the visual system estimates features of the distal environment (Knill and Richards 1996). The science illuminates diverse phenomena, including perceptual constancies and illusions.[3] As Burge (2010) notes, its explanatory generalizations routinely cite truth-conditions.[4] For instance, vision science explains how the human visual system deploys various cues – including binocular disparity and monocular linear perspective – to estimate *that* a perceived object has a certain depth. Any two cues may yield conflicting depth estimates. The science delineates algorithms, grounded in Bayesian decision theory, through which the visual system fuses conflicting estimates into a unified estimate *that* the object has a certain depth (Knill 2007). Explanatory generalizations of vision science share a crucial feature with folk psychology: they routinely individuate mental states through their truth-conditions.

b) *Empirical semantics* models linguistic comprehension as a speaker's attempt to pair an utterance with its truth-condition. Truth-conditional semantics, usually credited to Frege, illuminates numerous linguistic phenomena, including fine-grained inference patterns featuring quantifiers and other locutions. Recent integration with generative linguistics has substantially increased its explanatory power (Heim and Kratzer 1998).

---

[3] Millikan writes that explaining perceptual constancies is "… a problem of nearly unimaginable complexity that is still largely unsolved" (LBM, 67). Perceptual constancies are indeed complex. However, vision scientists have made substantial progress in illuminating them (Knill and Richards 1996).

[4] Burge speaks of "veridicality" rather than "truth." I use "truth" in a broad sense equivalent to Burge's "veridicality." Another wrinkle is that the explanatory generalizations of vision science individuate mental states not through their truth-conditions but through state-types that set truth-conditions *relative to context*. Truth-conditions vary depending on the context (e.g., which of two qualitatively indistinguishable objects the perceiver is currently perceiving). For discussion, see Burge (2010, 384–396). This wrinkle does not affect my argument, so I ignore it.

Thus, vision science and empirical semantics illustrate the explanatory benefits of truth-conditional individuation.

To what the extent does the truth-conditional explanatory paradigm generalize beyond human vision and linguistic comprehension? Does it fruitfully apply to non-human creatures? We must exercise caution here. As Dennett (1987, 23) observes, we can provide an intentional "explanation" for why a lectern remains stationary: it believes that it has the optimal location in the universe, and it wants to remain at the optimal location. Virtually all philosophers regard this description as a mere *façon de parler*. But why? Partly, I submit, because it yields no explanatory benefits. Truth-conditional attribution plays no role in explanatory generalizations that subsume the lectern, so we have no reason to attribute truth-conditions to the lectern. Truth-conditions play a crucial role in our best theories of human vision and linguistic understanding, so we should attribute truth-conditions to visual states and linguistic performances.

Truth-conditions must earn their explanatory keep. We should attribute them only if they yield explanatory dividends. Before applying the truth-conditional paradigm to non-human creatures, we should confirm that the paradigm offers genuine explanatory advantages.

With this background in mind, let us examine Millikan's account.

Millikan's teleosemantics

Millikan offers a "teleosemantic" theory, whose goal is to isolate naturalistically specifiable facts by virtue of which a state or event has truth-conditions. The basic idea is that a state has a certain truth-condition if that condition's being satisfied is required for the state to fulfill its "proper function," in a sense of "proper function" determined partly by evolutionary history. Her actual theory is quite complicated, but not in ways that affect my discussion. Millikan repeatedly illustrates her theory by citing the honeybee dance. For instance, she writes that the "interpreter mechanisms in the watching bees ... will not perform their full proper functions of aiding the process of nectar collection in accordance with a normal explanation unless the location of nectar corresponds correctly to the dance" (WQ, 91). She concludes that bee dances "... display the characteristic trait of the intentional; namely, they can be wrong or false. They can fail to correspond to a place where there is nectar. Should anything disturb the normal mapping between the shape of the dance and the location of nectar, this misalignment will, quite literally, lead the workers astray" (VOM, 97).

Millikan also illustrates her theory by discussing *bacterial magnetotaxis*. Magnetotactic bacteria contain inner magnets, *magnetosomes*, that cause the entire cell to orient along geomagnetic field lines. Thus, the bacteria "... behave like tiny, self-propelled magnetic compass needles" (Bazylinski and Frankel 2004). As a result, the bacterium moves downward toward less oxygenated regions of its habitat, which are also the regions in which it prospers.[5] Millikan concludes that the "proper function" of the magnetosomes is "... to

---

[5] See Bazylinski and Frankel (2004, 218–220) for a more accurate description.

effect that the bacterium moves into oxygen-free water" (WQ, 93). From her teleosemantics, she infers that bacterial states have truth-conditions: a given magnetosome orientation accurately represents the world just in case oxygen-poor water is located in the appropriate direction (Millikan 2000b, 400).

A natural objection to Millikan's analysis is that truth-conditions do not "earn their explanatory keep" in scientific theories of magnetotaxis. As Burge (2010, 300) notes, we can offer a detailed biochemical theory of magnetotaxis (Bazylinski and Frankel 2004). Our theory may mention the "biological function" of heading toward oxygen-poor areas. But it does not require that bacterial states are *accurate* or *inaccurate*, depending on distal conditions. Truth-conditional attribution contributes no explanatory force to a purely biochemical explanation. For instance, we gain no explanatory power by saying that magnetosome orientation is *inaccurate* in those cases where the reliable correlation with oxygen-poverty fails. Thus, the bacterium seems closer to Dennett's lectern than to humans. As applied to magnetotaxis, truth-conditional locutions are an expository flourish, not a serious contribution to good explanation.

Initially, Millikan seems on firmer ground regarding honeybee navigation and communication. It is tempting to say that a honeybee dance is *correct* or *accurate* just in case nectar is present at the appropriate location. It is tempting to describe a honeybee as communicating *that* nectar is present at some location (Tetzlaff and Rey 2009). But I agree with Burge (2010, 509–514) that we must tread carefully here. I do not *deny* that honeybee navigational states and dance performances have truth-conditions. I say only that, given current scientific knowledge, we have scant reason to believe that they do.

Consider dead reckoning. We can easily build a machine that dead reckons (Gallistel and King 2009, 198–203). Our hypothetical machine has three components: a speedometer, a compass, and a central processor that performs iterated vector addition. Nothing about a speedometer *in itself* or a compass *in itself* generates truth-conditions. *We* can use such devices so as to confer truth-conditions upon them, but the devices do not have "original intentionality." Nor do truth-conditions seem to emerge from coupling a speedometer and a compass to a processor that executes vector addition. We can offer a complete scientific account of vector addition without even mentioning distal conditions. The dead reckoning machine's computations yield states that reliably correlate with distal conditions (until enough noise accumulates to break the reliable correlation). The reliable distal correlations help explain why the machine facilitates successful navigation. Yet the machine's computations do not *in themselves* seem to involve truth-conditional content in any essential way. Nothing about dead reckoning *in itself* secures an explanatory role for truth-conditional individuation. (Cf. Burge 2010, 502–507.)

Honeybee navigation is much more sophisticated than our hypothetical dead reckoning machine. Even the honeybee "sky-compass" involves impressive computations grounded in the solar ephemeris function. Moreover, honeybee navigation deploys route following and cognitive maps, so it is vastly more sophisticated than any kind of dead reckoning. But do these increases in sophistication suffice to generate truth-conditions?

Consider the experimental results collected by Menzel and colleagues concerning VF, SF, and R bees. To explain those results, we can posit a "cognitive map," subject to certain mental operations, whose properties reliably correlate with the spatial distribution of distal objects and properties. What further explanatory power do we gain by attributing truth-conditions to the cognitive map? A key point here is that, on *any* plausible view, reliable correlation does not suffice for truth-conditions. Black clouds correlate with rain, and a high temperature correlates with illness, but neither phenomenon is truth-conditional (VOM, 31). So the mere fact that a honeybee cognitive map reliably correlates with distal states is not enough to show that the map is truth-apt. Truth-conditional attribution requires additional backing beyond mere reliable correlation. Where does one find such backing in the current scientific literature?

A similar worry applies to the waggle dance. As I emphasized, the mechanisms underlying dance production and reception remain mysterious. Perhaps our best theory of those mechanisms will ultimately treat the honeybee as pairing dances and truth-conditions, just as empirical semantics treats a human speaker as pairing sentences and truth-conditions. At present, this is mere speculation. We lack a developed theory of the mechanisms underlying dance production and reception, let alone a developed theory that assigns an essential role to truth-conditions. For instance, suppose a honeybee dance "represents" location x but that nectar is not located at x. Then the dance leads recruits astray in that they fly toward a location that holds no value for them. Perhaps, following Millikan, we should say that the waggle dance does not achieve the function for which it was evolutionarily selected: guiding other bees toward desirable locations. Yet why should we say that the waggle dance is *inaccurate*? Truth-conditional attribution is an expository flourish, not a serious contribution to good explanation.

We humans can confer truth-conditions upon the waggle dance. We can recognize that the dance is "inaccurate" *for us*, just as we recognize that a faulty speedometer is inaccurate *for us* or that magnetosome orientation is inaccurate *for us*. But the question is whether something about the bees themselves, as opposed to our own activity, renders the dance truth-apt. Do honeybees confer truth-conditions on their dances? Millikan's teleosemantics says that they do. My question is whether we have any independent reason, aside from Millikan's own theory, for endorsing that conclusion. In particular, are there any explanatory generalizations that individuate honeybee mental states and performances truth-conditionally?

The current scientific literature features no such generalizations. The literature emphasizes that map and waggle dance properties reliably correlate with distal properties. It emphasizes that the correlation serves an important biological function. It nowhere requires that the map or waggle dance is *inaccurate* in those cases where the reliable correlation fails, any more than current science requires that magnetosome orientation is *inaccurate* in those cases where the reliable correlation with oxygen-poverty fails.

Broadly representational language does play a significant role in the scientific literature on honeybee navigation and communication. Ethologists frequently write that the waggle dance "encodes spatial information," that aspects of the dance "represent" location, that the honeybee navigates by deploying mental "symbols," and so on. As Burge (2010, 492–518) notes, however, we can easily paraphrase this representational talk by citing reliable correlations, biological functions, and the like.

Gallistel's (1990, 1998) treatment of representation, which Millikan (VOM) cites sympathetically, illustrates the point. On Gallistel's theory, "[t]he brain is said to represent an aspect of the environment when there is a functioning isomorphism between some aspect of the environment and a brain process that adapts the animal's behavior to it" (Gallistel 1990, 15). The isomorphism obtains between the representing system (e.g., mental representations exploited during dead reckoning) and the represented system (e.g., velocity and displacement in physical space). The isomorphism is *structure-preserving*, since relations in the represented system (e.g., the relation between velocity and displacement in physical space) correspond to operations within the representing system (e.g., integration or vector summation). The isomorphism is *functioning* because it is "... used by the representing system in coping with – surviving and reproducing in – the represented system" (Gallistel 1998, 15).

Gallistel argues that dead reckoning and map-following involve "representation," in his proprietary sense. For instance, the honeybee exploits a systematic structure-preserving correspondence between its cognitive map and physical space. Similarly, Gallistel argues that the honeybee dance represents the distance and direction of relevant food sources: "[t]he correspondence between elements of the dance and the geometry of the terrain outside the hive is used to control the food-seeking flight of the recruited bees" (Gallistel 1998, 22).

Let us grant that Gallistel's theory isolates one legitimate sense in which honeybees "represent" their environment. The fact remains that Gallistel does not assign truth-conditions a significant role. Nothing in Gallistel's treatment requires that honeybee states and performances are prone to *error* if the world does not cooperate. Explanatory power resides solely in the "functioning isomorphism" between mind and world. There is no obvious reason why "functioning isomorphisms" must have truth-conditional content. For instance, magnetotaxis exploits an isomorphism between magnetosome orientation and direction of oxygen-poor water. So Gallistel's theory of representation, like Millikan's, entails that magnetosome orientation "represents" oxygen-poverty. Yet there is no clear reason, independent of Millikan's teleosemantics, for saying that magnetotaxis involves truth-conditions. The burden of proof lies with those who claim that functioning isomorphism suffices for truth-conditions.[6]

Differences in explanatory structure

Millikan might turn my argumentative strategy against me. She might deny that truth-conditions play a central explanatory role even in vision science or empirical semantics.

---

[6] Gallistel classifies beaconing as "nominal representation," i.e., "something that technically satisfies the definition of a numerical representation, but is not 'really' such" (Gallistel 1990, 27). Non-nominal representation requires a "... rich formal correspondence between processes and relations in the environment and operations the brain performs" (Gallistel 1990, 27), whereas the only relation preserved by nominal representation is identity. Presumably, Gallistel would also classify magnetotaxis as nominal representation. In contrast, honeybee navigation "non-nominally represents," because it supports structure-preserving operations such as vector addition. Thus, one might propose that "non-nominal representation" (in Gallistel's sense) suffices for truth-conditions even though "nominal representation" does not. This proposal would avoid the magnetosome counter-example. But it faces other hurdles. For instance, a simple dead reckoning machine "non-nominally represents," but I see no reason why it must generate truth-conditions.

She might propose that we can replace any intentional explanations offered by scientific psychology with explanations that cite reliable correlations, biological functions, functioning isomorphisms, or other non-truth-conditional notions.

I see little prospect for successfully developing this proposal. Vision science features numerous explanatory generalizations that type-identify mental states truth-conditionally. The generalizations derive from ordinary belief–desire explanation, by way of Bayesian decision theory. A typical model (Knill 2007) explains how the visual system estimates *that* an object has depth d by fusing an estimate *that* it has depth d' (based on binocular cues) with a possibly conflicting estimate *that* it has depth d'' (based on monocular cues). There is no obvious way to preserve the benefits of such a model while eschewing truth-conditions (Rescorla, forthcoming). Apparently, eliminating truth-conditions would require wholesale reconstruction of the science. Similarly, truth-conditions occupy a seemingly indispensable role within empirical semantics. In contrast, nothing like proper functions or functioning isomorphisms plays an explicit role within vision science or empirical semantics. For instance, a perceptual state studied by vision science may well *have* a biological function. But the explanatory generalizations of vision science do not mention that function. The generalizations type-identify mental states through their truth-conditions (e.g., estimating *that* an object has a certain depth), without any explicit mention of biological fitness, natural selection, functioning isomorphism, and so on.

As Burge (2010) emphasizes, we must distinguish two explanatory paradigms, subserved by distinct schemes for taxonomizing mental states. The first paradigm mentions reliable correlations, functioning isomorphisms, biological functions, and the like. The second paradigm mentions truth-conditions. One need only compare our current best science of magnetotaxis (Bazylinski and Frankel 2004) with our current best science of vision (Knill and Richards 1996) to appreciate how profoundly the two explanatory paradigms differ. There is no reason to think that we can reproduce the second paradigm's explanatory benefits within the first paradigm. There is no reason to think that we can gut vision science and empirical semantics of their core theoretical notions without explanatory loss. Philosophers who claim that we can owe us detailed reconstructions of the relevant scientific theories.

Millikan might respond that she uses "representation" and "truth-conditions" in a proprietary sense not answerable to pre-theoretic or traditional philosophical usage. If we grant that bacterial and honeybee states have the appropriate proper functions, then it follows by stipulation that those states have truth-conditions *in Millikan's sense*.

This response does not address my underlying objection: that Millikan elides disparities between the two explanatory paradigms distinguished above. The first paradigm traffics in notions such as functioning isomorphism, proper function, and so on. The second paradigm traffics in truth-conditions. By attributing truth-conditions even to bacteria, Millikan blurs the difference between these two paradigms. She thereby obscures crucial differences between humans and bacteria. She also distracts attention from a crucial *empirical* question: Which explanatory paradigm most appropriately applies to more difficult cases, such as the honeybee? We cannot overcome these worries by introducing a special stipulated usage on which bacteria have "truth-conditions." Such a usage merely reinforces the misleading impression that no significant difference separates the two explanatory paradigms.

Good terminology tracks underlying distinctions in explanatory structure. Well-chosen theoretical terms "carve nature its joints," rather than blurring important distinctions among explanatory paradigms. We should reserve truth-conditional locutions for those domains where they play a genuine explanatory role.

Psychological Structure

By questioning whether honeybee navigation and communication involve truth-conditional contents, I am not urging that we treat honeybees within an associationist or behaviorist framework. Any plausible view of honeybee cognition must recognize complex psychological structure that such frameworks do not accommodate.

As I have discussed, honeybee navigation and communication involve computation over mental states that "represent" (in Gallistel's sense) various distal properties, such as displacement and solar bearing. To take a particularly compelling example, there is decisive evidence that honeybees execute dead reckoning. Dead reckoning is a very simple navigational capacity, but it already illustrates the limitations of a purely associationist approach. During dead reckoning, the animal records its current displacement from a home location. Even very simple invertebrates can store this record in memory for relatively long periods. No one has the slightest idea how to model such memory storage within a purely associationist model (Gallistel and King 2009). The divergence from associationist psychology becomes particularly vivid when we consider the solar ephemeris function, which plays an integral role in the honeybee's sky-compass. To compute compass direction from current solar bearing, the bee must consult a stored memory that encodes the solar ephemeris function. It must perform a trigonometric computation of a kind totally alien to associationist psychology.

As I also emphasized above, there is evidence that honeybees store cognitive maps in memory. The mere suggestion is anathema to behaviorism or associationism.

Thus, by questioning whether honeybee navigation and communication involve truth-apt mental states, I am not questioning the broadly "cognitive" nature of those phenomena. Honeybee mental activity involves far more internal psychological structure than any behaviorist would countenance. Indeed, as I will argue in the next section, there is an important respect in which Millikan underestimates the extent of this psychological structure.

Pushmi-Pullyu Representations

Does honeybee cognition exhibit anything like belief-desire structure, as enshrined in ordinary folk psychology? Carruthers (2004) answers affirmatively. He holds that bees can execute a practical inference schematized roughly as follows:

BEL [nectar is 200 meters north of hive]
BEL [here is at hive]
DES [nectar]
MOVE [200 meters north]

Millikan rejects any such suggestion:

> Does the bee come to believe there is nectar at location L, desire to collect nectar, know that to collect nectar at L requires going to L, hence desire to go to L, and hence, no other desires being stronger at the moment, decide to go to L, and proceed accordingly? Surely not. The comprehending bee merely acquires an inner representation that is at the same time a picture, as it were, of the location of nectar (relative to its hive) and that guides the bee's direction of flight. The very same representation tells in one breath what is the case and what to do about it. (Millikan 2004a, 17–18)

On Millikan's picture, honeybee cognition does not divide neatly into "cognitive" and "conative," or "informational" and "motivational," states. Instead, it operates at a more primitive level that blends information and motivation together inextricably. In that respect, it differs profoundly from human cognition.

This picture of honeybee cognition recurs regularly in Millikan's work. In her later writings, she adopts the label *Pushmi-Pullyu Representation* (PPR). A PPR has both "descriptive" and "directive" content. It mediates directly between perception and behavior, so it is a more primitive form of representation than a purely descriptive representation (such as a belief) or a purely directive representation (such as a desire). As Millikan notes, a PPR resembles Gibsonian perception of an *affordance*, although Gibson himself was leery of "internal representations." Cognition defined entirely over PPRs is closely tied to perception and behavior, in a way that more sophisticated cognition is not: "[r]epresentations that are undifferentiated between indicative and imperative connect states of affairs directly to actions, to specific things to be done in the face of those states of affairs" (LTOBC, 99). Thus, a PPR generates a "perception–action" cycle, in which sensory input "directly" causes action. Millikan suggests, without asserting, that primitive animals display only pushmi-pullyu representation (Millikan 2004b, 18–19): "[o]ne possibility is that the simplest animals, at the level of insect, for example, may be governed almost entirely by a set of perception–action cycles arranged in a hierarchy that determines which shall take precedence over which." She contrasts the "... sort of inarticulate pushmi-pullyu comprehension the bee has" with "... articulate, well-differentiated, and uncommitted human beliefs and desires" (Millikan 2004b, 22).

How does Millikan's analysis of honeybee cognition fare against Carruthers'? Carruthers' specific formulation strikes me as problematic, for two reasons: first, it assumes that relevant honeybee mental states have intentional content; second, it depicts nectar location and honeybee location as recorded by discrete belief-states, not integrated into a holistic map-like structure. I find no warrant for either suggestion in the current science. Nevertheless, I believe that current science favors Carruthers' general approach. Current science supports a clean division between "informational" and "motivational" elements in honeybee cognition. Indeed, Menzel (2008) explicitly commends Carruthers' belief–desire analysis as a framework for studying honeybee cognition. Menzel does not mention anything like PPRs. His discussion is unusual only in being so explicit. Honeybee researchers routinely assume a sharp demarcation between memories and motivational states.

Research by Menzel and colleagues illustrates the point. As noted above, displaced SF bees first flew the route they would have flown if not displaced, then explored the environment to orient themselves, and finally flew *either* directly to the hive *or* to the hive by way of the feeder. Besides supporting the existence of a cognitive map, this phenomenon indicates a motivational element that varies independently from the map. As Menzel and Giurfa (2006, 28) put it, one operation at the bee's disposal must be "... a shift in motivation (fly toward the hive or toward the feeder)." The shift in motivation is independently manipulable from the cognitive map. Apparently, then, honeybee cognition features a primitive analogue to the separation between beliefs and desires.

Other results bolster this conclusion. As noted above, whether and how a forager dances depends in complex ways on various factors (such as profitability of the food source, danger of predation, hive nutritional status), as does the reaction of prospective recruits to the dance. How can we explain this complexity unless we posit honeybee spatial memories that influence behavior in conjunction with independently manipulable informational and motivational states?

Current science provides no indication of honeybee representations that blend descriptive and imperatival elements. Nothing in current science suggests that honeybee memories and motivational states are inextricably intermingled. Attributing imperative force to honeybee dances or to honeybee cognitive maps adds no explanatory force to the scientific theories canvassed above.

More seriously, Millikan's emphasis on PPRs obscures the complex psychological underpinnings of honeybee cognition. Her discussions suggest a fairly Gibsonian picture: honeybee navigation and communication involves perception–action cycles, rather than complex cognitive chains that connect to perception and action only at the peripheries. Millikan's picture conflicts with the results canvassed above. As she notes, a "... purely pushmi-pullyu animal" – a hypothetical animal whose only representations are PPRs – "is certainly capable of learning, but this learning is what psychologists call 'procedural learning'. It learns what to do after what, completion of each link in the chain producing perception of a new affordance, which guides production of the next link" (VOM, 185). Route following is an example of procedural learning. Perhaps one can also regard dead reckoning as a kind of procedural learning. Yet honeybees do not navigate solely by dead reckoning and route following. They also navigate by employing cognitive maps. Map-based navigation requires formation and revision of a cognitive map based on dead reckoning and sensory stimulations. It requires path planning with respect to the map, including vector addition. These mental computations are not reducible to direct links between perception and action. They are not purely procedural. Millikan's Gibsonian picture omits crucial cognitive activity underlying honeybee navigation and communication.

To be fair, scientists such as Collett and Collett (2002) embrace something like a Gibsonian picture. They try to explain honeybee navigation solely in terms of "procedural learning," such as route following, rather than "declarative learning," such as mastery of a cognitive map. It seems to me, however, that this position does not accommodate the experimental results of Menzel and his colleagues, which reveal a "landscape memory" distinct from mastery of sensorimotor routines.[7]

Millikan sometimes notes the developing scientific evidence for honeybee cognitive maps, but she does not substantially alter her approach: "... there is evidence that bees carry neural maps in their heads, but we do not have to assume this is so to see that the dances are representational. It could be that the watching bees responded directly to the dances by pivoting about to a certain direction and flying that way for a certain time" (LBM, 97). I find this passage misleading, because nothing resembling the honeybee dance could exist without the surrounding cognitive structure that Millikan dismisses as incidental. Honeybees make sophisticated use of dead reckoning, landmarks, the solar ephemeris function, and so on. They perform impressive computations, store memories for use at unknown future times, and exploit those memories during path planning. It is doubtful that any navigation–communication system remotely similar to the honeybee's could function as Millikan suggests: through "direct response" to dance signals. For instance, the hypothetical system proposed by Millikan would not incorporate computations involving a solar ephemeris function. Hence, it would not allow recruits to exploit a given waggle dance at a significantly later time of day (as honeybee recruits can do).

In another passage, Millikan tries to reconcile her PPR framework with the cognitive map hypothesis. She writes:

> Actually, there is evidence that the bee has a map in its head of its environment and that the dance induces it, first, to mark the nectar location on this map (Gallistel 1990). Still, assuming that the only use the bee ever makes of a mark for nectar on its inner map is flying to the marked position to collect the nectar, then the nectar on the bee's inner map is itself a PPR. And it seems reasonable to count a representation whose only immediate proper function is to produce an inner PPR as itself a PPR. (LBM, 174)

I see no reason for "... assuming that the only use the bee ever makes of a mark for nectar on its inner map is flying to the marked position to collect the nectar." I am not even sure that it makes sense to talk about the honeybee making use of a *particular* mark on the cognitive map. The bee makes use of the map *as a whole*. It uses the map to localize itself and to plan paths between arbitrary locations. As far as we know, the honeybee may also consult its map when deciding on a response to another bee's waggle dance. What use the bee makes of its map depends on its motivational state, which is independently manipulable from the map itself, and probably on other memories as well. Thus, this passage understates relevant cognitive structure.[8]

I conclude that Millikan's account downplays the cognitive complexity underlying honeybee navigation and communication. Those phenomena involve sophisticated patterns of

---

[7] Something close to Millikan's Gibsonian picture may apply to the navigation of some other insects, such as the Australian desert ant (Wehner et al. 2006).

[8] Also problematic is Millikan's claim that the waggle dance induces recruits to revise their cognitive maps. The waggle dance surely induces *some* relevant change in spatial memories of recruits. But the change may not involve a change in the cognitive map. For all we currently know, the recruits may simply acquire a sensorimotor vector that is not integrated with the cognitive map.

mental computation, patterns that we are only beginning to understand. Calling the patterns "perception–action" cycles misleadingly suggests absence of intervening cognitive processes.

Folk Psychology as an Explanatory Paradigm

To what extent should scientific theories of insect cognition replicate ordinary folk-psychological practice? I have proposed two respects in which Millikan answers this question unconvincingly. On the one hand, she overextends the folk-psychological practice of individuating mental states truth-conditionally. On the other hand, she understates the extent to which simple creatures instantiate something like the division between beliefs and desires. For these two reasons, Millikan provides a misleading picture of the overlap between human and non-human mental capacities. Needless to say, my two objections do not diminish the abundant insights into insect cognition offered by Millikan's work.[9]

---

[9] I am indebted to Kevin Falvey, Peter Graham, Dan Ryder, and Kenneth Williford for helpful feedback on this essay.

Reply to Rescorla

Michael Rescorla offers a very interesting summary of current research on the inner mechanisms that explain honeybee navigation. He has two stated purposes in doing so. First is to question the explanatory value of my claim that simple creatures such as bacteria and bees use representations, a central feature of representation being, as Rescorla emphasizes, possession of truth-conditions. Second is to examine my contribution to what he terms "insights into insect cognition."

The second concern I will say little about since I really did not intend any of the brief comments on bee dances that Rescorla has culled and set side by side (from six sources written over twenty years) to be contributions to insect cognition theory. The wider discussions in which these comments were embedded all concerned either the nature of representation generally or the natures of specific kinds of representations, the bees entering only as possible illustrations to help clarify general points. If the reader looks back at the text surrounding any of these passages, I think it should be clear that whether the bees themselves actually succeed in being instances of the phenomena I am discussing is in each case beside the point. On pushmi-pullyus I must mention, however, that Rescorla has overlooked my chapter-long discussion in VOM where many complexities are introduced, including, especially, much discussion of the kinds of restrictions on pushmi-pullyu function that he labels as "motivational." Purely pushmi-pullyu animals can, on my description, live extremely complicated lives that are exquisitely sensitive both to current environmental conditions and to the current needs of these animals. (When I wrote VOM I was, I think very reasonably, influenced by (ethologist) Randy Gallistel's idea that intricately complex "lattice-hierarchies" govern behaviors of many simple animals, which hierarchies, I suggested, might be explained in purely pushmi-pullyu terms.)

Turning to Rescorla's first concern, just what is the explanatory value of claiming that bee dances, or that the direction the bacterium's magnetosome points, are representations? To ease the tensions a little let me mention that originally, in LTOBC (1984), I did not call these simple devices "representations" but instead "intentional icons," reserving the term "representation" for those intentional icons that were designed to become involved in reidentifications and inference processes, paradigmatically perceptions, human beliefs, human language. But there being, as I thought, no point in fighting over words, I later followed Dretske, Fodor, and others, turning all of my intentional icons, by a verbal switch, into representations. I wonder if Rescorla (and Tyler Burge) might be set smiling again if I simply reverted to my earlier terminology in LTOBC?

But probably not. For the main thing that bothers Rescorla seems to be assigning truth-conditions to bee dances and magnetosome pointings, and on my account, even the simplest intentional icons do have truth or correctness conditions. So I must dig deeper.

Both Rescorla and Burge maintain that truth or correctness conditions are supposed to "explain" some things, and they want to know what ascribing truth-conditions to bee dances and magnetosome pointings explains. A problem for me is that I cannot find, in either man's work, any clear statement of what it is that they take truth-conditions to "explain" in the cases of those representation they do sanction, so I don't know what I am to be looking for in the case of bees and bacteria. It would, of course, help in trying to figure out what kinds of things truth-conditions might explain to know what truth-conditions are being taken to be. But again, I can find no hint either in Rescorla or in Burge on this matter. A good deal is said by Burge about how certain vision scientists describe the formation of visual representations and about what the content of these representations is, but what representations actually are and what it is for them to have truth-conditions is not addressed. Indeed, Burge hints strongly that representations and their correctness conditions are *sui generis*. Similarly, although Rescorla says that he is not yet sure whether or not bee dances are representations having truth-conditions, he is silent about what kind of evidence would settle the matter.

In LTOBC and later works I offered a theory about what truth-conditions are. The original and paradigm application of this theory was to sentences (LTOBC, ch. 6), then also to beliefs, desires and perceptions, and – as it turned out, though, this certainly was not a desideratum at the start – it applied also to simple, naturally designed signals such as bee dances and magnetosome pointings. My suggestion was that intentional icons (representations) fall within systems such that there is a semantic mapping between the set of possible icons in the system and the set of things these icons would represent (see my reply to Shea, this volume). Intentional icons have what I called "proper functions," these being "biological" in a broadly metaphorical sense, derived from perceptual tuning, from various kinds of learning or from cultural selection as well, of course, as from natural selection. These functions are sometimes multiple and often very general; for example, presumably one crucial proper function of human beliefs is to participate in inference processes. Intentional icons were interesting because they worked by running isomorphic to things in the world so as to guide behaviors or thoughts in a way that took account of those things. It was the principle by which they worked rather than their specific functions that characterized them. The argument was that language forms, thoughts, perceptions, and many simple things such as bee dances and magnetosomes (but obviously not lecterns) all exhibited the same fundamental principle (the use of mappings or isomorphisms) at work in performing whatever functions they had. Truth-conditions were determined by the icon–world relations that needed to be in place for the icons (non-accidentally) to do their proper jobs of guiding further icon production ("information processing") and/or (immediately or eventually) guiding behaviors useful to the organism in taking account of that world. In short, reference to intentional icons having truth-conditions was supposed to explain how a creature's behaviors could be controlled and coordinated with the world around it, at the same time accounting (with help from the teleological notion of function) for the difference between mere natural signs and genuine representations, between things that just happen and things that are working either rightly or wrongly – exactly the distinction that Burge seems to be after.

A relevant challenge to the above position would have to argue either that sentences, beliefs, and perceptions are not intentional icons, their truth-conditions being determined in other ways instead, or else that bees and magnetosome pointings do not fit my description of intentional icons. Rescorla has made neither challenge. He claims that "[t] he burden of proof lies with those who claim that functioning isomorphism suffices for truth-conditions," but he has not a word to say about any of the large corpus of my work that argues for exactly that position (substituting for "functioning isomorphisms," "isomorphisms when functioning normally and properly"). He does claim that "... nothing like proper functions or functioning isomorphisms plays an explicit role within vision science or empirical semantics. For instance, a perceptual state studied by vision science may well have a biological function. But the explanatory generalizations of vision science do not mention that function." But on the contrary, following in the Helmholtz tradition, Marr's classic modern theory of vision, along with many of its progeny, sets as its explicit goal to explain how the mechanisms of visual perception manage to perform what he takes to be their primary *function* of turning proximal stimulations of the retina, via a computational process involving intermediate representations, into representations of distal objects – a series of "proper functions" selected for by evolution and perceptual learning if ever there was one. A more general point, however, is that philosophical inquiry into the underlying nature of representation simply is not the job of the pure vision scientist – or of the semanticist – so of course we do not find them discussing these matters.

Rescorla says "[a]s Burge (2010) emphasizes, we must distinguish two explanatory paradigms, subserved by distinct schemes for taxonomizing mental states. The first paradigm mentions reliable correlations, functioning isomorphisms, biological functions, and the like. The second paradigm mentions truth-conditions.... Millikan elides disparities between the two explanatory paradigms distinguished above. The first paradigm traffics in notions such as functioning isomorphism, proper function, and so on. The second paradigm traffics in truth-conditions." But the point at issue is exactly whether these *are* different "paradigms." I have argued in detail that they are two parts of a single explanation of the phenomenon of intentionality, arguments that neither Burge nor Rescorla have addressed.

The many vision scientists who work at the intersection of perception and neuroscience are actively engaged in trying to understand how neurons manage to be the vehicles of visual representation – how their changes do somehow run isomorphic to the changes what they represent. And in the process of trying to understand how dead reckoning works in insects or how bees produce and use bee dances, ethologists run experiments in which they try to disrupt these activities, for this information yields evidence on how *correct* performances are achieved. That is, they look for conditions under which these insects will represent things wrongly. These researchers do not express themselves using the philosophers' jargon "truth-conditions" to be sure, but frankly, I don't recall any vision scientists using this jargon either.

In discussing how insects perform dead reckoning, Rescorla says "reliable distal correlations help explain why the machine facilitates successful navigation. Yet the machine's computations do not in themselves seem to involve truth-conditional content in any essential way. Nothing about dead reckoning in itself secures an explanatory role for truth-conditional individuation. (Cf. Burge 2010, 502–507.)" But these "reliable distal correlations" precisely are reliable satisfactions of truth-conditions on my analysis. Nor would it make sense to look for truth-conditional content *within* "the machine's computations ... themselves" if this means by examining only the inside operation of the machine. Dead reckoning is ("constitutively," if you fancy Burge's terminology) a kind of coordination of a "machine" (if you will) with its environment. "Contents" of any representations involved are determined by relations that would normally hold *between* these representations and the environment were the machine actually performing its (designed) dead-reckoning function. No isolated "machine" sitting quietly on a shelf by itself is performing dead reckoning. If this is not a correct view, then we need to be told what should take its place.

My worry is that what Burge and Rescorla think should take its place, in the case of *real* representation, is some kind of inner "non-physical" happening that within itself somehow projects to a possible state of affairs that it somehow envisions. For example, Rescorla complains that the devices responsible for such things as dead reckoning do not have "original intentionality." In the case of bees he says, "Perhaps our best theory of those mechanisms will ultimately treat the honeybee as pairing dances and truth-conditions, just as empirical semantics treats a human speaker as pairing sentences and truth-conditions." The image here seems to be that the human speaker's "pairing of sentences with truth-conditions" is some kind of inner act of understanding over and above the mere capacity to translate between sentences and inner representations. (Other acts of this kind would be, I suppose, inner pairings of one's beliefs with truth-conditions, inner pairings of one's desires with satisfaction conditions, and inner pairings of one's perceptions with correctness conditions?) It's hard to know how to answer this kind of argument other than to reassert that I am not a dualist.

5

Concepts

Useful for Thinking

LOUISE ANTONY

I want to talk about abilities, and about substance concepts, and about the relation between them. Millikan argues that substance concepts *are* abilities – abilities to reidentify objects. Or, to be more precise, she thinks that "Substance concepts … are, in part, abilities to reidentify their assigned substances" (OCCI, 51). She also says, in places, that concepts are "representations," but cautions us not to be misled by this term. In particular, we must not think that concepts are vocabulary items in some kind of internal language, some "language of thought." Obviously, if concepts are abilities, then they cannot be such things. Indeed, it's initially puzzling how something could be *both* an ability and a representation. (As it turns out, Millikan wants to blur the distinction between what is sometimes called "declarative knowledge" and "procedural knowledge" in a rather general way – the treatment of certain abilities *as* representations should fall out. But more on that later.)

I think Millikan is badly wrong about what concepts are. I think they are *exactly* what she thinks they are not, namely, vocabulary items in the Language of Thought (LOT). One important reason why she thinks that concepts must be identified with abilities is that she thinks we need to understand concepts – the paradigmatic instances of which, she and I agree, are naturally occurring parts of biological organisms – in terms of what they are *for*, what their *function* is. Now I do not, in general, accept the idea that the only or even the most fundamental way of studying something is via a study of its function (especially if "function" is understood to mean "teleofunction"), but I'm willing to let that point pass. I don't think Millikan has done justice to the range of functions concepts *in fact possess*. I'll argue in due course that her neglect of (what turn out to be) primarily mental functions reflects a behavioristic bias lurking beneath the surface of Millikan's work.

As I said, I also think she is wrong in much of what she has to say about abilities, and I think that the problems with her account of abilities stem from the fact that she constrains her account with the requirement that concepts must be a kind of ability. This requirement means that abilities have, somehow, got to have individuation conditions and "content conditions" that match those we intuitively assign to concepts. This requirement leads to strange results. Let me start with Millikan's account of abilities, and then say why I think the deficiencies in Millikan's account are connected to deficiencies in her account of concepts.

Abilities

Millikan argues that, because substance concepts *are* abilities ("in part, the abilities to reidentify their assigned substances"), the question of what determines the content of a concept ("How are these substances assigned?") is the same question as the question "What determines what a learned ability is an ability to do?" (I note that this question is posed only for *learned* abilities. That could be for two reasons: either concepts are to be identified only with abilities that have been learned, or it's presumed that the answer for native abilities is obvious. If the first is the case, then this would seem to imply that no concepts are innate. That would be a bad consequence, surely, both because no theory of concepts should rule out nativism *by definition*, and because there's lots of evidence that at least some of our concepts *are* unlearned. I'll have some things to say about native abilities in what follows.) In pursuing this question, Millikan considers whether abilities can be adequately characterized in terms of dispositions. She rejects two possible dispositional analyses:

(A) S has the ability to do $X =_{df}$. S is disposed to X in (statistically) normal conditions

and

(B) S has the ability to do $X =_{df}$. S would do X if S tried.

(A) is obviously a non-starter; (B) is rejected for two reasons:

1. The notion of "trying" seems to involve *intending*, which seems to involve *concepts*. This would make for a vicious circle, given Millikan's purposes. This seems to me to just be part of the reason why it's a bad idea to identify concepts with abilities. But let that pass, since Millikan's second reason is the one she regards as the more important.
2. Trying and subsequently succeeding is insufficient for successful manifestation of ability, but the conditions we need to add turn out to render the whole definition of "ability" in terms of dispositions vacuous.

I don't quite follow the reasoning here, so I'll give you the argument in Millikan's own words: "The only true disposition here is the disposition to succeed if one tries under certain conditions, namely the ones one attempts to recognize ..." as the conditions under which trying would likely yield success. Since these attempts involve trying to recognize those conditions "... *under that very description*, namely *as* certain conditions *under which one can succeed*," "... the description of this disposition appears to be empty" (OCCI, 59).

Since neither (A) nor (B) works, Millikan says, we are still seeking "... the relation of an ability to the conditions of its successful exercise" (OCCI, 59). Through consideration of an example – a person who learned to word-process on a now utterly defunct word-processing program – she comes to the conclusion that this relation that we are looking for is "historical."

> In general, the conditions under which *any* ability will manifest itself are the conditions under which it was historically designed as an ability. These are the conditions under which it was learned, or conditions in which it was naturally selected for. (OCCI, 61)

In the case of learned abilities, the relevant conditions will be a matter of the personal history of the individual:

> Each person's ability to do A rests on a disposition defined through their particular past. Each person has a disposition to do A if they try to do A under the conditions that accounted for their own past successes in doing A. If they have no such disposition, of course, then they have lost the ability to do A. (OCCI, 62)

This all seems to me to be quite wrong, whether we're talking about learned abilities or native abilities. First of all, let's consider some learned abilities. I have the ability to sing the theme song to *The Mr. Ed Show*. This is obviously a learned ability. I know exactly how I learned it. I learned it by watching *The Mr. Ed Show* every week for several years. (At pretty much the same time, I learned how to drive my parents crazy.) Does it follow, is it part of what it is for me to have this ability, that I am disposed to sing the song, if I try, under the conditions in which I learned to sing it – that is, when the show is on? I don't see that it does follow, or that it is part of my ability. It may be that I would sing it in those circumstances if I tried, but there doesn't seem to me to be anything special about those, as opposed to any number of other circumstances, such that my doing it in those circumstances demonstrates my ability. What seems important is just that there be *some* circumstances in which I will sing it, if I try. The history of my acquisition of the ability seems just irrelevant.

Indeed, it could even turn out that those circumstances happen to be among the relatively small number of circumstances such that I *wouldn't* succeed if I tried in them. I usually can sing the song, I explain, but something about seeing that ridiculous horse supposedly talking is so distracting that I just can't do it while the show is on. Such an eventuality would not, it seems to me, in any way impugn my ability, in general, to sing the song. (Millikan and I agree that failure to do X on some occasions is not enough to withdraw credit for the ability to do X, period.)

The historical account that Millikan is offering does not even appear to fit her own examples very well. Consider the ability she says she has to kill cats. Let's assume that this is a learned ability. And let's consider some ways that someone might have learned how to kill cats (I don't know how Millikan did it). One way someone might have learned this is by learning how to kill dogs – the skills transfer pretty readily, let's suppose. In that case, there may have been "conditions that accounted for [would-be cat-killers'] own past successes" in killing *dogs*. But if the ability to kill cats is simply a transference of the ability to kill dogs, then it looks like the proprietary conditions for manifesting the cat-killing ability would have to involve dogs, but no cats. It would then follow that one could manifest one's ability to kill cats by trying to do it when only dogs are around. This is strange in the extreme.

Alternatively, one might have learned how to kill cats by reading about it in a book. The book might have described the process extremely perspicuously, so that one knows exactly what to do: it may have an instruction like, "soak a cloth with chloroform and hold it over the animal's nose for two minutes." In this case, it looks like the ability to kill cats has been acquired *without any cat ever having been harmed at all*. (Perhaps one is a well-trained, but extraordinarily fortunate veterinarian, who has never been called upon to put down a cat.) The point is, there are lots of ways of acquiring abilities, and some of those ways may *never* involve the *exercise* of the ability one learns, hence never provide the kind of distinguished set of conditions for testing the ability that Millikan has in mind.

Things are even worse, it seems to me, for unlearned abilities. Consider again my ability to sing the *Mr. Ed* song. Clearly, I learned how to sing this particular song. But unless I'm greatly mistaken about my youth, I never learned how to sing, *simpliciter*. Nobody taught me how to reproduce melodies or rhythms; it's just something I found myself able to do. So singing is, for me at least, a *native* ability. In that case, the distinguished circumstances, the ones in which I must be disposed to succeed in singing if I try, are, I presume, the conditions in which singing was selected. But there's an obvious problem: What reason is there for supposing that singing was selected *for* at all? Millikan can hardly claim to know that singing was selected for, much less what the conditions were within which it was selected. One might plausibly claim that singing is but a sub-ability of the larger ability to *vocalize*, and that vocalization *was* selected for. But this would do Millikan no good. The Normal conditions for communicating vocally just are not the same as the conditions under which we would particularly expect singers to be successful if they tried to sing.

To take a starker example, consider another ability of mine – the ability to crack my toes. [Insert toe-cracking demonstration here.] Now this is clearly something I am *able* to do – after all, I just did it. But there are no particularly propitious conditions for manifesting this ability. *Or*, if some are better than others (it's easier against hard-soled shoes), it's certainly not because they have something to do with the conditions under which my toe cracking was selected for – because toe cracking was not selected for at all!

Let's review. Millikan wants to characterize concepts in terms of *abilities*. For that, she needs a principled taxonomy of abilities, one that will answer such questions as "Is the ability to kill cats the same ability as or a different ability from the ability to kill dogs?" Standard dispositional accounts of abilities seem to Millikan to founder on the difficulty of characterizing a distinguished set of conditions – the conditions under which trying and succeeding would constitute possession of the ability. Millikan is accustomed to appealing to Normal conditions in order to select a distinguished set of circumstances for determining what functions are; she thinks it will work for abilities, too. But what I think I've shown in the foregoing is that an appeal to evolutionary history does not deliver, or is not guaranteed to deliver, the *right* distinguished conditions. The etiology of an ability is irrelevant to the nature of the ability it is; that's why I can possess the ability to sing the *Mr. Ed* song even if I cannot manifest that ability in the same circumstances as the ones in which I learned it, and why I can possess the ability to kill cats even if I have never ever killed one (or indeed, killed any small mammal), and why I can possess the ability to crack my toes even if that ability never had any effect whatsoever on my or any of my ancestors' levels of fitness.

Let us look at Millikan's proposal in a slightly different light. This will help us see a different way in which Millikan's proposal for getting conceptual content out of abilities is inadequate. In the literature on action theory, it used to be standard to distinguish *ability* from *opportunity*. Roughly, "ability" pointed to an *internal* condition of success, and *opportunity* to an external one. Admittedly, the distinction is not altogether clear: If I'm tied up, do I have the ability to run away, lacking only opportunity? But it doesn't have to be perfectly clear, because it doesn't have to do the kind of heavy lifting Millikan needs done. Millikan wants, in effect, to reconstruct *opportunity* in terms of her own notion of Normal conditions. But the examples above, and many more, show that this doesn't work. *Opportunity* is just a much broader and less ruly notion than Normal conditions. In particular, opportunity can exist even when there are no Normal conditions on the horizon.

Why does Millikan need the narrower conception? She needs to have some *distinguished* set of circumstances associated with each ability, because abilities, to be concepts, have to have intentional contents. There has to be a determinate – and univocal – answer to the question of what a mechanism has the ability to do. But as I've just argued, there is no reason to think that abilities and mechanisms (or structures) line up in this neat fashion. While it seems plausible that my vocal apparatus was selected for its role in oral communication, it still, once in place, also enables me to sing. The particular structure of my toes may have been selected for its role in enabling me to walk, or it may just be an allometric byproduct of selected-for finger structure; either way, having it gives me the ability to make noise with my toes.

In light of these considerations, I suggest that there is a way of viewing abilities that is different from Millikan's. It is one that, by the way, takes account of much of her criticism of conditional analyses of "is able." It says that an ability always involves some kind of categorical property, one that is the ground of a complex of dispositions, a property possession of which *enables* a creature (or device) to do something, provided there is opportunity to do it. We could even say, without too much damage to intuitions, that the ability *is* this categorical grounding. That might make my ability to sing the same thing as my ability to talk. Alternatively, if you don't want to make this identification, we would say that I am able to sing and talk *in virtue of* the same physiological structure. (Maybe the neurology involved is different. If so, then they wouldn't be the same abilities on the identification alternative.)

Looking at abilities this way, it always makes sense to ask what properties of a thing are responsible for its abilities, to ask *in virtue of what* can this person do X? Often, when we are talking about overt behavioral abilities, like swimming or standing on your head in the commuter train, the answer will involve properties like having muscles with a certain degree of strength and having neural pathways of a certain sort. But abilities that involve cognition (and I leave it open whether, for creatures like ourselves, this includes almost everything we are able to do) will have as their categorical grounds our possessing concepts. It is in virtue of our having concepts that we are, *inter alia*, able to reidentify objects.

If we actually identify abilities with their categorical grounds, we can agree with Millikan that concepts are kinds of abilities, but we no longer need to answer the question what determines the content of a concept *by* trying to figure out what an ability is an ability to do. Also, we needn't privilege the ability to reidentify objects in our investigation of the nature of concepts. In fact it will be neither necessary nor desirable to find some privileged ability concepts afford, some fundamental or primary function of concepts. Instead, it will become necessary to survey *all* the things that having a concept enables us to do. An adequate account of concepts must explain *everything* that having concepts enables us to do.

Concepts

When I speak of Mr. Ed, you may and you may not know who or what I am talking about. (You may know exactly what I am talking about and wish that you didn't.) You may have any of a large variety of attitudes toward Mr. Ed. You may *wonder* who or what Mr. Ed is. You may know that *The Mr. Ed Show* was a popular television show in the early 1960s, and the eponymous main character was a talking horse (played by a palomino named "Bamboo Harvester"). You may now remember who Mr. Ed is. You may hate Mr. Ed. You may now remember that you hate Mr. Ed.

The point is, you all have the ability to *think* about Mr. Ed, and his TV show, and you had both these abilities even if and when you didn't have the ability to *recognize* Mr. Ed. You have the ability to think about Mr. Ed in virtue of having a *concept* of Mr. Ed. Some of you, in fact, used your concept of Mr. Ed to *wonder* who the hell Mr. Ed was. Those of you in this position got your concept of Mr. Ed from reading about him just now. You didn't have, in Millikan's terms, much of a conception of Mr. Ed. She and I agree that it's a mistake to confuse concepts and conceptions. But you were able to think about him, and now that I've told you who he is (or was, alas, he's no longer with us), your conception of him has been enlarged.

So if someone put a gun to my head, and forced me to say what I thought the *function* of concepts was, I'd say this: concepts are for thinking about things with. That's it. If one wanted to identify a concept with a single function, it ought to be this. But as I've been arguing, if we approach both the issues of what an ability is and what a concept is without prejudice, there is no need to pick one function apart from all others. What we ought to want to know is what sort of thing a concept is that it can be used to think with, *inter alia*. What I wanted to illustrate with the Mr. Ed example is two interesting features of concepts: first, that they can be – in some yet-to-be-explicated sense – cognitively minimal, and second, that they are "separable." To say that they are cognitively minimal is basically to say again that they are distinct from conceptions; they are not clusters of beliefs, they are the stuff beliefs are made of. This much Millikan agrees with (I believe). To say that they are separable is to say that they can occur – be tokened, if you will – in the *absence* of their extensions. This makes them extremely handy. You can use them to remember things, to imagine things, or just to refer to things, without having to get hold of the thing itself. Concepts are our internal proxies for objects and substances. This is a feature of concepts that I think Millikan sorely neglects.

Concepts also, of course, make up beliefs, and beliefs are involved in the acquisition, storage, and utilization of information. So part of understanding what concepts are is understanding what beliefs are. And here are a couple of things that we know about beliefs:

1. Beliefs can figure in inferences, both theoretical (where the conclusion of the inference is another belief) and practical (where the "conclusion" is an action). In the latter case, they interact with desires.
2. Beliefs display opacity – I can believe that Brian Warner is a nice young man without believing that Marilyn Manson is a nice young man, despite the one's being the other.

Time is short, so I'll cut to the chase. The best overall account of these (as well as other) features of belief, and propositional attitudes generally, is the Computational Representational Theory of Mind (CRTM) according to which the mind does its thinking in an internal medium of representation, the Language of Thought (LOT). Beliefs on this view are sentences in the LOT, and concepts are vocabulary items in the LOT. Concepts and beliefs have both syntactic properties and semantic properties. The mind's computational machinery is sensitive to the lexical and syntactic properties of strings in the LOT, and the machinery is built ("designed by nature" if you like) to effect state transitions that mirror rational relations among the semantic values of the LOT strings. The CRTM explains how the rationally relevant properties of thought contents can be *causally* relevant to the operation of a physical mechanism. This goes some way toward accounting for the role of belief and desire in the production of human action.

But it is the opacity phenomena that I want to focus on. On the LOT view, the difference between believing that BW is a good kid and believing MM is a good kid is explained by the *lexical* difference between two mental sentences. Lexical elements serve, in effect, as "modes of presentation." (Fred Adams has also made this suggestion.) Now you'd think that there's a lot Millikan would like about this picture: it enables us to distinguish concepts from conceptions (because it distinguishes parts of beliefs from beliefs); it does not confuse our means of identifying a thing with what is identified (because a concept is not, *per se*, a method of identification – one needs perception and memory and other beliefs for that); and it does not commit one of the central sins of which Millikan convicts mode-of-presentation views – it does not project structural features of extensions onto features of the vehicle for thinking about those extensions. That's one of the great advantages of *discursive* representation over other forms, like pictorial representation, as Millikan herself points out. But, of course, there's another almost immediate consequence of the LOT view, and that is the possibility of false identity judgments. Nothing about either the world or our internal cognitive economy guarantees, on the CRTM view, that distinct lexical items in the LOT designate discrete things in the world. LOT does not "externalize difference" in Millikan's terms. This is the thing Millikan wants to deny is possible.

Millikan claims to have a general argument against all forms of mode-of-presentation views, including LOT. This argument comes in two stages: first, she argues against Strawsonian modes of presentation, and then second, she attempts to show that *all* modes-of-presentation views are actually versions of the Strawsonian view. I'm willing to spot

Millikan the first part (for one thing, I'm not very clear how Strawson's view works), but I dispute the second part.[1]

LOT is, according to Millikan, a version of the "duplicates" model of mental representation, according to which I represent an object as the same by repeating or duplicating a mental symbol that stands for that object. So far so good. But then Millikan goes on to say that an identity judgment, on such a model, could only amount to the melding together of what had previously been two distinct mental symbol types: one could only come to realize, e.g., that Cicero is Tully, by putting "... all the Cicero and Tully information into sentences using the same mental name ..." (OCCI, 160). That is, the identity conditions for mental symbol types have to be functional; if what were two symbol types come to have the same functional profile, then, Millikan argues, they cease to be two different types. As she says, "Whatever the individual mind/brain treats as the same mental word again *is* the same word again" (OCCI, 166–167). Millikan concedes that there appears to be one kind of "duplicates" system that doesn't have this effect, and that is "the equals model of marking" (OCCI, 137). On this view – which I'm happy to say is the LOT view – an informative identity judgment consists in two type-distinct mental symbols related by an "equals" marker, a mental representation of the identity relation. But in the end, Millikan argues, this model still collapses into the old duplicates view.

I think there are two important errors in Millikan's arguments for this position. One argument involves the general theoretical point that formal axiom systems are intertranslatable with rule systems, without loss of computational power, whereas *rules* are ineliminable. This last is because there must be some principles guiding transitions from one line of the demonstration to another, or, if we are talking about physical mechanisms, principles in accordance with which the mechanism's state transitions occur. Call the principles involved "implicit rules." Then we can put the point this way: systems involving explicitly represented rules are fungible with systems involving only implicit rules, and hence are in a sense eliminable. But implicit rules are ineliminable.[2]

So far so good. But Millikan goes wrong in thinking, as she seems to, that it follows from this general theoretical point that there is no fact of the matter whether any particular physical mechanism is performing its inferences by means of explicitly declared rules, or by means of "implicitly represented" rules. Millikan describes a physical device that "performs" universal instantiation and *modus ponens* by first matching the shape of a piece of putty, and then reshaping it. (Properly shaped putty – "representing" that a is A – falls through one hole, and then is deformed (transformed into putty that "represents" that a is B) to go through a second.[3] The whole system, then, has "inferred" from a is A to a is B,

---

[1] I cannot really address Millikan's claim that LOT is a version of Strawsonian modes of presentation since, as I said, I don't understand what these are supposed to be. At the same time, Millikan misunderstands LOT, and so the version of it that she likens to Strawson's is not a version that I would defend in any case.

[2] For further elaboration of this point, see my "In Praise of Loose Talk: Three Ways of Following a Rule." (Antony, ms).

[3] I would count this as a form of "rule-following" only in the weakest sense I would allow: such a process might count as an "intelligible causal" process – a process that mimics an intelligent inference under an interpretation of the informational contents of its input and output states. But it is not a "rational causal" process precisely because it does not involve structured representations. See "In Praise of Loose Talk" (Antony, ms).

without any explicit representation of the needed inference rule "For all x, if x is an A, then infer that x is a B.") But it's also quite obvious that this device is *not* utilizing a representation of "All As are Bs." There's no candidate element of the system that is playing the causal role that would need to be played in order for us to denominate that element a representation *of* that generalization. So the theoretical point is idle: that there is no difference *in principle* between axioms and rules of inference hardly entails that one cannot mark the distinction in the case of the operation of a particular machine. It is hardly true, as Millikan asserts, that the latter difference is "altogether chimerical" (OCCI, 165).

Note that, in Millikan's example, lots of features of the "representations" that are potentially rationally relevant are invisible to the machine. In her example, the putty has propositional content – that *a* is A. But her machine is not sensitive to this propositional structure. It is not sensitive to the fact that the putty is "about" individual *a*, rather than individual *b*. One could mark that difference on the putty in all sorts of ways, say by coloring putty that's about *a* differently from putty that's about *b*, but it will not help the machine discriminate the two. So this machine, for all that it can do, will be unable to perform inferences that depend on realizing that we're talking about one thing rather than another. It is then an empirical constraint on theories of a mechanism that we explain how it can make any discriminations – and inferences based on those discriminations – that it manifestly makes.

Here is where the CRTM comes into its own. The point of positing a LOT, together with a syntactic engine sensitive to lexical and syntactic difference, is to account for the extremely general inferential powers of the mind. Not just any computing device can do what the human mind can do; so not just any way of realizing any of the various cognitive functions the human mind carries out will do. Again, it may be true *pretheoretically* that, as Millikan says, "Ways that various individuals' inferencing systems are put together can themselves be representations ..." (OCCI, 164), but that doesn't mean that there are not additional constraints that will decide the question. Ways of representing, when we are trying to account for specific phenomena, are not necessarily fungible.

So I don't think Millikan has the basis she thinks she has for thinking that any system that treats identity judgments as explicit declarations can be simply reinterpreted as one that "computes" identity by means of some rule governing the treatment of certain kinds of symbols. But there is a second error at work.

Millikan says repeatedly that for a mental symbol to be type-identical with another mental symbol is for it to be "treated the same" (e.g., OCCI, 166–167). But she seriously equivocates on "treat the same." One way in which the mind could treat two symbols the same is if the mental machinery were insensitive to physical differences between the two tokens. In that case, despite the fact that the symbols involve differently colored neurons, let's say, the machine treats them as the same, much as letters of the alphabet are treated as the same by readers even if they are written in different colors or fonts. This just means that there is a dimension of difference along which tokens can vary to which the machine is insensitive. But there is a second way that we might speak of the mind's "treating symbols as the same": it could be that the system as a whole produces the same downstream *consequences* by means of these two symbols. The mind might do this, for example, if it learns an identity. If it learns that Marilyn Manson is Brian Warner, it then may (as Millikan suggestively puts it) pool information stored under the two different symbolic markers. But there's no reason to deny that this can be done without the machine's losing its ability to discriminate between the two symbols.

The equivocation at work here is, ironically, one that Millikan herself warns us against: between the *identity of vehicles* (what the cognitive machinery treats as the same symbols) and the *identity of contents* (what the person, for any of a variety of reasons, comes to recognize as the same thing). It's a kind of use/mention equivocation, and it's pervasive in Millikan's book.

Consider: as an ordinary speaker of English, I can recognize that Cicero is Tully, and still distinguish the *symbols* "Cicero" and "Tully." (This has nothing to do, by the way, as Millikan suggests it does, with public conventions of use.) I then can perform two distinct practical inferences:

> A. Cicero is fun at parties
> I want to invite fun people.
> _____
> [Straightaway I invite Cicero]
>
> B. Tully is fun at parties.
> I want to invite fun people.
> _____
> [Straightaway I invite Tully]

As a rational agent who knows that Cicero is Tully, I'm indifferent to which inference is performed. Either way, I get the guy at my party. That hardly means that there is not – much less that there *could not* – be a difference between those two inferences. I have treated Cicero and Tully as the same, as well I ought, but I have not treated "Cicero" and "Tully" as the same.

I have a diagnosis as to why Millikan is insensitive to this difference. The key comes in the following quotation:

> ... [W]hat substance concepts are initially *for* is grasping, which requires marking sameness in substances. But since there is no difference between marking sameness and fixing identity beliefs, it follows that there are no representations that *are free from the possibility of empirical error*. (OCCI, 173)

What is it, I ask, for a single *concept* to either "be correct" or be "in error?" Think of Mr. Ed. Now, is what you did either right or wrong? The question makes no sense. All that can be right or wrong is a judgment, a complete thought about Mr. Ed: "That's Mr. Ed" or "Mr. Ed was a beautiful horse." The fact is that there are no correct or incorrect tokenings of substance concepts – not even from a biofunctional point of view. I'm not misusing my concept of Mr. Ed if I use it to think about him in his absence, nor am I misusing it when I think a false thought about him, including a false identificatory thought. It's *because* my concept is being used properly – as a way of thinking of Mr. Ed – that I am *able* to think false thoughts about him.

This shows what's confused in Millikan's notion of an "equivocal" concept. Equivocation, according to Millikan, occurs "when more than one substance has become tracked under one concept" (OCCI, 97–98). This is the sort of thing, Millikan says, that happens all the time: "Each semester when I acquire a new class of freshmen I go through it again, making embarrassing mistakes because I have got Johnny and William or Susan and Jane mixed together in my mind" (OCCI, 98). But if Millikan has an *equivocal* concept – call it SUSANJANE – she has a concept that refers to Susan on some occasions, and to Jane on others. In that case, it is difficult to see how Millikan can ever make a mistake. To see this, let's suppose that there is something true of Susan and false of Jane. Susan, but not Jane, made a brilliant comment in the first class meeting. Jane, in fact, was paying no attention whatsoever to class. Suppose that by the end of the day Millikan has lost the ability to discriminate Susan from Jane. Later that evening, reflecting on her teaching experiences, Millikan thinks, "SUSANJANE promises to be an excellent student." Now either Millikan thought truly of Susan that she is a promising student, or falsely of Jane that she is a promising student. But the factors that determine concept extension for Millikan – the source of the information – determine that in this case, "SUSANJANE" refers to Susan. If Millikan had thought, instead, "SUSANJANE looks to be hopeless," she would be deploying information about Jane, and hence "SUSANJANE" would refer to her. The information source, while disciplining the tracking, and hence the extension of the concept, simultaneously keeps Millikan accurate. Millikan cannot get what's needed to think something false; her equivocal concept will just shift reference as needed.

The idea that a concept is an ability only further obscures the main point. It facilitates a confusion between the mere tokening of a concept and the utilization of a concept to think a complete thought. Now Millikan addresses this concern. She says:

> There are lots of ways to do things right rather than wrong without making claims.... You don't make claims when you stand up to walk just because it's possible you could trip and fall. (OCCI, 172)

But this example doesn't do what she needs it to do. When we speak of something's being done "right" there *is* an implicit appeal to something propositional. What makes my walking an instance of something being done "right" is my *intention* to locomote. Had I intended to imitate a clumsy person instead, exactly the same set of motor movements would have counted as my having done something *wrong*. Normativity with respect to abilities takes us back to the normativity of intentionality, and thence to the truth or falsity of belief.

Conclusion

All in all, Millikan's initial identification of concepts with abilities, which I've argued causes her a lot of trouble, is the step at which the crucial question is begged, and that crucial question is this: Are there or aren't there mental intermediaries that enable us to behave in the way we do? In saying that concepts are abilities to *do* things – where the doing is overt and behavioral – Millikan puts herself in the company of behaviorists or behavioristically oriented psychologists and ethologists like Gibson, theorists who hold that behavior is a function solely of environmental variables, that the environment offers us "affordances" and such like. CRTM, on the other hand, contends that this is an empirically inadequate model for the explanation of human (and much non-human animal) behavior. We advocates of the CRTM point to inference and opacity, *inter alia*, as phenomena that militate in favor of the view that human behavior – at least much of it – is a function of both our environment and our representations of it. Once that point is accepted, I see no reason not to embrace the consequence that false identity judgments are possible. Thoughts may be confused, but not in the way Millikan thinks.

Reply to Antony

My good friend Louise Antony's feisty chapter deserves a feisty reply (and she is quite capable of taking it as well as giving it), so I will not mince words. Her misunderstandings of my position are deep and numerous. I will address what seem to be the most important of them.

The second paragraph of the first chapter of OCCI (the book from which Antony is working) finishes with the words "I will claim that the task of substance concepts [the only kind discussed in OCCI] is to enable us to reidentify substances through diverse media and under diverse conditions, and to enable us over time to accumulate practical skills and theoretical knowledge about these substances and to use what we have learned" (OCCI, 2). Notice that several kinds of abilities are mentioned here, not just the one that Antony discusses. I am happy to add more detail for Antony here, including, for example, that imaginative thought is a use of concepts and so forth. The next paragraph of OCCI reads:

> There is another tradition that treats a theory of concepts as part of a theory of cognition by taking a concept to be a mental word. If one takes it that what makes a mental feature, or a brain feature, into a mental word is its function, then this usage of "concept" is not incompatible with my usage here. Indeed, during the first part of this book I will rely rather heavily on the image of a substance concept as corresponding to something like a mental word (while plotting subsequently to demolish much that has usually accompanied this vision). But if a substance concept is thought of as a mental word, it must constantly be borne in mind that the category "mental word for a substance," like the category "tool for scraping paint," is a function category. My claims will concern the function that defines this category. If a mental word for a substance is to serve a certain function, the cognitive systems that use it must have certain abilities. It is onto these abilities that I will turn the spotlight, often speaking of a substance concept simply as *being* an ability. (OCCI, 2)

What often accompanies the mental sentence model of mental representation – this accompaniment being the part to which I object – is the assumption that sameness versus difference in representational vehicle type must be what represents sameness versus difference in content, that is, in reference/extension. Ignoring this problem (to which I will turn in a moment), my view is, exactly, that a substance concept is an ability to use a mental word – to iterate and reiterate it in response to incoming information about the same thing, thus using it for purposes of collecting practical and theoretical knowledge, and using it in thinking where this thinking may result in action. In saying that concepts are mental words, surely Antony does not wish to deny that mental words count as mental words because of how they get tokened and how they get used?

Mental words are representations, and given a (partly) consumer-based teleosemantics, this entails that they not only have functions (Normal effects) but that these functions (participating in mental sentences that participate in inference processes sometimes producing action) help to determine representational content. Especially important, and very much emphasized in OCCI, they figure as middle terms in inference processes. Indeed, I ultimately describe the act of reidentifying as being, exactly, a preparation for moves akin to mediate inference. *Of course* concepts are, as Antony's title puts it, "useful for thinking."

There are a couple of places where Antony says she agrees with me, but unfortunately they both involve misunderstanding of my use of the word "conception." I introduce this term in OCCI, 10–11 as different from, though not unrelated to, the more common notion of "conception" as a person's knowledge of a thing. My "conceptions" are ways of identifying a thing. Various kinds of knowledge of a thing can sometimes facilitate identification. But conceptions include, more fundamentally, ways of recognizing through perception, *including perception through words* (also through inductive inference). Having spent a chapter (OCCI, ch. 6; but see also Millikan ms) explaining how being able to recognize a word for a thing in context often provides enough of a conception of that thing to ground a genuine concept of it, I was happy to acquire from Antony a conception allowing me to have – it actually was new to me – a concept of Mr. Ed. I acquired it from reading her very first sentence naming him, given my ability to recognize that name, in context, as she repeated it.

It is good that Antony is puzzled about Strawsonian modes of presentation since there are none, especially not in Strawson. But if she means, rather, than she doesn't understand Strawson on identity judgments, let me recommend Strawson's *Subject and Predicate in Logic and Grammar* (1974) to her and all others, especially those who may otherwise think that Jerry Fodor invented Strawson's idea on identity judgments in his book *LOT 2* (2010). Let me then say a word about identity judgments and about why I believe that the idea that thoughts are like sentences can be misleading.

In language, that word tokens in different sentences have the same referent is typically indicated by using the same orthographic type over in each sentence, the same combination of sounds or letters. But in OCCI, I explained that there are lots of other ways referential identity might be marked in a representational system, lots of relations other than physical property identity that might be used for that purpose – indeed, there undoubtedly are, given what we know about brains. Just as the content of a representation is determined in part as a function of its Normal use, *identity* of content *between* representations (as recognized, for example, in taking something to be a middle term during mediate inference) is determined by how these representations would Normally be used together. If two different propositional representational vehicles would Normally be used together in a way that requires certain parts or aspects of these to have the same referent/extension, then whether or not these vehicular parts or aspects are alike in any physical way, they represent the same to that cognitive system. These parts or aspects are, for that system, tokens of the same mental type. To believe that one thing is the same as another – this might be the result, for instance, of believing a sentence that asserts identity – is the same as to identify these things in thought, thus turning their mental representational vehicles into vehicles of the same meaning type. Thinking in terms of mental sentences, hence of mental word types and tokens, prejudges the matter of identity marking in a very confusing way. It makes us suppose that when an identity judgment is made, a new mental identity sentence must enter and then permanently reside in the brain with a different representation type at each of its ends, rather than that a change is made in concepts, in the way identifications are made – a change in the representational system itself.

Confused or equivocal concepts, which so confuse Antony, are what I have put in place of language-of-thought false identity beliefs. The central claim of OCCI is that not all cognitive error consists in affirming false propositions. There can be serious error *prior* to belief. Our concept-learning systems, the systems designed to fashion our conceptual abilities, can sometimes fail to design adequate concepts. Equivocation and also emptiness are sometimes accidentally built right into to our basic representational/conceptual systems as a result of faulty learning. Antony wishes to deny this, but what are her reasons?

The most challenging question that Antony raises, I think, is why I said that a person's ability to do A consists in the disposition to succeed in doing A if they try *under the conditions in which they* (or their species) *learned to do A*. Actually, I didn't say quite that. I said "... under the conditions that accounted for their ... past successes in doing A" (OCCI, 62). In a footnote I said that these conditions were what in LTOBC I called "Normal conditions" for exercise of that ability, that is, conditions needed to explain how exercises of the mechanisms responsible for that disposition had succeeded in those past cases responsible for their having been maintained or selected for. Presumably that the Mr. Ed show was on was never such a condition for Antony's singing the Mr. Ed song, although the presence of oxygen in the air, relatively normal air pressure, and certain features of her healthy vocal tract and general physiology undoubtedly were.

Much more important, however, we need to begin with a correct description of what the ability involved actually is. Antony's current ability should not be described, say, as the ability to sing the Mr. Ed song in the key of G while soaping in the shower in a brown bathroom in Amherst at 7:15 a.m. on Tuesday. Her ability is much more general than that. It is, probably, an ability to sing any song within her vocal range that she remembers, to remember songs that she has learned, and to learn songs that are not too difficult by hearing them enough times. As for the ability to sing songs within her vocal range, that she woke up one morning as a child with this ability without ever having practiced or even tried before to produce different pitches with her vocal chords seems implausible. Surely she had at least observed/learned how to make her voice go up and down. One of the most unique abilities that humans have, surely developed through natural selection, is the general ability not just to remember and repeat what they have discovered, either by accident or through repeated trials, how to do, but also to recombine things that they have discovered how to do into novel and often long sequences in order to achieve newly adopted ends. As for Antony's ability to learn and remember a song, this would also seem to be but an instance of a much more general ability humans have to learn and remember all kinds of sequences. But whether Antony's disposition to remember songs was due to natural selection, or whether it is a spandrel, a mere accidental byproduct of design for other abilities, still that she can use the mechanisms supporting this disposition, along with dispositions for her voice to go up and down upon contracting certain muscles, for the purpose of singing the Mr. Ed song is as much an ability as the ability purposefully to use her ears and nose for supporting her eyeglasses. To exercise this ability she needs oxygen, no one's hand over her mouth, not to be in free fall parachuting at the time (inside joke), no lung paralysis, and so forth. She needs what Antony later calls "enabling conditions," conditions that are needed for the mechanisms that express these abilities to do their work. Presumably, these are conditions which were fulfilled and operative when these mechanisms were selected.

The main point about abilities that I was trying to make is nicely illustrated by a problem in the suggestion Antony makes on the nature of abilities. She suggests that an ability is "... the ground of a complex of dispositions, a property possession of which enables a creature (or device) to do something, provided there is opportunity to do it." First we need to unpack "enables a creature to." This cannot mean "affords the creature the ability to," for that would produce circularity. Nor can it mean "disposes the creature to," for in that case letting go of the creature in mid-air will provide opportunity for it to exhibit its ability to fall. We need somehow to prevent people from having "the ability" to fall or to trip or to drown, given that those who do so do indeed have mechanisms that enabled these doings/happenings when there was an opportunity. This could be accomplished, of course, by quietly packing into "enabled" and "opportunity" that, necessarily, these people were trying to do just these things at these times. But bringing all into the open, what we really seem to need is something like "makes it so that the creature will succeed in doing it if it tries." And we would also need to take "if it tries" very generally as, say, "if it should become its biological or psychological purpose to do it." Thus the first point here is the teleological point. Abilities have to do with purposes.

But having added purpose, there is also a second problem with Antony's analysis. For people can succeed in what they intend or try by accident (a point familiar in the literature on intentional action). If in trying to stop the oncoming car I step out with my hitch-hiking thumb up, thus slipping on ice into the car's path, thus stopping it, this does not show that I had the ability to stop the car. The driver might have paid no attention to my thumb; he just didn't want to run over me. But it would satisfy Antony's definition. It's only if that's how you've learned or been designed to do a thing that it's no accident that you succeed in doing it, and the proper results of abilities *are not accidents*.

6

Properties Over Substance

RICHARD FUMERTON

In *On Clear and Confused Ideas* Millikan presents an interesting, original, and thought-provoking exploration of some of the most fundamental issues in ontology and the philosophy of mind. One of her particularly striking claims is that substance concepts are ontologically and epistemologically prior to other concepts, in particular the concept of property. In this chapter I critically examine that position.

Before we are in a position to adjudicate disputes concerning the respective priority of a given sort of concept, it would be helpful to have a clearer understanding of just what is involved in possessing a concept. For Millikan, possession of a concept is intimately connected with the ability to recognize and, particularly, *reidentify* a thing or a kind of thing. In characterizing the possession of a substance concept, Millikan insists on a distinction between one's *conception* of a substance and one's *concept* of the substance. You and I might recognize and reidentify an individual in quite different ways while we nevertheless possess precisely the same concept of the thing in virtue of the fact that we *are* identifying and reidentifying the same thing. Of course, with respect to both individuals and kinds we need some criteria for determining when we have successfully reidentified the same individual or the same kind. Crudely put, according to Millikan, two things are the same kind of substance when they share a common underlying structure that causally explains a great many of their more superficial characteristics, an underlying structure that allows one to make successfully certain sorts of inferences or projections based on relatively few or even a single observation of a member of the kind.

Because her characterization of substance is so closely linked to recognition of an individual or kind about which one can successfully make certain inferences, Millikan sometimes suggests that the category of substance is an *epistemological* category (OCCI, 26). On the other hand, she also makes clear that the existence of substances and the question of whether a given object falls under a substance category is fundamentally a matter that is quite independent of conscious beings, their representations, and their capacity to make inferences. As I indicated above, her more *metaphysical* characterization of substance seems to rely heavily on the notion of an underlying structure that causally explains similarities between things at a time and over time. While the existence of such underlying structures may *explain* epistemic success in projection, the underlying structures are presumably what is essential to the existence of a substance kind. While I don't agree with some of the views sketched above (in particular the liberal criterion for sameness of concept), I won't challenge them in the context of this chapter.

As noted earlier, Millikan argues that the concept of substance is more fundamental than the concept of property. This claim about conceptual priority can be understood in a number of different ways. We should distinguish all of the following kinds of claim:

1   The concept of X is more fundamental than the concept of Y in virtue of the fact that we always or typically acquire the former before we acquire the latter. So, for example, many philosophers would argue, *contra* the British empiricists, that our concept of the external world is prior in this sense to the concept of subjective appearance. It's not, perhaps, that we can't develop a concept of being appeared to a certain way, but it is (the argument sometimes goes) a skill that it is rather difficult to acquire – the sort of skill, perhaps, that a painter works hard to acquire.

2   The concept of X is more fundamental than the concept of Y in virtue of the fact that the concept of Y can be analyzed, in part, into the concept of X. So, for example, almost all moral philosophers would concede that the concept of being intrinsically good is more fundamental in this sense than the concept of being instrumentally good. We can define something's being instrumentally good in part by reference to its causal power to produce that which is intrinsically good.

3   The concept of X is more fundamental than the concept of Y in virtue of the epistemic fact that we can recognize Xs much more easily that we can recognize Ys. Again, critics of radical empiricism have often claimed that the radical empiricist is just wrong in suggesting that we must base our knowledge of physical objects on our knowledge of subjective sensation. As a matter of contingent fact (the argument goes) we have a much easier time recognizing familiar facts about our physical environment that we do recognizing subtle changes in the character of sensory experience.

4   The concept of X is more fundamental than the concept of Y in virtue of the linguistic fact that we typically develop a language to describe the fact that something is an X before we develop a language to describe the fact that something is a Y. To be sure, 4 may collapse into 1 if we assume that concept possession is inextricably tied to language. But it is an understatement to suggest that in the history of philosophy that claim is more than a little controversial. So a philosopher might claim that we have a grasp of subjective difference in experience long before we have the capacity to describe such differences. An adverbial theorist, for example, had probably better admit that we had a language to describe physical objects long before we were able to talk about the fact that we are appeared to redly-and-roundly!

So in what senses, if any, is the concept of substance more fundamental than the concept of property? Put another way, in what senses, if any, is our ability to recognize or represent substance more fundamental that our ability to recognize and represent properties?

Before answering this question, we might understandably worry about just what exactly the difference is between the categories of substance and property. So there is the relatively uncontroversial fact that we use and understand *nouns* like "water," "gold," "human," "George Walker Bush," "Paris," "mother," "father," and so on. We also use and understand *adjectives* like "yellow," "round," "wet," "generous," "smooth," and so on. I suppose we could proceed as neutrally as possible by stipulating that the first list of nouns is the list of terms picking out substances, and the second list of adjectives is the list of terms picking out properties. There is, however, a danger that we are making too much out of mere grammatical difference. Philosophers have not been shy about simply converting common nouns into adjectives in their attempts to render logically perspicuous the content of various claims. Indeed, contemporary predicate logic is largely insensitive to the linguistic distinction noted above. The sentence "Gold is yellow" is virtually always translated into predicate logic as the following quantified claim: $(x)(Gx \supset Yx)$, where "G" represents the property of being gold and "Y" represents the property of being yellow (more linguistically, where "G" and "Y" are treated as predicate expressions). The bound variable takes as its values the various *particulars* to which the predicates apply.

Now one should no more draw exotic ontological conclusions from the particular idiosyncratic features of contemporary predicate logic than one should draw exotic ontological conclusions from contingent features of the grammar of natural language. There certainly is a grammatical distinction between the statement "This is gold" and the statement "This is yellow." The former completes the copula "is" with a predicate noun, the latter completes the copula with a predicate adjective. Perhaps one should worry about uncritically treating "gold" and "yellow" as both predicates. But those who think that common nouns are typically used in assertions to make general claims about things exemplifying certain properties are surely right to be relatively indifferent to the grammatical difference noted above. As I shall argue below, if they are right in their understanding of common nouns, claims about the relative priority of our understanding of substances over properties receive no support from certain empirical claims about our acquisition of common nouns before adjectives.

Even if it is a fact that we could do perfectly well without common nouns (translating sentences containing them into sentences containing only adjectives), there remains, of course, the fact that our language contains not only common nouns, but *proper* nouns, demonstratives, and indexicals. And it is, of course, far more difficult to replace their use in sentences with translations employing adjectives. To be sure, philosophers have tried. Russell famously suggested that the key to understanding informative identity claims is to translate all such claims into quantified claims about the unique co-exemplification of various *properties*. Proper names flanking informative identity claims *disappear* in the correct analysis. And even demonstratives and indexicals flanking informative identity claims disappear in the logically perspicuous translations of these claims. While the descriptivist's position has fallen on hard times these days, it remains a puzzling feature of contemporary philosophy of language that the increasingly popular direct reference theorists and causal theorists have a horrible time duplicating Russell's success in explaining referential opacity. I have argued elsewhere (Fumerton 1989) that the descriptivist can always "steal" whatever is plausible in direct reference theories while retaining all of the virtues of their descriptivism, not the least of which is the descriptivist's ability to give a plausible account of informative identity claims.

The Russellian descriptivist project was not, by itself, intended to show that we do not need the category of substance or particular. While ordinary names disappear in the proposed reduction, there remain bound variables whose values could well be construed as substances or particulars. Still, it seems obvious that the success of a Russellian reductionist program might well cast serious doubts on claims that our ability to *represent* substances is not parasitic upon our ability to represent *properties*.

Let's return then to Millikan's claim that our concept of substance is more fundamental than our concept of property. As I indicated earlier, I found the claim intriguing precisely because I have always thought exactly the opposite. I have always thought that our concept of property is both conceptually and epistemically prior to our concept of both individual and substance. Like most of the British empiricists, it seems to me almost obvious that we have a clearer grasp of properties than we have of the objects that exemplify them. And my reasons are primarily phenomenological. So-called "bare" particulars or Aristotelian substances have always struck me as ontologically "spooky" – phenomenologically inaccessible, necessary, if at all, only for purely dialectical reasons. Like many radical empiricists, I am so attracted to the view that we have a clearer understanding of property than we have of the bearers of properties that I am sometimes even drawn to a perhaps quixotic attempt to reduce individuals to "bundles" of "particular" properties.

Millikan is not, of course, alone in thinking that the concept of property is, in some sense, more problematic than the concept of substance or particular. Quine famously suggests that we identify our ontological commitments with our willingness to view an entity as taking the value of a bound variable. When I assert that there is an x that is F, I commit myself to the existence of whatever takes the value of "x." But, according to Quine, I apparently get an ontological free lunch with my use of the predicate expression "F." This too has always struck me as a baffling view given that the predicate expression is obviously just as relevant to representing the content of the quantified statement as is the bound variable.

The Quinean "prejudice" against properties finds expression in the closely related and equally bewildering linguistic nominalism that seeks to identify the meaning of a predicate like "red" with its extension either in the actual world or through possible worlds – with the class of actual or the class of possible red things. A novice to philosophy can't help but wonder what the relevant *principle* of class membership is. And when told that a thing enters the class of red things in virtue of the expression "red" applying to it, that novice will typically think you are making a bad joke. The philosophically unsophisticated know, just as you and I know, that we must look beyond language to find something in the world – the *property* – that makes red things red. But most of this is merely rhetorical (albeit rhetorically effective) hand waving. Can we muster arguments for or against the relative priority of our understanding of property over our understanding of substance?

In advancing her thesis Millikan does point to some intriguing empirical evidence regarding language acquisition. In particular, she appeals to evidence suggesting that we acquire the ability to use concrete nouns and terms for kinds earlier than we acquire the ability to use adjectives, where adjectives are construed as the kind of term that picks out properties (OCCI, 69). If common nouns or proper nouns do pick out something other than properties, this alleged empirical fact might trivially entail that substance concepts are more fundamental than predicate concepts in sense 4 defined above. But would this linguistic fact constitute even *prima facie* evidence for the conclusion that one *thinks* of substances before one thinks of properties (that a thesis of conceptual priority in sense 1 distinguished above might be true), that one can *recognize* substances before one can recognize properties (the epistemic priority defined in 3), or that one cannot *define* the concept of substance in terms of the concept of property (the conceptual priority defined in 2)? It's not obvious what the connection between the alleged linguistic fact and these other claims would be. And furthermore, we have already had occasion to question whether the use of common nouns is anything but a grammatically disguised way of talking about things having certain properties. If in an "ideal" language one can philosophically treat common nouns as if they were predicate expressions, then why shouldn't we think of learning how to use common nouns as really just a way of learning how to use predicate expressions?

Why do I find so odd the suggestion that our concept of substance is more fundamental than our concept of property? Well first, and foremost, it seems almost obviously true that we identify and reidentify individuals and kinds of things only *through* properties or characteristics those individuals and kinds exemplify. The baby recognizes its mother. I'd be willing to bet my house, however, that if we cleverly disguised the mother the baby wouldn't recognize her (though we might, of course, need to disguise not just her visual appearance, but her voice and scent, for example). And that is surely *prima facie* evidence that the baby's recognition of the person as her mother is entirely parasitic upon the baby's recognizing *at some level* properties or characteristics that the mother has. Similarly I can recognize a dollar bill, but even a pretty bad counterfeiter could present me with a cleverly disguised piece of paper that looks to me just like a dollar bill and that I would, consequently, mistake for a dollar bill. And that surely suggests that my ability to recognize something as a dollar bill crucially relies on my recognition of the properties or characteristics that thing has. The same is true for virtually all recognition of both particular things and things being of a certain kind. *The only reidentification that doesn't seem obviously to rely on the recognition of properties exemplified by that which is reidentified is recognition of properties themselves.* We need to avoid a vicious regress of recognition just as we need to avoid more generally a vicious regress of justification.

Now Millikan might well take issue with the rather primitive argument presented above for the epistemic and representational priority of property over substance. She might well respond that while both initial identification and reidentification of individuals and kinds would be impossible without a *causal* role played by the properties exemplified by those individuals and kinds, it doesn't follow that we need to *represent* or *discover* the exemplification of those properties in recognizing the individuals. Brain states, after all, play a critical causal role in our ability to represent anything, but it hardly follows that we need to represent brain states prior to our acquiring the ability to represent other sorts of things. That suggests that we might add a causal dimension to our list of ways in which one kind of thing might be more fundamental than another. Properties might be more fundamental than substance in the sense that property exemplification plays a fundamental role in the *causal* explanation of how people identify and reidentify both particulars and kinds of things. Indeed, on certain traditional conceptions of explanation, that thesis might be necessarily true.

A similar sort of distinction is often invoked by epistemologists in rejecting the kind of foundationalism that insists that empirical knowledge rests on knowledge (and hence representation) of subjective mental states. Most of these philosophers do not (certainly need not) reject the claim that there is such a thing as sensation, or even that sensations cause us to believe what we do about the world around us. In rejecting classical foundationalism, they need only deny that propositions describing such sensations serve as the premises from which we must infer those other propositions believed.

Still, it is difficult for me to see what a representation of substance would be without representation of properties. On some popular externalist accounts of representational content (accounts which I take to be mistaken), one represents X as being of a certain kind in part in virtue of a causal connection between the things of that kind and the kind of internal state that is the representation of X. The crudest (and wildly implausible) view maintains that I can represent X as being a certain kind of thing only if I personally have causally interacted in the right sort of way with at least one member of the kind. A still crude, but *slightly* more plausible view, allows me to "build" the concept of a kind F out of simpler concepts that satisfy the relevant causal constraints.

The correct philosophical analysis of causation is hardly a matter upon which philosophers agree. Nevertheless, the most plausible approaches to understanding causation, I would argue, are all generality theories. They take the relata of the causal connection to be something like the occurrence of a state of affairs, or the occurrence of an event, where the critical constituents of the state of affairs or event relevant to the obtaining of the causal connection are *properties*. I don't think that one can even state a coherent regularity theory of causation without thinking of the relevant regularities in terms of the constant correlation of property exemplification. And if one turns to necessary connections between universals to understand causal connection, it is the exemplification of the relevant universals/properties that yields the causal connection. The upshot is that what gets represented on a causal theory of representation is either a property or something that essentially involves a property. And if that is true it is hard to see how representation of substances or particulars can be prior to or more fundamental than the representation of properties.

There are, of course, other externalist accounts of how our language and thought reaches out to the world. So Kripke and his followers famously argue that we can introduce an expression into our language through a reference-fixing definite description. We can run across a person with a certain sort of dementia and "dub" the cause of these symptoms Lewey-Body disease. On the theory (if ever there were a truly magical theory of reference this would be it) one can "discard" the definite description after introducing a name with it and subsequently use the name as a "directly referring" expression for the kind of disease in question (assuming that there actually is just one disease responsible for the symptoms). I've argued elsewhere (Fumerton 1989) that one can't really make sense of reference-fixed rigid designators (though one can make sense of conventions governing the interpretation of modal operators used in connection with names introduced with definite descriptions). But whether or not I'm right about this, it seems that there can be little solace in this conception of how we successfully acquire competence using names for the philosopher who wants to think of our grasp of things or natural kinds as prior to our grasp of properties. The definite descriptions that fix the reference of names for individuals and kinds critically involve the use of predicate expressions, understanding of which is critical to the definite description performing its reference-fixing function. Of course the predicates contained in the reference-fixing definite description might themselves have their reference fixed by still other definite descriptions, but again regress lurks. And the only way the regress ends is with a definite description whose predicate expression is understood "directly." Again, it looks as if we will eventually need representation of properties to secure meaningful use of names introduced using definite descriptions.

But if all this is true, what are we to make of the linguistic data to which Millikan refers? Let's assume that it is true that children typically learn to use nouns before they use adjectives. We can't just ignore this data and insist that nevertheless the grasp of properties is more fundamental than the grasp of individuals or natural kinds. Even if talk of a kind is best understood as talk about things having certain properties, why is it that we acquire skill at using the nouns before we acquire skill at using the adjectives?

In addressing this question, it is important to realize that the friend of properties can make all sorts of distinctions between how one notices, grasps, represents, or comes to discover the presence of properties. On most plausible views of properties, we take notice of only a fraction of the relational and non-relational properties that are exemplified by various objects. And of this fraction, even a smaller number have the distinction of being represented by concepts. And a yet smaller fraction still get linguistically represented by adjectives that work their way into our language. What does seem entirely plausible to me is the suggestion that we tend not to notice properties individually but, rather, notice "congeries" or complexes of properties. It is simply the fallacy of division to suppose that because we notice, and even represent conceptually or linguistically, a complex conjunction of properties, that therefore we must have noticed or represented conceptually or linguistically the individual properties that make up the complex.

But while this hypothesis is surely an abstract possibility, why should anyone think that it is true? Well, this is rank armchair empirical speculation, but it surely would be plausible to suppose that there are enormous evolutionary advantages to being able to recognize the complex set of properties associated with, say, a dangerous animal like a lion. Being yellow, by itself, tends to be of little interest to an organism trying to survive. Being yellow, fast moving, and having large teeth and a mane might be something to which it would be good to have a quick reaction. Indeed, it might be downright dangerous for an organism to be too interested in recognizing isolated, individual properties even if coupled with the capacity to "put those representations together" to form representations of various complex properties. Recognition had often better be fast if it is to serve any sort of useful purpose, and losing the forest for the trees could easily spell disaster.

If even some of what I have just speculated is true, we have a relatively straightforward explanation for why it is easier for us to learn the meaning of nouns than it is to learn the meaning of adjectives, and that despite the fact that we have no access of any sort to individuals or kinds except through access to properties. Many adjectives function in language to isolate single properties. To put it as neutrally as possible, proper names and common nouns have associated with them networks of non-accidentally correlated property exemplifications. Because we recognize patterns earlier than we recognize the constituents of the patterns, we have an easier time making associations between the use of names and common nouns and the complexes of properties with which they are associated, than we do making the associations between adjectives and more "isolated'" properties. The making of these associations is, of course, critical to grasping the meaning of linguistic expressions. Put somewhat cryptically, we don't have a more fundamental grasp of particulars or substances than we do of properties. We just have a more fundamental grasp of proper*ties* than we do of property.

It is possible that the view I have put forth above is only terminologically different from Millikan's view. One can simply stipulate that a substance concept is the concept of connected properties, while a property concept is the concept of an isolated property. On such an understanding of substance, I have no quarrel with the suggestion that there might be a number of senses in which our concept of substance (connected properties) is more fundamental than our concept of (individual) property. In particular, I would have no quarrel with the suggestion that the concept of connected properties is more fundamental than the concept of individual properties in senses 1, 3, and 4 distinguished above. There remains the obvious fact that the concept of property is more fundamental than the concept of connected properties in sense 2. Complexes still decompose into simples.

Reply to Fumerton

Richard Fumerton is right that my category *substance* is an epistemological category only in the sense that the ontological structure that makes something a substance is what explains how it can support knowledge. But he oversimplifies in a couple of ways. First, the point of emphasizing that substance is in its way an epistemological category is that ontological structures that are substances can be quite different in various *other* ways. In chapter 2 of OCCI, I explain:

> What makes a substance a substance is that it can be appropriated by cognition for the grounded, not accidental, running of inductions, or projecting of invariants. This will be possible in different cases for very different reasons, due to very different sorts of causes, which is, of course, exactly what interests me about substances. It is their variety, considered from other ontological perspectives, that makes it easy to overlook their similarity relative to the projects of cognition. (OCCI, 26–27)

I discuss, in particular, three very different kinds of structures that can serve this purpose, namely individuals, eternal kinds, and historical kinds, giving each its own detailed discussion. These cannot be collapsed together in the way Fumerton suggests.

Second, if you and I use quite different ways to identify the same substance we do not "possess precisely the same concept of the thing." In the first chapter of OCCI, I explain:

> … consider a child and an organic chemist. Each has an ability to identify sugar and collect knowledge about it. Does it follow that there is a concept that they both have, hence that they have "the same concept"? In one sense they do, for each has the ability, one more fallibly, the other less fallibly, to identify sugar, and each knows some kinds of information that might be collected about sugar. But in another sense they do not have "the same concept." The chemist has much more sophisticated and reliable means at her disposal for identifying sugar and knows to ask much more sophisticated questions about sugar than the child. (OCCI, 11)

The notion "same concept" is equivocal. Usually, however, it is sensible to individuate substance concepts merely by their referents/extensions, ignoring conceptions, because it would almost never be true that two people had exactly the same conceptions of a substance, and no particular set or subset of conceptions is definitional of a word for, or of "the concept" of, any substance. The important point is that there really is no sensible meaning for the expression "*the* concept of a substance" (for detail, see Millikan 2011).

Fumerton asks, "What exactly [is] the difference ... between the categories of substance and property?" That was the topic of LTOBC chapter 16, where I discussed Leibniz's law and its inverse for property identity. Starting with Strawson's idea that the basic distinction between "general concepts" and "spatio-temporal particulars" was that general concepts come in contrary ranges, the members of which compete to be exemplified by each particular (Strawson 1974), I generalized, to draw a more global distinction between 'substances' and 'properties'. Properties come in determinable ranges relative to various substance categories, such that only one determinate in each determinable range can characterize each substance in that category. But substance and property are relative categories. A substance is the same as itself (cf., Leibniz' law) relative only to a certain set of contrary spaces and not others, and a property has contraries relative only to certain categories of substance and not others. Some things, such as real kinds of various sorts (say, *gold* and *human*), are substances relative to certain property ranges but also properties relative to certain substance kinds (*gold* relative to chunks of matter, *human* relative to individuals). Even such things as *white* and *cube* are substances relative to certain properties; for example, white *shows up well in dim light*, a cube *will not balance on edge easily*, and so forth. "The same thing can be a property relative to certain substances and also a substance relative to certain properties" (OCCI, 27).[1]

Substances, then, are not ontologically prior to properties but are correlative to them. But "epistemological" priority seems to interest Fumerton more. He wonders whether one must employ concepts of properties in having concepts of substances. "... [I]t is difficult for me to see what a representation of substance would be without representation of properties," and "[t]here remains the obvious fact that the concept of property is more fundamental than the concept of connected properties in sense 2 [that is, 'in virtue of the fact that the concept of Y can be analyzed, in part, into the concept of X']. Complexes still decompose into simples." This reminds me a little of Wittgenstein in *Philosophical Investigations*:

> Then does someone who says that the broom is in the corner really mean: the broomstick is there, and so is the brush, and the broomstick is fixed in the brush?(¶60; Wittgenstein 2009, 33ᵉ–34ᵉ)

If the thing has parts and it is essential to its being this thing that it has these parts, won't the concept of it have to specify these parts? If a substance has certain properties and this is essential to its being that substance, won't the concept of the substance have to include concepts of these properties?

---

[1] These claims about ontology in LTOBC involved considerable argument, and were tied closely to the theory of negation and the epistemology of concepts developed there.

Suppose you take it that to think of an individual or of a kind is to think of some fact or set of facts about it that distinguish it from all other things. How else, you may ask, could it be the case that you were thinking of that individual or kind, and not of some other instead? In the case of individuals, this may involve thinking of properties that no other individual has. In the case of kinds, it may involve thinking of properties that define the limits of the class. If that is where you begin, you will probably agree with Fumerton. But I have departed from this view in several ways.

First, I have tried to show why real kinds, the sorts that are "substances" in my sense, are not *classes*, any more than individuals are single-member classes. Their parts (members) are not collected together by the mere fact of common properties or, say, densely overlapping properties, thinking of which would determine the kind. Real kinds are structured more like individuals and are identified, and thought of, more in the way individuals are (OCCI). Second, I have argued that to have a concept of a substance is merely to posses a mechanism that has been designed (through learning backed by evolution) to be able, in actual practice (that is, in one's own actual world, indeed, in one's own peculiar restricted environment), to recognize a variety of manifestations of that thing as they impinge on one's senses. Such abilities are highly fallible. What one's identificatory dispositions will actually effect given a current local environment is not always what they were designed to accomplish. Thus our current dispositions to apply an empirical concept do not, in some simple way, determine its object. Thinking of things clearly, univocally, has as much of an element of luck to it as does any other fairly routine activity taking place in our cluttered and contingently arranged world.

In this light, the question of "epistemological priority" concerns whether one must have and use concepts of properties in order to *identify* substances. Must one employ concepts of a substance's properties during the process that originates with the arrival of energies manifesting the substance at one's sensory surfaces and ends with identification of the substance manifested ... ends, say (it's a helpful model), by tokening a mental term for the substance? Before tackling this question directly, one more point needs to be clear.

I describe two kinds of substance concepts, calling them "practical" and "theoretical." Theoretical substance concepts are developed in the context of the use of intentional attitudes having subject–predicate structure. Possessing a theoretical substance concept does involve understanding what some of the property contrary spaces are that are correlative to that substance. It involves having concepts of properties in some of those determinable ranges. However, understanding that a certain theoretical substance has its identity relative, in part, to these and those determinable ranges is only knowing some questions it would be sensible to ask about it, not knowing the answers. I can know that John must have some definite height and weight and place of birth without knowing what his height or weight or place of birth is. In having a theoretical substance concept it is not implied by this that any particular predicate concepts are involved.

That said, how could it be, as I have claimed, that one might recognize a substance owing to the fact that certain of its properties are manifesting themselves to one's senses but without having to employ concepts of those properties?

Dogs can generally recognize their masters by their smell, or by their looks, or by their voices. Human infants can generally recognize their mothers in these same ways. Now suppose we ask exactly what kinds of property concepts might they have to employ in order to accomplish this recognition. Consider what your concepts of smells are like. Are smells properties, or are they things that have properties? Just what properties does the smell of oregano have? If you think you have concepts of these properties, enough to describe this smell uniquely, is it plausible that a dog or a three month infant has concepts of properties of this kind, concepts of the properties of smells? (It is always important, of course, not to confuse concepts with mere capacities to discriminate. Bacteria and earthworms have capacities to discriminate.)

Suppose then that instead of focusing on properties of smells we take a smell to be itself a property. Would it be necessary that the dog or the infant respond to the chemicals wafting through its nostrils first with a concept of a smell, and only later with a concept of its master or mother? Or could it respond directly with a concept of, with a recognition of, its master or its mother?

It would help, of course, in answering this, to have on hand a theory of what a concept is. On my view, a concept involves a (fallible) ability to token the same mental representation again in response to manifestations of the same thing. The content of a (descriptive) representation, on my view, is determined (not by the causes of its tokening, but) by what it needs to correspond to if its interpreters or users are to be able to perform their designed functions in a Normal way. Concepts are of what they are of because they have uses that require them to correspond, in the context of complete (i.e., truth evaluable) representations, to these things. Any cognitive response that the infant might have to a smell prior to its causing a recognition of Mama would have to have a use within the infant's cognitive or behavioral activities that required it to correspond, specifically, to that smell. But no such use has been proposed for this postulated intermediary. It's use is only to cause a tokening of *Mama*. What it has to correspond to in order to serve its functions Normally is merely Mama.

If you consider recognizing Mama by the look of her face or by her voice, I think you will see that the same kind of considerations apply. And I think they apply also to your own ways of recognizing your friends. For example, try to say exactly what general property concepts you have to apply to get from seeing your best friend's face to identifying him or her. Moreover, I think they apply to most of the ways that you actually identify water, dogs, and, for example, fear and anger. (There is more detail on all this in my reply to Nussbaum, this volume.)

7

Millikan's Historical Kinds

MOHAN MATTHEN

Ruth Millikan introduced history into the theory of functions, and so changed the game of naturalism. The central insight of her approach to functions lay in her original use of the notion of copying or reproduction. It seemed to her that if functions are based on reproduction, they look backward in time to the earlier items from which they were copied, and to the circumstances under which the copies of some things were preserved while copies of other things were destroyed. Thus, she reasons, biological functions are historical. This was not clearly evident in other analyses of function.

The historicity in Millikan's account is contested – it is not universally accepted that functions should be understood historically – as is the *kind* of historicity she postulates. However that may be, it is now (thanks in the main to her) thought to be a live option to hold that historicity has a place in the theory of universals. Straightforwardly, Millikan's account historicizes apparently functional categories such as types of bodily organs; for example, the heart. For if, as many think, the category *heart* is defined by a certain functional role in the circulatory system, then Millikan's analysis would make the history and emergence of this role important. (This is an area in which a different kind of history has sometimes been urged: see, for example, Matthen 2000.) More surprisingly, Millikan argued that her account historicizes non-linguistic meaning – she defines the representational content of perceptual and other mental states by their indicator-function. Finally, the notion of reproduction can be recruited to help with the definition of biological kinds such as species. For many think that these biological kinds are reproductive families defined by a unique originator. All of these universals turn out to have historically determined content in Millikan's system.

Like many pioneers, Ruth does not always cleanse herself of the past from which she seeks to liberate the rest of us. She bursts through the walls that enclose earlier paradigms, but fragments of brick and bits of mortar cling to her. It remains for minor figures to dust her off. In an earlier article (Matthen 2006), I tried to play such a minor figure: I suggested how her treatment of non-linguistic meaning could benefit from greater consumer orientation.[1] Here, I shall attempt to play that role again. I will suggest how her "historical kinds" can be made more historical. I shall also suggest that in one way, she goes too far with the historicity of biological kinds.

Introduction: Russell's Natural Kinds

The "tradition of natural kinds" (as Ian Hacking (1991) calls it) was initiated in modern philosophy by William Whewell, who defined a kind as "… a class denoted by a common name about which there is the possibility of general, intelligible and consistent, and probably true assertions" (Hacking 2007, 216). In this tradition, kinds are thought of not as definable properties, but as bases for generalization and induction – "Any one can make true assertions about dogs, but who can define a dog?" said Whewell (quoted by Hacking).

A paragraph from Bertrand Russell (1948) illustrates this:

> The existence of natural kinds underlies most pre-scientific generalizations, such as "dogs bark" or "wood floats." The essence of a "natural kind" is that it is a class of objects all of which possess a number of properties that are not known to be logically interconnected. [As Ian Hacking (1991, 112) notices, Russell misplaced the 'all' in the preceding sentence: he meant the "class of *all* objects which …"] Dogs bark and growl and wag their tails, while cats mew and purr and lick themselves. We do not know why all members of an animal species should share so many common qualities, but we observe that they do, and base our expectations on what we observe. We should be amazed if a cat began to bark. (Russell 1948, 335)

Russell here follows both Whewell and his (Russell's) godfather John Stuart Mill's way of capturing the significance of "classes," or predicables, such as *gold* and *tiger*.[2] In Frege's theory of Concept and Object and Russell's Theory of Types, these predicables are like *white*: that is, they subsume multiple individuals under a single class. (Equivalently, they are functions from individual objects to truth-values.) But there is a significant difference between these natural kinds and *white*. Mill wrote:

> White things are not distinguished by any common properties, except whiteness. But a hundred generations have not exhausted the common properties of animals or plants, of sulpher [sic] or phosphorus; nor do we suppose them to be exhaustible, but proceed to new observations and experiments, in the full confidence of discovering new properties which were by no means implied in those we previously knew. (Quoted by Hacking 1991, 117)

---

[1] My own work has often paralleled, sometimes diverged, from Ruth's, and so I mention some of my writings in this context, though I am not fond of self-citation. I apologize for the volume of advertising in this chapter.

[2] Two notes concerning terminology. First, I shall italicize terms when I intend them to refer to predicables or classes; when they are in ordinary script, I am using them predicatively. Thus, *tiger* is a kind, but Benji is a tiger. Second, I'll say that kinds are "classes." While it is not entirely clear from the above quotation, Russell thought that natural kinds are properties, not sets. However, I want here to distinguish them from *properties*, in view of the Mill–Russell thesis that kinds are associated with *many* properties.

As defined by Russell, natural kinds are not classifications like *white* that are based on a bounded commonality. They are rather associated with "a number of properties," some of which may not be known. Let us say, anticipating a bit, that they are unbounded "property clusters."

Russell writes:

> The bearing of all this on induction is of considerable importance. If you are dealing with a property which is likely to be characteristic of a natural kind, you can generalize fairly safely after very few instances. Do seals bark? After hearing half a dozen do so, you confidently answer "yes," because you are persuaded in advance that either all seals bark or none do. (Russell 1948, 336)

Why do these properties come in clusters? We do not know, says Russell, and for this reason, "... the doctrine of natural kinds, though useful in establishing such pre-scientific inductions as 'dogs bark' and 'cats mew', is only an approximate and transitional assumption on the road towards fundamental laws of a different kind" (Russell 1948, 461–462).

Why do natural kinds support induction? This is one traditional puzzle. Another puzzle concerns natural kind terms. Corresponding to the difference between *tiger* and *white*, there are differences between the terms that denote them:

1   *User Ignorance*: If I do not know the basis for distinguishing white from non-white things, I don't possess the concept *white* and do not understand the word 'white.' Much the same is true of concepts such as *Italian*, *dollar*, and (perhaps) *democracy*. The same is *not* true of *gold* or *tiger*. I can possess, use, and perhaps even understand the concept *tiger* even though I do not know how to distinguish tigers from other things. I may, for example, confuse tigers with leopards, but nevertheless understand the meaning of "Tigers are striped" – this is the meaning of Whewell's aphorism about dogs, quoted at the start of this section. Even experts can be wrong about the identifying marks of tigers (or of gold). They may think that tigers are always orange and black, though in fact some are white. Experts may even once have said *a priori* that nothing white is a tiger, and they would have been wrong. Nevertheless, the concept *tiger* was, even in such a time, in the possession of humans.
2   *Variety*: Finally, members of kinds such as *tiger* may not even share a distinguishing set of properties, bounded or unbounded.

With biological kinds in mind, Russell (1948, 461) writes:

> A natural kind is like what in topology is called a neighbourhood.... Cats, for example, are like a star cluster: they are not all in one intensional place, but most of them are crowded together close to an intensional centre. Assuming evolution, there must have been outlying members so aberrant that we should hardly know whether to regard them as part of the cluster or not.

The current consensus with regard to natural kinds derives from a proposal by Richard Boyd (1991). Boyd's claim is that natural kinds are "homeostatic property clusters" or HPCs. According to Boyd, these classes satisfy Russell's characterization because "homeostatic" causal influences push their members toward conformity with each other.

With *gold*, atomic structure plays this homeostatic role. That is, pieces of gold are like each other because they all have the same atomic structure. The case of *tiger* is interestingly different. A long process of selection has adapted tigers to the environmental niche that they uniquely occupy. This process has resulted in their sharing the properties by virtue of which they are so adapted. For example, they are typically brownish-orange with black stripes, because this provides good camouflage in a forest. As well, new tigers are born by a process of reproduction. This means that they will resemble already existing well-adapted tigers. Hence, they too will share the properties in question. Selection and reproduction are the homeostatic mechanisms here: tigers are like each other because they are shaped by these mechanisms. However, neither adaptation nor reproduction ensures perfect conformity. For though selection has pushed tigers toward their orange-and-black coloration, it nevertheless permits considerable variation in the actual pattern that individual tigers display. Moreover, reproduction is not perfect. Some tigers are albino because of a mutation. The net result is that tigers share a lot of properties, and in accordance with Mill's description above, we may not know all the properties they share. But they also display a certain amount of variety, in accordance with condition (2) above.[3]

In ordinary language, a term like 'tiger' is used as a predicable: "Benji is a tiger," we say, thus predicating *tiger* of Benji.[4] Assuming that we are using the term 'tiger' here as a natural kind term, HPC theory holds that we are in effect relying on a hypothesis, namely that tigers are subject to homeostatic mechanisms that cause them to be associated with a property cluster. We rely, in other words, on an implicit theory that there are "homeostatic influences" such as shared structure, or selection and reproduction, which induce an open-ended set of similarities that we can discover. (This implicit theory constitutes what Boyd calls the "accommodation demands" of our use of natural kind terms.)

We should note, in passing, that HPC theory is not, as yet, sufficiently well defined. HPC theorists such as Wilson et al. (2007) say that natural kinds are defined by "causally

---

[3] HPC theory recognizes both the superficial property clusters that members of kinds share (sometimes imperfectly) and the underlying homeostatic mechanisms that explain these clusters. Should kinds be defined/ individuated in terms of the former or the latter? According to Boyd, the answer is both: "Species are defined, according to the HPC conception, by those shared [phenotypic] properties *and* by the mechanisms ... which sustain their homeostasis" (Boyd 1999, 81). But according to Wilson et al. (2007), the burden falls on the homeostatic mechanisms. These authors, however, make the interesting claim that the species *category*, i.e., the class of species, is an HPC kind. In arguing for this claim, they follow Ereshefsky (1992) in saying that species instantiate an imperfectly shared property cluster, but do not show how an underlying homeostasis keeps these together. So perhaps there are three views (two of which are held by these authors in the same paper): namely that superficial property clusters constitute HPCs – this last attitude seems to reduce to Wittgensteinian family resemblance theory.

[4] Ghiselin (1974) and Hull (1978) argue that species are individuals. My small amendment (Matthen 1998) is that species *populations* are individuals – species themselves are collections of populations – while *belonging to a population* is a predicable. In 'Benji is a tiger' the term is being used in the second way, i.e., to assert that Benji belongs to a tiger population. In this sentence 'tiger' does *not* denote a whole of which Benji is said to be a part. (Ghiselin and Hull take 'Benji is a tiger' to mean that Benji is *part of* the individual denoted by 'tiger.' This implies that the copula 'is a' denotes the relationship of parts to wholes. This would be unprecedented: "My hand is a me" would be a parallel, and it is evidently inadmissible.)

basic" properties, and assert that these properties must be identified empirically, rather than *a priori*. But they do not identify the empirical criterion by which the following cases should be judged:

a) Sub-classes of natural kinds may or may not be natural kinds on the HPC account. Hacking (1991, 112) noted that the class of white cats is a natural kind under Russell's characterization (as well as Whewell's and Mill's). Is it so under the HPC account? There may be a homeostatic influence that has kept them white – reproduction. This uncertainty is not happy for HPC theory. Hacking thinks, rightly, that white cats should definitely not be a natural kind. But HPC theory leaves the question open, and offers no guidance on how to decide.

b) Again, it is far from clear what counts as an admissible homeostatic influence. To adapt an example from Elliott Sober (1984), let's say that Snooty Hall (a famous school) accepts only the children of parents with a household income of $1 million per year or more. The admission policy is homeostatic. It ensures that all the children are from well-off families, and thus supports an indefinite number of generalizations, some still unknown, about their apparel, mode of transportation, health, and so on. Is attendance at Snooty Hall a natural kind? If it is not, is it not *only* because Snooty Hall's admission policy is not a part of "nature?" (Be careful how you answer: is assortative mating natural or unnatural? Among peacocks? Among humans?)

Let us set these doubts aside, and continue with the exposition of the theory. With terms like 'white' we know the associated property – if we didn't, we would not understand the term. But in the case of natural kind terms such as 'tiger,' we do not necessarily know how to identify the things that belong to the class. In this case, understanding the term consists of (a) knowing (by acquaintance or by report) some tigers, and (b) knowing that these tigers belong to an HPC class (satisfying some further conditions of specificity and type). We use the term to denote the HPC class that we thus posit. This open-ended understanding accounts for why the term is epistemically unbounded. We do not know what properties are homogenized by the homeostatic influences underlying the class (thus the User Ignorance above).

The hypothesis of homeostasis could be wrong: racists make the mistake – not their only one, of course – of assuming that certain human subgroups are natural kinds. When they use a term like 'Muslim' they use it as if it denotes a natural kind. But, as Mill pointed out, there is no property that Christians have and Muslims lack except as follows from their being members of their respective religions (Hacking 1991, 118). *Muslim* is therefore not a natural kind in the intended sense. One does *not* understand the term correctly if one does not know that the distinguishing mark of a Muslim is that s/he is a member of a particular religion, and that one is *not* entitled to assume that Muslims have unknown characteristics that follow upon their membership in this kind.

So goes the consensus, setting aside the sectarian skirmishes sketched in notes 2–4. Millikan accepts a version of it.

Is Biological Homeostasis Historical?

According to Millikan, biological taxa are *historical* HPCs:

> The members of these kinds are like one another because of certain historical relations they bear to one another... ... (1) ... something akin to reproduction or copying has produced all the various kind members from one another or from the same models ... (2) ... the various kind members have been produced in or in response to the very same ... historical environment. (Millikan 1999a, 54–55)

In the case of natural kinds such as *gold*, the properties of instances come about because of "an underlying structure common to members of the kind" (Millikan 1999a, 49). The homeostatic influence that ensures conformity between one bit of gold and another is that each of these bits has the same underlying atomic structure. But in biological kinds, homeostasis operates through relationships among members of the kind, and relationships that those members bear to their environment. It is these relationships that ensure that "the kind does not do as Achilles' horse did and 'run off in all directions'; but remains relatively stable in its properties, maintaining its integrity as a kind" (Millikan 1999a, 55).

Put in this way, however, it is not entirely clear why Millikan says that biological kinds are *historical*. (Boyd 1999 denies, by the way, that biological HPCs are historical, though he acknowledges that selection and reproduction are homeostatic.) The difference between *gold* and *tiger* seems to lie, as Millikan describes it, in the relationality of the homeostatic influences operating in the latter. Pieces of gold are like each other because gold always has the same atomic structure. Thus, the similarity of one piece of gold to another is a matter merely of each having a certain intrinsic structure. The case is different for tigers. They are like each other because of how they stand with respect to other things, including each other. Or so Millikan argues. But where does history enter the picture?

Millikan answers this question by invoking the spatiotemporal locatedness of homeostatic relationships in biological kinds – selection and reproduction happen in time. She contrasts the "historical relations" that biological organisms bear to one another with the "eternal essence" that pieces of gold have in common. This is unconvincing.

First of all, the fact that this piece of gold has atomic number 79 is a temporal, not an eternal, fact – it has atomic number 79 *now* (and at every other moment of time), which is why its solid form is yellow, heavy, malleable, and non-reactive *now* (and always). Gold does not have atomic number 79 outside of time. (What is more, it could be argued that gold came into being, with atomic number 79, some time after the Big Bang.) Conversely, in the sense that gold *is* eternal – i.e., in the sense that the kind *gold* existed, uninstantiated, before the Big Bang – so too is the kind, *tiger*. Any organism *thus* adapted to *this* kind of environment would participate in the *tiger* property cluster whenever and wherever it existed. (Again, I am speaking on behalf of HPC theorists: I don't agree with this way of understanding *tiger*.)

Secondly, Aristotle's species were relational in pretty much the same way as Millikan's. Aristotle too knew that biological organisms reproduce, and believed that every organism belongs to the same species as its parents. He could perhaps have been persuaded to adopt the Biological Species Concept. Moreover, Aristotle held that every animal's essence is to function well in its environment. So his way of characterizing species has something to do with reproduction, function, and adaptation. But any relationality that this entails should not be taken as making Aristotle a proponent of historical kinds. In fact, when philosophers of biology say that biological essences are historical, they are precisely repudiating Aristotle's view of these essences. To say that taxa, or biological kinds, are historical is, at a minimum, to say that they are not fixed in time (as Aristotle thought), but that they change or evolve, and display variation besides. But Millikan conspicuously avoids all talk of evolution or species-change in her argument for this particular claim of historicity. (It is actually something of a gloss on her account to say that she speaks of selection.)

Millikan is out on a limb talking about historicity in this context. Is she even right about relationality? There is another issue that intrudes here. What exactly does it mean to say that relations among organisms and environment bring it about that tigers are the same as one another? Let's consider selection first. Elliott Sober (1984) makes an important distinction along the following lines. Selection comes about by individuals of certain types perishing. Let's suppose tigers once came in two types: stripy and non-stripy. Prospective prey spot non-stripy tigers easily; thus, they manage to elude these tigers, which, as a consequence, starve and die. The tigers that remain are almost all stripy.

Now, one can ask two questions:

A   Did selection bring it about that most tigers are stripy?
B   Regarding any individual tiger, did selection bring it about that *it* is stripy?

The answer to question (A) is clearly "yes." But the answer to (B) is arguably "no."[5] Any stripy tiger descends from an ancestor who was stripy: the elimination of non-stripy tigers has nothing to do with why *it* is stripy (as opposed to why most extant tigers are stripy).

In light of Sober's distinction, certain of Millikan's turns of phrase become quite significant. She says that selection maintains the integrity of a *kind*; that it prevents the *kind* from running off in all directions. Her view is *not* that selection ensures that any particular tiger is stripy, but rather that selection ensures that the kind *tiger* is more or less uniform with regard to stripiness.[6] Thus, she implicitly acknowledges that the reason why any individual tiger is stripy is that it is descended from (or that it originates by mutation from) a line of stripy tigers, and not because of selection. Let us say that selection explains trait-distributions *non-distributively*: it explains why the distribution has a certain shape, but it does not explain why any particular individual has the traits it has. This distinction will become significant later.

---

[5]   Sober gives a "no" answer. I argued for a "yes" answer in Matthen (1999). But my reasons for contesting Sober are not relevant as yet. I'll return to them later.
[6]   Wilson et al. (2007, 198) seem to miss this point. They write that "... biological individuals often are as they are and behave as they do because of the relations in which they stand.... Finches may tend to have beaks of a certain size and shape because of selective regimes their ancestors faced in the deep past...."

Now, Millikan says that reproduction too is a historical relation: so it is open to her to say that the reproduction is the locus of historicity in distributive explanations of generalizations over biological kinds. However, this too is dubious. It's certainly true that Benji the tiger is stripy because he is copied from his stripy tiger parents – and that the same is true of every other tiger (unless they originated stripiness through a mutation). However, in any such case, there is a more proximate cause of Benji's stripiness – the genes that Benji inherited from his parents. The parents cause him to have his genes, and the genes cause him to be stripy. So the most proximate, most direct, most relevant cause of Benji's stripiness is something *intrinsic* – is genes. If the same genes could have been implanted in him, it wouldn't have mattered whether his parents were stripy or not. (Technically, the genes screen the parents off as causes.)

At this point, one half of the difference between *gold* and *tiger* has just about disappeared. The kind *gold* consists of those things that have atomic number 79. The kind *gold* is associated with a property cluster because each of the things that it comprises has this structure, which is a cause of the property cluster. Similarly the kind *tiger* consists of those things that have a certain genotype (or disjunction of genotypes). This kind is associated with a property cluster because the genotype is a cause of the property cluster. Selection only shows us why some other genotypes and phenotypes were eliminated. It shows us why there are no non-stripy tigers. It doesn't tell us why *these* tigers (all of them) are stripy, or why *those* tigers – the ones who were eliminated without issue – were non-stripy.

I said that one half of the difference between *gold* and *tiger* has disappeared. But one half remains. For in the story we have told, the environment eliminated non-stripy tigers; this implies that there were once non-stripy tigers, and would be now, except that these unfortunate animals perished. There is no analogue of this in the *gold* case. To be gold is to have atomic number 79. There were never any non-79-atomic number bits of gold. No such thing was ever gold, nor could it have been. This is an intriguing difference, and it gives relationality (and historicity) a residual role in biological explanation. But as we shall see in the following section, it is not enough. A wily Australian spoiler waits his turn in the wings.

Let me summarize the results of this section. Millikan speaks of two sources of homeostasis within biological kinds, selection and reproduction (or at least she seems to endorse Boyd in this matter). She claims that these mechanisms are relational. We have found that selection does explain certain uniformities across kinds, but it does so non-distributively. Reproduction, on the other hand, explains distributively, but it is trumped by a non-relational explanation.

Intrinsic Properties Redux

The argument of the preceding section constitutes a major disappointment for relational natural kinds *à la* Millikan. In the terms that Russell sets, HPC theory is a success. It schematically explains why most tigers are stripy – the homeostatic mechanisms make it so – and thus it makes successful induction over HPC kinds not just a matter of chance. On the other hand, it was always a hope that HPC kinds would be explanatory. However, the Sober argument rehearsed in the previous section indicates a difficulty. Selection shows us why all tigers are stripy, but it does not explain why Benji is stripy. Even with the Theory of Natural Selection firmly in place, the reason why Benji is stripy has something to do with causal interactions between his genes or other intrinsic properties and his environment.

This is where Michael Devitt's (2008) brilliant and pugnacious attempt to revive biological "essentialism" enters the picture. Devitt takes on board the traditional commitment of natural kind theory to explanation. But he distinguishes between *historical* explanation, which involves selection and other occurrences in the past, and *structural* explanation, which involves explanation in terms of concurrent structures intrinsic to an organism.

> Some intrinsic underlying property of each Indian rhino causes it, in its environment, to grow just one horn. A different such property of each African rhino causes it, in its environment, to grow two horns. The intrinsic difference explains the physiological difference. If we put together each intrinsic underlying property that similarly explains a similar generalization about a species, then we have the intrinsic part of its essence. (Devitt 2008, 352)

Devitt correctly insists that structural explanation is different from historical explanation. "Regardless of the history of its coming to be true, in virtue of what is it now true?" he asks – history cannot answer *this* question, he rightly says.[7] The arguments of the preceding section bolster his conclusion. Selection is historical, but it does not explain the properties of any individual. Parentage and reproduction are historical antecedents of an individual's properties, but structurally explaining these properties is more fundamental.

Devitt assumes that any individual tiger's stripes are explained not by just any property, but specifically by it being a tiger.[8] (We are assuming here that stripiness is a characteristic property of tigers; as we shall see in a moment, Devitt needs to maintain that only characteristic properties of a kind are explained by the kind.) Now, presumably being-a-tiger is the tiger essence. But we have just shown that the relational properties both of individual tigers and of the tiger kind are not fundamental with respect to explanation. Thus, Devitt concludes that species kinds have intrinsic essences, and that these structurally explain characteristic species properties. To me, this further argument seems specious. Is it not possible that tigers were once non-stripy? If so, whatever explains their stripiness is not the essence of *tiger*. So if being-a-tiger explains stripiness, it must explain it in some other way than structurally. I'll put this point aside for the moment. (See, however, the fourth section, below.)

The question now arises: What happens when there is variation across a species? Consider:

1   Mohan is 5 feet 11 inches tall and Shaq is 7 feet 1 inch.

---

[7] Some take Devitt to be asserting that structural explanation is more fundamental than historical. In a personal communication, he denies this. He says that he is not dealing with historical explanation – so in the end, he might well agree with everything I say below.

[8] Devitt writes: "Matthen points out that 'many biologists seem committed to the idea that something is striped *because* it is a tiger'" (1998, 115). This is a bit cheeky, since the 'because' I had in mind was historical, not structural. Nevertheless, I take the point. Any tiger's stripiness must be explained in terms of an intrinsic property. So if you want to show what makes each and every tiger stripy, you need to find a common intrinsic property.

Though (1) is evidently a concomitant of within-species variability, it should give Devitt no difficulty at all. For though Mohan and Shaq are both humans, their widely divergent heights are not characteristic species properties and are not explained in terms of both Mohan and Shaq being human. Devitt is committed to characteristic generalizations being distributively explained in terms of intrinsic species properties. He would presumably deny that divergences could be so explained. From his point of view, then, Mohan's height will be partially explained structurally by some intrinsic property that he does not share with Shaq. Since Mohan and Shaq are both humans, this will not be a species property. However, their both falling within a certain height-range characteristic of humans *would* be so explained.

It should be noted *en passant* that relational HPC theory is also quite comfortable with (1). After all, HPC theory knows all about imperfect homeostasis. HPC theory will explain the divergence of height in (1) by (a) the failure of selection to homogenize height. Presumably, this failure of homeostasis is connected to (b) the viability of the two lineages that led respectively to Mohan and to Shaq. But (b) is not a species property. So here, once again, the explanation will involve factors special to these individuals, and not shared by them as members of the species.

Population Structure

Mill and Russell thought that natural kinds support induction. Following in that tradition, Millikan, Boyd, and now Devitt put a lot of weight on uniformities across species. All think that the explanatory task of systematic biology is to explain characteristics across biological taxa. What I want to argue now is that there are certain *differences* among organisms that are also maintained by species structure. (See Ereshefsky and Matthen 2005 and Matthen 2009 for a fuller argument than I shall present here. The main novelty of the present argument is the discussion of the contrast between (1) above and (2) below.)

Here are two closely related phenomena of the sort that I want to consider.

The Sexes: Most sexually reproducing species have two types of organism, males and females. (Some sexual species are self-fertilizing.) Though these types of organism might be very similar to one another in many respects, there is a crucial difference with regard to their reproductive role that is maintained by natural selection and species structure.

Sexual Dimorphism: Females have an interest in mating with fit males. In some species, males develop a trait that has no function other than to indicate their fitness, or lack of deficiencies. In some cases, the indicat*ed* deficiency is ecological. The indicat*or* trait is not a deficiency of this type. However, it is a sign of such a deficiency. For example, in many birds, males do not display their characteristic bright color in the presence of worms or blood-infections (Hamilton and Zuk 1982). Thus, a female will be interested in choosing brightly colored mates, and males have an interest in developing traits that advertise their fitness. In other cases, the proper expression of the trait – for example, a complex courting behavior or a suite of decorative features – depends on a host of other fitness determining genes. In these cases, an easily perceptible trait (or traits) indicates fitness, and evolves to draw more and more on fitness (Rowe and Houle 1996). In both these sorts of situation, females scrutinize prospective mates for the presence of these traits. Thus, males (but not females) end up with exaggerated traits that distinguish them sharply from females. Females, on the other hand, evolve exaggerated preferences for indicator traits.

In the sort of case that Millikan has in mind, all or most members of a species have a characteristic. This is explained by some underlying property that each has. We have argued that relations between organisms do not best explain these characteristics. By contrast, relations between organisms are essential to explaining the above phenomena. In these cases, we appeal to relations among the members of a species to explain a *difference* among them. Thus:

2   Males of species S have characteristic C and females of S lack C.
3   (2) is explained by mate choice behavior in S.

Cases like this constitute the heart of Population Structure Theory (Ereshefsky and Matthen 2005). PST is meant to be an alternative to HPC theory. In PST as in HPC theory, there are underlying influences that explain the properties of individuals. However, in PST, some of these influences are difference-making or *heterotic*, not similarity-maintaining or homeostatic. In such cases, the species S historically explains why a certain organism has characteristic C, despite the fact that not every member of S has C.

We saw that Devitt's structural explanation theory and Boyd's HPC theory have no difficulty dealing with (1). What about (2), which mentions a property that males have and females lack? Here, Devitt and HPC theory would presumably take a similar line. Devitt would insist, rightly, that there must be a structural explanation of why any male or any female became colored in the way they are. But, he will say, this explanation will fracture: it will not appeal to species characteristics, much less to species essences, but to characteristics peculiar to each sex. In fact, Devitt would be hard pressed to distinguish case (1) from case (2): both involve properties not shared by every member of the species in question, and so they involve explanatory intrinsic properties that are not species properties, but rather subgroup properties.

The same goes for HPC theory. Wilson et al. (2007) explicitly insist that cases like (2) are the same in kind as cases like (1), and hence pose no problem for HPC theory:

> ... [I]n light of the intrinsic heterogeneity of biological kinds, a feature highlighted by our earlier appeal to the *natural flexibility* of the HPC view, it is difficult to see how particular forms of such heterogeneity – various polymorphisms – could undermine that view. (Wilson et al. 2007, 211; emphasis in original)

Pace these authors, dimorphisms due to sexual reproduction and sexual selection have to be treated differently from mere variation. The reason is *not* that sexual dimorphism is a "clustering." It is rather that sexual dimorphisms arise irreducibly from relations between the males and females of a species. The case of sexual conflict, which has been intensely studied only during the last couple of decades, is particularly revealing. (See Arnqvist and Rowe 2005, chs. 1–2, for a review.) Sexual conflict arises because males and females invest unequally in offspring, for instance because of the unequal size of their gametes. Thus, the optimal mating frequency for one sex is different from that of the other – typically females invest more in a gamete, and therefore mate more selectively. In such asymmetric situations, the male's fitness is best served by having offspring with as many females as possible and not caring too much about the risk of unfit progeny. On the other hand, the female, who might produce only one offspring per mating cycle, cannot afford to lose her offspring.[9] Consequently, it is in the male's interest to induce a female to mate with him, but at the same time, it is in the interest of the female to choose the fittest male, and hence to postpone mating until she is ready to choose. Both strategies are costly; that is, they risk or incur reductions of fitness.

Here is an example adapted from Holland and Rice (1998). (Note how this example is different from that of fitness-indication described earlier.) Suppose that a fish feeds on red berries and has hence evolved visual sensitivity to red spots.

> ... [The] preexisting sensory bias of females selects males to evolve an initial, rudimentary display trait [a red spot] that enhances their attractiveness to females.... These overly attractive males induce females to mate in a suboptimal manner (e.g., too often, less-than-ideal time or place). This counter-selects females to evolve resistance to (i.e., decreased attraction), rather than preference for, the male display trait, for example, a higher requisite stimulatory threshold to induce her mating response.... Males are now selected to evolve a more extreme display trait to overcome the increased receiver threshold (by receiver we mean the signal receptor(s) and all associated neurological processing of the display signal), and cyclic antagonistic coevolution ensues. (Holland and Rice 1998, 1)

In this case, there is antagonistic selection for a larger vivid red spot in males, and for lower visual sensitivity for red spots in females. This selection is balanced by interactions with the environment – the red spot may make the male more subject to predation; the reduced visual sensitivity may make the female less good at finding food. These diverse pressures eventually find an equilibrium that is quite different from what one would expect on the basis of survivorship benefits alone.

This kind of case illustrates what sort of entity a species population is. (In my view, a species is a collection of populations such that there is a non-zero probability that an organism from one can participate in gene-flow to the other – see Matthen 2009 for details.) Organisms do not just interact with the environment and evolve so as most efficiently to survive and reproduce in it. They also interact with each other. Sexual reproduction, which initially evolved because of environmental advantages, creates subgroups

---

[9] Though anisogamy, gametes of unequal size, is always a factor, the males of many species, including humans, devote considerable resources to parental care. This evens up parental investment. Mate choice is presumably more mutual in such species, though the characteristics used by males for sexual display might be different from those used by females.

in a population. Each subgroup now constitutes a part of the environmental problem that the other subgroup must solve. Since each subgroup is responding to the other one, each subgroup has a different puzzle to solve. Yet, because they use each other to reproduce, they are not independently evolving groups. The two different problems faced by the subgroups have to be understood against the background of the integration of the group and the environment they face in common.

In a case like this, I would argue, we have:

4   Males of species S have a red spot and high visual sensitivity to red spots, and females of S lack red spots and have low visual sensitivity to red spots.

And in accordance with Devitt's argument, we have

5   Males of species S share an intrinsic character IC that explains the male-characteristics mentioned in 4, and females of S share an intrinsic character IC' that explains the female-characteristics there mentioned.

The explanations alluded to in (5) will not be species-based, but in light of my earlier arguments, we will also have:

6   Males and females of species S jointly participate in mating behavior common to their species that historically explains why males and females have the divergent characteristics mentioned in (4).

So though there are structural explanations in terms of divergent intrinsic characteristics, there are historical explanations in terms of shared relational characteristics.

Two small points:

a)   Are these relational explanations non-distributive? Sober's argument would indicate that they are. I am dubious. In my view, populations are concrete collections or pluralities (i.e., not just sets) of interbreeding organisms. The conditions that prevail within such concrete collections influence which mating relations will take place, and thus which individuals will be born (see Matthen 1999). However, this is not a particularly important point here. The important point for me here is that there is a species-based explanation of difference.

b)   Does my evocation of populations entail the existence of irreducibly population-level causes? I do not think so. The relations that exist between subgroups of a population change the adaptive landscape for individual members of populations, but they do not force the view that selection, for instance, acts on populations. Selection is simply a bias in births, matings, and deaths of individuals induced by the adaptive landscape.

Millikan rightly spoke of "historical relations" that members of a population bear to one another. But the relations she had in mind – reproduction and adaptation to the "external" environment – are uniform throughout the species population. Given Devitt's arguments recounted above, her historical relations seemed to be preempted, as far as explanation goes, by intrinsic species properties. The fractured problems that I have introduced cannot be preempted in the same manner. But once one has seen how population relations influence evolution, it seems to me that one will resist the preemption of historical explanation even in cases where the outcome is uniform. The point that I want to make is that population structure explains trait distribution, whether that distribution is more or less uniform, or sharply divergent.

Populations are collections of individuals within which there are, as I said earlier, both similarity-making or homeostatic influences and difference-making or heterotic ones. The problem for HPC theory does not arise from a failure of uniformity at the shallow level – as Wilson et al. (2007) say, this theory is designed to cope with such failures of uniformity. The problem arises from the fracturing of homeostasis at the explanatory level. Because homeostasis is broken up in this way, the small-sample generalizations noted by Russell are more complicated than he thought. Do seals bark? You would be rash to conclude that they do after observing half a dozen males.

Conclusion: Are Species Duplicable?

In the preceding two sections, I have attempted first to take relationality away from Millikan and then to give it back. I argued that reproduction and environmental adaptedness are not reasons enough to make species relational entities, but that the divergence of secondary sexual characteristics is. My conclusion does not, however, touch on another point – are species *historical* entities in the sense of being individuated by their history? For this is what Karen Neander (1996) asserts (see below), with Millikan's (1996b) assent.

In her 1999 paper, "Historical Kinds and the Special Sciences" (Millikan 1999a), Millikan drew a parallel between biological taxa and kinds like the 1969 Plymouth Valiant. Something is a 69PV because it has a certain historical origin; similarly, something is a tiger because it has a certain historical origin. A car just like a 69PV but not copied from the Chrysler Corporation plan is not a 69PV. Similarly, if a stroke of lightning were spontaneously to create an animal just like a tiger, it would not be a tiger, Millikan says, since as she puts it, its ontogeny and phylogeny are wrong. This strong conclusion does not follow from anything we have said so far, or from anything else that Millikan (1996b, 1999a) says.

First of all, it is unclear to me that the analogy with the 69PV throws much light on the biological case. Clearly, the 69PV is a historical kind. It is so not just because its members are reproduced from a template created at the Chrysler Corporation at a certain time, but also because they were branded as such. Thus:

a) Imagine that some Brazilian engineers designed a car, which turned out to be an exact replica of the 69PV. This car would not be a 69PV. This is a bit like the Swamptiger example, and one might be tempted to export the negative result. But now consider:

b) Imagine that Chrysler had sold a slightly modified version in Australia under the marquee, Dodge Dingo. The Australian car would not be a 69PV.
c) Imagine an Australian enthusiast painstakingly modified his Dodge Dingo in such a way that it exactly matched the 69PV, down to the PV plates and branding. His product would not be a 69PV.
d) Imagine that Chrysler then copied the Australian modification back to the PV late in the 1969 model year. This version would be a 69PV despite having been copied from the Australian Dodge Dingo, and not directly from the original 69PV plans. Note that it is an exact copy of something that is *not* a 69PV.
e) Finally, imagine that some Brazilian knock-off artists had copied the 69PV exactly and sold it under that name. The result would not be a 69PV.

In b–e, the copying relation works differently from that in biological taxa. So I do not think that we should use the 69PV case as a source of intuitions about biological taxa.

What historical properties mark tigers off from other things? The process of speciation suggests that tigers are those animals that belong to populations that descend from an originally speciated tiger population, which has not itself speciated. Here, reproductive isolation can be taken as a criterion for speciation: so, the mark of speciation in the original (tiger) population could be that its members ceased naturally to mate with prototigers – i.e., with animals of the type with which the tigers' ancestors did naturally mate. This gives the original tiger population a unique historical role in the origin of the species. But does this historical role *define* the species? What would happen if an ancestor population gave rise to two descendant populations each of which speciates, i.e., each becomes reproductively isolated from the common ancestor? Suppose that the two descendants happen naturally to interbreed with each other. Would the two descendant populations then be properly regarded as conspecific? If one were a group of tigers would the other one also be? See Figure 7.1.

Karen Neander writes as if this question were completely settled:

> How best to define "species" is still highly controversial, but all the major contemporary schools in evolutionary biology agree that conspecifics must be united by descent. Species are, on all accounts, unified segments of the phylogenetic tree; the debate between the schools concerns how that segment is to be demarcated. (Neander 1996, 119)

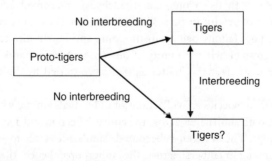

Figure 7.1 Is this a case of independent species origins?

And she adds in a note:

> According to Cladistics, for example, a species starts at one speciation event and ends at the next, whereas, according to Phenetics, parent species can survive the budding of a new species. (Neander 1996, note 2)

It seems to me that this is far too strong a statement. It does not follow from the broad historical approach that Millikan favors – namely that biological kinds have trait-distributions that are ecologically shaped, even when one adds the notion that the ecology in question includes relationships within a population. For Millikan's idea tells us nothing about the identity of kinds in cases of parallel evolution.

A moment ago I imagined an example of two speciated descendant populations that turn out to be inter-fertile. The example is not wholly invented. Schluter and Nagel (1995) give two examples of this sort of process in fish among which there is assortative mating by size. One of these is the Pacific sockeye salmon, which gives rise to several freshwater populations, each isolated from the others. In each such freshwater population, the smaller sized varieties rapidly become isolated from larger varieties and thus speciate. The other example is that of the three-spined stickleback which have independent populations in various places, including Japan and British Columbia. In each of these habitats, assortative mating by size and by habitat (shallow vs. deep water; fresh water vs. oceanic) has resulted independently in speciation. The question is whether to regard the newly evolved populations in the separate locations as conspecific. Clearly, they are not "united by descent," since there are two separate speciation events in two separate locations, and the fish from separate locations do not have the same ancestors of the same species.

Schluter and Nagel suggest that this is a case of "parallel speciation." In a later study (Rundle et al. 2000), in which Schluter and Nagel are co-authors, parallel speciation is characterized as "… a special form of parallel evolution in which traits that determine reproductive isolation evolve repeatedly in independent, closely related populations as a byproduct of adaptation to different environments." "The outcome …," Rundle et al. say, is "reproductive compatibility between populations that inhabit similar environments and reproductive isolation between populations that inhabit different environments" (Rundle et al. 2000, 306). And yet another study (McKinnon et al. 2004) shows that the Japanese and British Columbia sticklebacks interbreed assortatively by size rather than by place of origin. In these cases, it does not seem unreasonable to say that the interbreeding populations belong to the same species. In any case, conspecificity does not seem to go so clearly with unity of descent as Neander and Millikan suppose. Nor do these scientists treat the matter as if it were settled by definition. Caution is desirable.

To summarize then: species are collections of individuals among which there are relations that produce both similarities across the entire collection and also sharp dissimilarities across subgroups. The between-subgroup dissimilarities trace to and are maintained by interactions among members across the subgroups; hence they are historically explained by heterotic selection within a population. HPC theory and Millikan's account are myopic in their paying attention only to homeostasis – this is a mistake that traces back to the idea that natural kinds, including biological kinds, support induction. Finally, Millikan is right to think that there is a sense in which biological kinds are historical. However, this sense of historicity does not exclude the possibility that species can originate independently. They may not be historical with regard to tracing back to a unique origination event.

Reply to Matthen

Mohan Matthen is primarily interested in how species should be delineated. This is a problem that I haven't ever addressed, but I have used species as examples of what I term "historical kinds" in a number of places. As I see it, all of the *actual* groups that have been proposed in recent times as species have been historical kinds in my sense (as I will try to show), but there are far too many historical kinds, embedded within one another, crossing over one another and so forth, to be used, alone, to delineate species. Richard Boyd and I once had an exchange over the purely logical possibility of a pair of species that were identical but not related to one another such as to form a single historical kind. Boyd said that on his homeostatic property cluster view, such a pair of species would be considered as one (Boyd 1999; Millikan 1999a, 1999b).[1] I have given reason to suppose that a kind term invented to cover one of these species (say, earth weasels) would not have the other (say, twin earth weasels) in its extension (Millikan 2010, para. 7). But that in itself does not tell us how species should be defined; indeed, I suspect the species problem has no set answer. I would like to explain, however, why I take extant species terms to pick out historical kinds.

Historical kinds are kinds whose members are clumped together in logical space – kinds whose members form property clusters or "topological neighborhoods" in Russell's sense – for a special kind of reason. They are not merely classes, certainly not disjunctive classes (Matthen: "… the kind tiger consists of those things that have a certain genotype (or disjunction of genotypes)"). Rather, it is, in part, because of a network of causal relations among the actual members of the kind that the members bear many similarities to one another.

> The members of these kinds are like one another because of certain historical relations [i.e., causal-order rather than logical-order relations] they bear *to one another* (that is the essence) rather than by having an eternal essence in common. It is not just that each exhibits the properties of the kind for the same … reason [contrast Boyd's homeostatic property clusters]. Rather, each exhibits the properties of the kind because other members of that same historical kind exhibit them. Inductions made from one member of the kind to another

---

[1] Boyd mistakenly took my description of historical kinds to be an attempt to correct or supplant his homeostatic property cluster kinds.

are grounded because there is a certain historical link between the members of the kind that causes the members to be like one another. (Millikan 1999a, 54, referenced by Matthen)

"[T]he [historical kind] principle explains the likeness between members, not, in the first instance, the properties themselves. (To explain why a photocopy is like the original is not to explain why either has the properties it has. I can know why the photocopy is like the original without knowing what properties either has)" (Millikan 1999b, 100). Putting things graphically, suppose that new twin earth is exactly like earth, even the water being H_2O. The tigers on new twin earth are clumped together in logical space because a network of causal relations among them has determined that they are much like one another in many respects. But no network of causal relations has determined new twin earth tigers to be like earth tigers, so earth tigers and new twin earth tigers are not members of the same historical kind. Members of historical kinds are tied down by their actual historical relations to other actual individuals in the same kind. Compare being a Rockefeller, being a mason, being a McDonald's restaurant, being a US senator. None of these properties are found on new twin earth.

If species are historical kinds then – *contra* Devitt and, as I understand him, also Matthen – being a member of the tiger species does not explain why Benji has the characteristic tiger properties that he has, either logically or causally. Not logically, because being a tiger is not a matter of being in a certain class; any distinguishing tiger property may be absent in some tigers. Not causally, because the ways Benji was caused to have his tiger properties is merely what *defines* him as a tiger; the causes come first, being a tiger comes logically after. Benji has four legs and stripes because of his genes and the environment in which they developed. He has those genes because they were copied from his parents' genes. His parents also had four legs and were striped because of those genes and the environment in which they developed. Those things together are what explain why Benji and his parents have the same number of legs and the same coat pattern. This, along with the same kind of explanation for many other similarities between Benji and his parents, determines them to be members of the same historical kind, which kind we call by the name "tiger." Benji's stripes would be determined or caused by his being a tiger only if tiger were the sort of kind that is a class, defined either directly by the properties of its members or indirectly by the causes of the properties of its members. But to assume that all kinds are classes is to beg the question against historical kinds, to beg the question, for example, against species considered as like big scattered "individuals."

Matthen says that "the kind tiger consists of those things that have a certain genotype (or disjunction of genotypes)." The difficulty with this idea is that the genotypes in the collection that produce tigers *count* as disjuncts in this collection only because they are genotypes that have produced actual tigers, which are the actual tigers having previously been determined by membership in a common historical kind. There is no independent reason why *just these* genotypes should be included. In fact, certainly in most sexually reproducing species, every gene distinctive of the species has alleles, so that there are an awful lot of alternative genotype kinds to be considered. Also, new minor mutations occur constantly. What's in the tiger gene pool this week may differ from what will be there – though perhaps very temporarily – next week. There is nothing to collect the disjunction of genotypes together except the historical clump that is the actual causally linked species members.

Matthen (and also Boyd) consider cases in which a species sprouts a new species in a certain kind of environment on one occasion, then later sprouts almost exactly the same kind of new species in the same kind of environment on a second occasion, there being no contact between the two sprouted species. (This kind of occurrence is actual, not merely logically possible.) Do these constitute one and the same species? I don't know. (I don't think Heaven knows ... or cares.) But members of these two sprouted species will have a great many properties in common due in part to common ancestry. And just as Benji and his parents both have stripes in part because of the environments their genes developed in, these sprouted species will have properties in common because of the similar environments in which easy mutations of these genes found themselves. (New genes don't come from nowhere but evolve from previous genes.) The two species are alike in part because linked by certain causal relations between them, so together they constitute a single historical kind. Separately, they each constitute a different historical kind as well. (Similarly, each of the sexes within a sexual species forms a historical kind nested within the historical kind that is the species itself.)

But the example has a lot of merit, for it reminds us that the distinction between eternal kinds and historical kinds is not dichotomous. In talking of the historical-kind unity of a species, I have previously talked of continuing selective pressures in an ongoing environment. The idea was that the environment was ongoing because of natural conservation laws; the environment itself was an ongoing historical entity or (parts of it) a repeating one. But in this example, the environments in which the two species sprout off are not supposed to be historically related. A merely eternal element has entered the picture. In general, the exact relations among eternal and historical kinds and various crossbreeds of these needs much more investigation than I have given it (a point my graduate students have urged on me). Between cases like the chemical substances on the eternal side and Gothic churches and McDonald's restaurants on the historical side, there are probably many hybrid cases. The resulting epistemological questions – questions about how inductions within such kinds are supported – especially need to be investigated.

Matthen is surely right that natural selection, just as such, is not a cause of the similarity among members of a species taken distributively. Natural selection, by weeding out members of a species that fail to have certain traits, does remove competition for resources, for mates and so forth, allowing members that do have those traits a better chance of passing them on. Over time this is certainly a very significant causal factor in proliferating these traits. But the direct way that natural selection helps to produce historical kinds all having many traits in common is by throwing away most contrary traits that arise, leaving behind only a very limited variety. This is a collective cause, not a distributive one.

8

Millikan, Realism, and Sameness

CRAWFORD L. ELDER

The subtitle of Millikan's ground-breaking *Language, Thought, and Other Biological Categories* (1984) is *New Foundations for Realism*. Does Millikan's work – whether in that book or afterwards – really provide such foundations? I shall argue here that the answer is "not quite – but very nearly." Realism about objects, I shall argue, requires realism about two forms of sameness. The first is sameness in natural kind – the sort of sameness that can bind numerically distinct objects together. The second is numerical sameness across time – the sort of sameness embodied in an object's persisting. Millikan does give a realist account of kind-sameness, I shall argue. Indeed she gives the only extant account that truly deserves to be called "realist." But her account of an object's persisting, I shall argue, does not stably differ from an anti-realist account. That is important because realism about persistence, as I shall be saying, is the key to realism about objects. Realism about the other sameness, about kind-sameness, supports realism about objects only if one is willing to hold that the conditions on kind-membership double as persistence conditions for the members of the kind – and this Millikan is unwilling to hold. But she could and should hold exactly that, I shall argue. Robust and powerful foundations for realism are lurking just beneath the surface of Millikan's work.

I

To bring out what is distinctive about Millikan's view on kind-sameness, I want first to set forth a very widely accepted view with which Millikan's view sharply contrasts. This view almost surely has the status of "the received view," though I will not attempt to substantiate that claim here. It is not in the end a realist view about kind-sameness – even though

---

Millikan and Her Critics, First Edition. Edited by Dan Ryder, Justine Kingsbury, and Kenneth Williford.
© 2013 John Wiley & Sons, Inc. Published 2013 by John Wiley & Sons, Inc.

it does incorporate some realist-sounding claims – and this *is* a claim that I will try to substantiate here. But for the first sections of this chapter, I will leave largely in the background the question of whether this widely accepted view is a realist one. The way in which it falls short of true realism becomes far clearer once one has seen the alternative that Millikan puts forth.

On this widespread view – I shall call it "conceptualism," using the label offered by one of its proponents – kind-membership is a language-centered affair. What it is for an object to belong to a natural kind is for it to satisfy one of the sortals that we take as picking out a natural kind (Thomasson 2007a, 56, 157). What it is for a portion of matter to belong to a particular natural kind of matter is for it to satisfy a mass-noun that we take as designating a natural kind of matter (Thomasson 2007a, 41). The satisfaction conditions for such a sortal or matter-name are fixed by a description that we associate with the sortal or matter-name (Thomasson 2007a, 39–40).

What is this associated description like? According to some conceptualists, the associated description for at least some sortals spotlights a list of qualitative features, of properties. This might be so, for example, with "marriage license" or "jokes" – or with non-count nouns such as "money" (cf. Putnam 1975, 159–160). The descriptions associated with many other sortals, according to all conceptualists, incorporate indexical elements, and in this way provide a "blank check" for the world to fill in. These descriptions, in other words, point us to certain standard examples of the kind in question, and say that the members of the kind are all those objects (or samples of stuff) that are similar in certain specified respects to the standard examples (see Fumerton 1989). Thus the description that fixes the satisfaction conditions for "water" – to pick a familiar example – might be "whatever is the same in respect of molecular structure as *these* particular samples of stuff," in other words whatever bears to the indexed samples Putnam's relation "same$_L$" (Putnam 1975, 141). Perhaps the description that fixes the satisfaction conditions for "chair" is "whatever artifacts are the same, in respect of the human function that they serve, as *these* particular objects." In all such cases the description leaves unspecified just *which* the qualitative features are, in virtue of which other members (or samples) of the kind bear the specified relation of resemblance to the indexed examples. That is left as a blank check for the world to fill in. Determining how the world fills in the blank check may require hard empirical work, as in the case of "water"; in other cases it may require relatively simple reflections on how different designs can meet the demands posed by human anatomy, as in the case of "chair."

In any case, ostension alone cannot fix satisfaction conditions for the sortals and matter-names that end up referring to natural kinds. That is the lesson that conceptualists draw from "the quâ problem" (Devitt and Sterelny 1999, 80–91; Papineau 1979, 158–168). Suppose that, in a baptismal utterance, I point to some teak chairs and say "by 'teeks' I shall mean whatever is like these," not specifying which sort of likeness is crucial. Does "teeks" refer to a kind of wood; to illegally imported goods; to chairs, or perhaps teak chairs; or to furniture? It has no determinate reference, conceptualists say, until the baptizer communicates which *sort* of features define the kind in question (Thomasson 2007b, 55).

The consequence of this view is that we have interesting *a priori* knowledge about the properties that characterize the members of any natural kind (Thomasson 2007a, ch. 3; Sidelle 1989, chs. 2, 3). Where the associated description consists wholly of a list of qualitative features, we will know *a priori* that members of the kind in question all have just those features. Where the associated description issues a "blank check," we will know *a priori* that members of the kind are alike with respect to certain *sorts* of properties – we have "template knowledge," as one might put it, about that kind. This *a priori* knowledge will moreover be knowledge about properties that, in any possible world, the members of the kind in question possess. We will know *a priori* that necessarily, if x is a member of the kind in question, it possesses just *these* properties, or some one property of just *this* sort, of just *that* sort, etc.

Millikan's view differs from the conceptualist view in respect of both metaphysics and semantics. Metaphysics: for Millikan, what it is for an array of objects to belong to a common natural kind – and what it is for some samples of matter to do so – is not a function of the satisfaction conditions for terms in our language. It is a wholly mind-independent affair. It is for these objects (or samples) to possess in common the properties in a certain cluster, and to do so because of the operation of a common causal factor (Millikan OCCI, ch. 2; Millikan LBM, ch. 6). Semantics: what it is for a sortal or matter-name to have the satisfaction conditions that it does has nothing to do with the description that we associate with it. Indeed it *could* not have anything to do with *the* description that we associate with that sortal or matter-name. For it follows from Millikan's views on the metaphysics of kind-membership – together with her naturalist picture of language-development – that, in general, *there is no such thing* as *the* description that we associate with a given sortal or matter-name (Millikan 2010, §2). Different speakers will typically respond to different properties characteristic of a given natural kind, in applying a given sortal or matter-name. Indeed, any individual speaker typically will, on different occasions, token a given sortal or matter-name in response to different features (OCCI, 7–8, 72; WQ, 92–94; Millikan 2010, §6). Speakers may achieve proficiency at reidentifying a given kind well before they have a conscious grasp of the properties to which they are responding (OCCI, 77–80). When speakers do construct descriptions of what a given kind is like they will have learned these descriptions empirically, not *a priori*; and typically, the descriptions they use will hold true of the kind in question only often, not across all actual circumstances, and almost surely not across all possible circumstances, in all possible worlds (OCCI, 7–8, 72). I shall begin by explaining how these anti-rationalist consequences follow from Millikan's metaphysics of kind-membership. Only afterwards will I turn to set forth Millikan's positive account of what does fix the satisfaction conditions for our sortals and matter-names.

II

Suppose, then, that the kinds of nature obtain independently of our having names for them or thoughts about them. Then at the time that natural selection was fashioning in our hominid ancestors the capacities to think about the world's kinds, and to name the world's kinds, our ancestors already were surrounded by objects and samples belonging to these kinds. These objects and samples were similar to their kind-mates in respect of many properties – often enough, in respect of properties crucial to our ancestors' survival. For this reason it would have been crucial for our ancestors to be competent, and useful for them to be skilled, at reidentifying nature's kinds – at judging (and saying) that this object now before me is the same in kind as that object that I (or a member of my clan) earlier examined (OCCI, 5). That is because not every property of an object or a sample that it is important for an agent to take note of is immediately obvious on every occasion of observation (OCCI, 5–6). Some such properties manifest themselves only from the right angle or in the right lighting, or only in certain circumstances or phases of the existence of the object that has them. So it was crucial for our ancestors to be able to bring to bear information gleaned from earlier inspections of members of a given kind to their dealings with members currently encountered. Our ancestors needed to be able to reidentify nature's kinds.

But the very fact that gave rise to this need – that not all the properties characteristic of a given natural kind are immediately apparent upon every encounter with a member of that kind – makes it likely that different individuals, within a population of our ancestors, would have learned to cue their reidentifications of a given kind to different features and marks (OCCI, 7–8). Some might have cued their identifications to color and shape, others to taste and sound, others to tracks and behaviors. Even an individual speaker would be likely, on different occasions, to cue his reidentification to different features. (To use one of Millikan's examples, think of how many ways one may use to tell that an object before us is a lemon: one can tell this by color and shape, by shape and texture, by smell, or by taste (WQ, 94).) Once our ancestors managed to communicate their several reidentifications of a given kind to one another – once they had language, and sortals by which they could communicate to one another the whereabouts and properties of members of a given kind – this divergence in recipes for reidentifying a given kind would have worked out to the individuals' mutual advantage. For language would have ensured when *one* member of a population identified a given kind as present again, here-and-now, *all* members of that population would have shared in that reidentification (cf. Elder 2004, 177–181). The capacity of each to reidentify and learn about a given natural kind would have been greatly amplified by the divergence in techniques that individual members employed. Our ancestors profited from the fact that there was no such thing as *the* description that all – or even each individually – associated with a given sortal. Even now, it is likely that different speakers associate different descriptions with sortals such as "lemon" or "vole" or "hickory tree," or with a matter-name such as "water" or "gold"; more precisely, that speakers cue different tokenings of these terms to different sensory cues. Such divergence in techniques for kind-reidentification works out for our benefit too.

It is important to note that if our capacities for reidentifying nature's kinds ultimately derive from abilities installed in us by natural selection, it is likely that the specific techniques by which we implement these capacities are fallible. The features I rely on, for reidentifying a particular kind of animal or fruit or tree, or a particular kind of liquid, may in *unusual* circumstances characterize the members of a *different* kind as well. They may even be features that do not universally turn up in members of the kind that I take their presence to indicate. The reason for this is that natural selection is typically content to install into organisms capacities that work only often, not invariably. The beaver's tail-slapping behavior is a capacity for warning other members of the lodge of impending danger, but it works only often; beavers sometimes slap their tails when no real danger is present. The human eye-blink reflex is a capacity for preventing entry into the eye of foreign matter, but it can fail to work when debris is blown toward the eyes by a strong wind, and it can also be triggered even when no foreign matter is about to enter the eye, for example by a rapid movement toward the face. Just so, one should expect that the capacity for reidentifying nature's kinds, fashioned by attuning capacities installed in us by natural selection, gets implemented in fallible ways (WQ, 88–92). Thus not only is there no such thing as *the* description that we associate with "vole" or with "hickory" or with "water": it also is likely that *your* description or *my* description applies only *often* to voles and to hickories. Certainly there is little reason to expect that anyone's description spotlights features that characterize voles and hickories in all possible worlds.

But this very point may seem to raise daunting challenges concerning the semantics of the sortals and matter-names by which we speak of nature's kinds. If different individuals cue their tokenings of a given sortal to different marks, why does it not follow that that sortal is simply ambiguous? Then too, if a single individual cues his tokenings of "vole" or "hickory" or "gold" to marks that can sometimes be presented by members of a *different* natural kind, why does it not follow that these terms, in that individual's idiolect, present "the disjunction problem"[1] – that that individual's term "vole" has as its extension voles-or-moles, or that his "gold" refers to gold-or-iron pyrites? Worse, if an individual cues his tokenings of a given sortal to marks that members of some natural kind do not universally present, why does it not follow that the extension of that person's sortal is some subset of the natural kind?

But worries like these presuppose that what grounds the semantic values of our various sortals and matter-names is something that lies causally upstream of tokenings of those sortals and names in sentences. That is precisely what Millikan's semantics disputes (WQ, 85–94; OCCI, 72). In order for my tokens of the sound-parcel "lemons" (or "gold") to be co-referential with yours, there need be no sameness between the description of lemons that I carry in my head and the description of lemons carried in your head, and there need be no sameness between my dispositions and your dispositions to respond to sensory cues by tokening "lemon" (Millikan 2010, §10). What does need to be the same is the explanation of how, when you and I both token "lemon" in sentences expressive of our beliefs and our desires, those sentences succeed in doing what they are supposed to do. "What they are supposed to do" here gets cashed out in terms of selectional history. Sortals that have won a place in our linguistic repertoire typically have done so because of the ways that tokens of these sortals have enabled speakers and hearers to construct mental maps of their surroundings.

This reference to selectional history is the beginning of a long story: Millikan's view on the semantics of public-language sentences, and of the mental states which such sentences express and inculcate in hearers, is complex, and is the topic of other chapters in this collection. Here I will limit myself to a brief and partial sketch, one that focuses on

---

[1] Fodor (1990, ch. 3; 1991, 293–296); Neander (1995); Millikan (1991, 158–163).

beliefs and desires. Millikan holds that beliefs and desires are, at least in many respects, like inner sentences (LTOBC, ch. 8; VOM, ch. 19; but cf. OCCI, chs. 10 and 11). I shall sketch how the elements in these inner sentences – our *inner* terms, so to speak, for natural kinds and for particular matters – acquire the semantic values that they have. It will then be fairly clear how the account extends to the semantic values taken by public-language sortals and matter-names in the sentences that express, communicate, and engender those beliefs and desires.

The heart of the account is the idea that natural selection has fashioned in each of us programs or devices for producing, in our brains, the inner representations (tokens) that are individual beliefs and desires. I will focus, for brevity, on the device that fashions individual beliefs. Any creature capable of flexible and varied behavior needs to have a capacity for adjusting its behavior to its actual circumstances. The creature needs to have a system or device that will adjust the timing of its behavior, or the spatial location and orientation of its behavior, or the velocity and vigor with which the behavior is carried out – or any of countless other features that characterize behavior – in such a way that its behavior will interact with environing circumstances in such a way as to yield satisfaction of its desires – and hence, in the cases when its desires themselves are well formed, satisfaction of its needs. The creature needs a system or device that will construct a mental map of its circumstances (WQ, 77–81; cf. VOM, ch. 14). The proper function of any particular belief is to help govern behavior by contributing to such a map, a map that can be enlisted in the satisfaction of whatever desires the creature happens to form. Since a given belief can get enlisted in the implementation of all manner of desires, there will be no limit to the ways in which a given belief may get its host agent to behave. Nevertheless there will be something constant and characteristic about the way that any one belief steers its host's behaviors. This will have to do not with *how* the agent acts but with *why*, in getting the agent to act in the way he does, the belief does what it is supposed to do. (More precisely, it will have to do with how the belief-fashioning device, *by* having fashioned that particular belief, does what *it* is supposed to do.) Any individual belief is supposed to attune action to circumstances in such a way that one desire or another in the agent gets satisfied. When a belief does this – whatever the desire that may thus get implemented – it does so because the attuned action intersects causally with some one environing state of affairs. That one state of affairs is then the belief's semantic value, its content: it is what the belief is supposed to correspond to. Semantic value, then, is after all fixed by a causal connection. But the connection is causally *downstream* from the belief (WQ, 85–92; cf. Elder 1998, §I). A given belief causes the agent to behave in some way such that its behavior will be caused to be successful (i.e., to result in desire-satisfaction) by some one state of affairs in the world.

How might natural selection have managed to create a device that can fashion indefinitely many different beliefs, beliefs that can interact to steer behavior in ways that yield satisfaction for indefinitely many different desires? It can have done so only if there is some *system* to the way in which this device fashions individual beliefs (WQ, 90). For the device to have been simple enough for natural selection to find it, and powerful enough for it to do the job that (we are supposing) won the favor of natural selection, there has to have been a *system* to how the beliefs that it fashioned mapped environing circumstances.

By far the likeliest guess, Millikan holds, is that the device was so constructed as to produce individual beliefs – individual inner sentences in the brain – that display compositionality. The inner representations whose function it is to map states of affairs involving the same element in the world – the same kind, the same sort of matter, the same individual, the same locale – must have had in common some recurring element, that therefore served as an inner name for that kind or matter or individual. The sentences whose function it is to depict different elements in the world as having the same feature must have resembled one another in respect of a different sort of feature, a feature that therefore amounted to a recurring predicate.[2]

Different creatures equipped with such a belief-fashioning device might well be disposed to token inner sentences featuring co-referring inner names – in particular, names for the same natural kind, or for the same particular matter – in response to different sensory cues. The names would be co-referential nevertheless, because, when those sentences functioned in the way they are supposed to – in the way that is called for by the design of the belief-fashioning device – they attune their hosts' behaviors to states of affairs involving the same element in the world. *Public*-language sentences *expressive* of those beliefs would feature sortals and matter-names that are co-referential, and for largely the same reason.

III

Millikan's view on sameness in natural kind, then, is robustly realist: kind-membership is language-independent, and we have no interesting *a priori* knowledge about nature's kinds. Soon I will argue that Millikan's view on numerical sameness across time – that is, on persistence – does not qualify as realist. But first let me say why this would matter – why, that is, realism about persistence does seem to be required for realism about objects.

Realism about objects, if it is to merit its name, must attribute mind-independent existence to objects that are neither abeternal nor eternal. These are objects that begin to exist at a certain time, that continue to exist across certain changes – even if only such trivial changes as change in location or change in age – and that pass out of existence at a later time. I contend that in order to attribute mind-independent existence to such perishable objects, one must attribute to them mind-independent courses of existence, careers, that mind-independently begin and end where they do, and that mind-independently span such changes as the object undergoes. One can indeed imagine a version of realism that made no such claim. This version would say that each perishable object exists mind-independently, but that the beginning of its existence, the continuing of its existence across changes, and the ending of its existence are all phenomena that obtain only in virtue of our ways of thinking or talking. But such a position would seem miserably unstable. Each perishable object would enjoy mind-independent existence, but there would be no mind-independent fact of the matter as to *when* it exists, no mind-independent *span*

---

[2] This oversimplifies just a bit, since a system can mark sameness otherwise than by resemblance among signs for the same (OCCI, chs. 10, 11). But for brevity of exposition, the oversimplification can be allowed to stand.

that its existence takes up. As it exists mind-independently, any such perishable object would be without temporal location – and hence would be quite disturbingly incomplete. Meinong believed that there are – in some sense of "there are"! – just such "incomplete objects" (Findlay 1963, ch. 6). There are persons who were not born on any particular day, golden mountains that do not have any particular height or weight, ivory spheres that have no particular diameter. But it is nearly impossible not to regard such incompleteness as the mark, precisely, of a fictional object. Just so with perishable objects that, supposedly, exist mind-independently, but do not mind-independently exist some*when*.

But realism about the *courses* of existence traced out by perishable objects is less easy to defend than is commonly realized. A large part of the reason is that it seems so very plausible that the continued existence of any such object both requires, and is ensured by, the continuous presence in that object of certain characteristic properties. (I will call this "the qualitative assumption.") An object passes out of existence, it seems, just when it loses certain crucial properties; it will have possessed these same properties over the course of its existence; its existence began when, at a certain location in space and time, these properties came jointly to be present. At the same time, many other properties that characterize the object, at one point or another of its existence, are properties that the object *can* lose, and many are properties that in fact the object *does* lose, replacing them with contrary properties. Realism about courses of existence therefore seems to require a mind-independent difference between properties essential to a given object, and properties that are merely accidental.

But what does this contrast exactly amount to? What is it for certain properties to be essential to a given object, while others are merely accidental? Many of the answers that actually get offered to this question appear to entail that this distinction in modal status is not after all mind-independent. Consider, for example, the position that sometimes is called "conventionalism."[3] This position holds that what it is for certain properties to be essential to a particular object just is for them to be treated as crucial by our "conventions of individuation," the conventions we follow in making judgments of the form "this is the same object still continuing to exist," "the object that formerly existed here exists no longer." This position says more than just that there is an *extensional equivalence* between the properties essential to a given object and the properties of that object that are spotlighted by our conventions for affirming persistence. The *latter* position *is* indeed compatible with the idea that there is a mind-independent difference between the modal status *essential* and the modal status *accidental*: our conventions *reflect* this difference, this latter position could say, and do not *constitute* it. But then we would still be awaiting an answer on what the distinction in modal status does consist in. "Conventionalism" properly so called is the stronger position that the properties essential to a given object are so *solely* in virtue of our conventions for judging persistence – that independently of our conventions, there simply is no such phenomenon as an individual object's persisting. This view

---

[3] The most careful presentations of the conventionalist position, in my opinion, are Sidelle (1989, in particular, 50–58), Sidelle (1998), and Thomasson (2007a, in particular, 57–59). But the most influential endorsement of conventionalism is probably Hilary Putnam's attack on "Self-Identifying Objects" (Putnam 1981, ch. 3), or on "ready-made objects" (Putnam 1983). Another widely read endorsement of conventionalism is Jubien (1993).

seems quite obviously to block realism about courses of existence. (For discussion of some tricky details, see Elder 2006.)

There is a more familiar and traditional answer on what it is for certain properties to be essential to a given object, while others are merely accidental. This is that the properties essential to a given object are those by virtue of which it belongs to the natural kind that it does – those properties individually necessary, and jointly sufficient, for membership in that natural kind. But then one must ask: What underlies the fact that such-and-such properties *are* individually necessary, and jointly sufficient, for membership in thus-and-such natural kind? On the conceptualist account of natural kinds, the answer – the whole answer – is that these properties are embroiled in the description that we associate with the sortal (or matter-name) by which we refer to that kind. Perhaps our description lists the properties by name. Perhaps it merely indicates which sorts of properties are crucial for kind-membership, and signs a blank check for empirical research to fill in with particular property-names. But even in the latter case, it is we who sign the check. It is we who determine which sorts of properties are crucial to membership in the kind in question. *That* certain properties end up as membership conditions for a given natural kind is up to us. And then if *what it is* for a property to be essential to a given object is for it to figure in the membership conditions for the kind to which that object belongs, essential status is likewise up to us.

But Millikan's account of natural kinds, we have seen, is robustly realist. Can we couple it with the traditional idea that the properties essential to an individual object are those by virtue of which it belongs to the particular natural kind that it does belong to, and end up with a realist account of the difference between properties essential to that object and properties that are merely accidental? To put it differently: Can we hold that the membership conditions for Millikan-kinds double as persistence conditions for the members of those kinds, and so end up with a realist account of the persistence conditions that characterize those members – and thereby with realism about courses of existence?

Surprisingly, the answer appears to be no. Clearly no one object can be characterized by divergent courses of existence. But it can well happen, Millikan holds, that one object belongs to two or more natural kinds (OCCI, 30; LTOBC, 293). For on Millikan's view all that it takes, for a given class of objects to compose a natural kind, is that they are characterized by a recurrent package of properties, and are so because of the operation of a recurrent causal ground. In this sense, human beings arguably compose a natural kind. But then so do human adolescents and (human) diabetes sufferers. That is because human adolescents share a common (transient) form of brain organization, and as a result are non-accidentally characterized by a common propensity to underestimate risks in behavior; they share hormonal disturbances, and as a result non-accidentally have a propensity for acne. Diabetes sufferers non-accidentally share a pattern of elevated blood sugar, and a host of attendant health risks. One and the same entity, then, can belong to three different natural kinds.

Indeed Millikan's view appears to saddle us with all the cases of "coinciding objects" that appear in the literature on material constitution. Suppose that a particular ice cube is composed wholly of H_2O. Then it is some H_2O; it belongs to the chemical kind H_2O. But ice, *frozen* H_2O, will be for Millikan a natural kind in its own right: frozen H_2O is in all portions non-accidentally characterized by a certain texture, a certain specific weight, etc. Or suppose that a certain statue is composed wholly of gold. Then that statue is some gold, a portion of gold. But what if *statues* too compose a Millikan-kind? Then if it is the case that membership conditions for Millikan-kinds double as persistence conditions for the members of those kinds, our statue will have both the persistence conditions for a statue and the persistence conditions for some gold. The ice cube and the diabetic adolescent will likewise be characterized by discrepant persistence conditions.

Here then is the reason why Millikan's views do not unambiguously provide "new foundations for realism." Realism about objects requires realism about courses of existence, and hence realism about persistence conditions; yet on Millikan's views about natural kinds, it can happen that a given object belongs to more than one natural kind. And so what grounds an object's persistence conditions – what grounds the status, *as* essential, of the properties essential to that object – apparently cannot be its kind-membership.

At first blush, there seem to be four ways in which we might amend Millikan's views so as to render them consistent with realism about persistence. (1) We might deny that some of the natural kinds that Millikan recognizes really are natural kinds – we might tighten the requirements for a class to qualify as a natural kind. (2) We might hold that only *some* natural kinds are such that their membership conditions double as persistence conditions for their members. (3) We might hold that membership in a given natural kind in *no* case sets up persistence conditions for the members of that kind. (4) We might hold that one and the same object can belong to different natural kinds, and *can* in a sense have divergent persistence conditions. This seems impossible because it seems to entail that divergent *actual* spans of persistence may hold true of a single object. But no such entailment holds, says this position. For the only real objects that there are do not persist at all: they are momentary object-stages. A *sentence* saying that such an object persists across a certain span can be true or false, and it is in this sense that such an object can be said to have persistence conditions. The truth-maker for any such sentence will be relations that hold among (suitably propertied) momentary object-stages. The particular relations that matter, for the truth of a particular claim of persistence, will depend upon the conversational context – upon which sortal is being used to pick out the momentary object in question, in other words which kind it is, among those to which that object does belong, that is conversationally salient. Different contexts can make different such kinds salient, and it is in this sense that different persistence conditions can attach to that single object.

Position 1 is the position that I myself favor. I will discuss it last, and briefly. Position 2 can fairly quickly be dismissed as incompatible with realism about persistence, and I will discuss it first. Positions 3 and 4 have adherents in the contemporary literature. Position 3 is treated as a serious option in Rea (2002). Position 4 will quickly be recognized as the "stage theory" advocated by Ted Sider and Katherine Hawley; interestingly, it is very close to the position that Millikan attributes to the "devil's advocate" in chapter 17 of LTOBC, a position that Millikan herself takes quite seriously. I will argue that Position 3 raises gratuitous and severe worries, and should be rejected. Position 4, I will argue, fairly clearly is incompatible with realism about persistence, since it entails that one and the same real entity can truly be said to persist over different spans of existence, relative to different conversational contexts. But the amendment of that position that Millikan offers in chapter 17 of LTOBC does not help. Position 1 will emerge as the only real option.

IV

Might one hold – as Position 2 does – that the membership conditions for *some* natural kinds double as persistence conditions for the members of those kinds, while *other* natural kinds have membership conditions that do *not* amount to persistence conditions for their members? The suggestion has a certain appeal. Perhaps Position 2 would enable us to claim that a human adolescent could cease to belong to the kind *adolescent* without ceasing to exist, while also claiming that once that same human adolescent ceases to belong to the kind *human being*, he really does exist no longer. But there is a serious problem with Position 2. If it has traditionally seemed plausible that an object could not lose the properties by virtue of which it belongs to its natural kind without ceasing to exist, that is not because it has seemed plausible that there is a mere extensional equivalence between the properties that are membership conditions and the properties that are the object's persistence conditions. Rather it has seemed plausible that an object must retain certain of its properties to go on existing just *on account of the fact* that it needs those properties to belong to its natural kind. To put it differently, it has seemed plausible that, for any object in nature, membership in the natural kind to which it belongs is a life-and-death matter. Position 2 needs to say that the properties required for membership in *certain* natural kinds (but not others) are, additionally, existence requirements for the members of those kinds – but it cannot say that they are existence requirements *just in virtue of* being required for those objects to belong to a natural kind that they do belong to. For in that case, *whenever* certain properties are required for any object to belong to a natural kind that it does belong to, those same properties would have to amount to persistence conditions for that object. No, Position 2 needs to say that something more must be added, to the mere fact that certain properties are required for an object to belong to a given natural kind, for those properties to qualify, additionally, as persistence conditions for the object. Certain natural kinds (and not others) have somehow to be dignified as "career-defining" – as being such that membership in them is a life-and-death matter. What extra factor could confer this dignity – could make membership conditions amount also to persistence conditions? The only plausible answers would depict this extra factor as being supplied by us. It would have to be our practices or conventions that render membership in certain natural kinds, and not others, career-defining. But then the courses of existence that do obtain in nature – the careers, in other words – would not be mind-independent.

Position 3 offers a radically different response to the problem that Millikan's position appears to face. Perhaps one and the same object can belong to two (or more) natural kinds, without inheriting incompatible persistence conditions, precisely because membership in a particular natural kind *never* sets up persistence conditions for the objects that belong to that kind. Perhaps, in other words, membership in a natural kind is *never* a life-and-death matter for the members. Perhaps this fact structures the semantic content of sentences that assign particular objects to particular natural kinds: perhaps when we say of an object that it "is a dog" or "is an electron," or that is "is some water," we are using a sortal or matter-name in only what Rea calls "a classificatory sense," and are really only claiming that some entity *now qualifies as* a dog or an electron or as some water, while remaining uncommitted as to the persistence conditions for that entity (Rea 2000, §II).

But Position 3 suffers from two problems, a lesser and a greater. The lesser is that it appears to reject what above I called "the qualitative assumption." If membership in a particular natural kind never amounts to a life-and-death matter for the objects that belong to the kind, it follows – barring special extra premises – that any object can depart from any natural kind to which it does belong, while still continuing to exist. What now qualifies as "a dog" can cease to be a dog at all, while still existing; what qualifies as "some water" or "some gold" can cease to be water or gold, but still *be*. The continued existence of such an entity seems not to require that it retain *any* particular properties. That violates the commonsense idea that continued existence, for a perishable entity, *does* require retention of at least certain properties. Even so, the position is not without philosophical precedent. Aristotle at one point worries that he is committed to a sort of matter that can come to have, and can later lose, the properties characteristic of water or earth: a parcel of this sort of matter would be a "this," but not a "this such" (*Metaphysics* Z, 1029a3–36). Commentators on Aristotle call this sort of matter "prime matter." It is matter to which *no* properties are essential. Position 3 invites us to entertain the thought that wherever we discern an object (or portion of matter) that belongs to a particular natural kind, we really are looking at a parcel of prime matter that currently is qualified, accidentally, by the properties characteristic of that natural kind (Rea 2002, 104; cf. 132, 134, 159–160).

The greater problem is that as soon as we do entertain the thought that nature is really populated by parcels of prime matter, we are required to envision massive, widespread coincidence, for which there can be no explanation. Let me approach this point by speaking metaphorically. If nature is populated only by parcels of prime matter, it can address, to the entities that populate it, only hypothetical imperatives. It can say only, "Parcels of matter!: to the extent that you elect to belong to one natural kind or another, you must display the properties characteristic of that kind, and you must respond to outside influences in the ways characteristic of that kind; in certain circumstances, you must depart from that kind altogether. If you elect to belong to the kind *dog*, you must appear furry and four-legged; you must respond to being kicked by yelping; if you should be fried by a bolt of lightning, you must depart from the kind *dog* altogether. If you elect to be *some water*, you must appear clear and potable; you must boil if heated to 212°F or higher; and if subjected to hydrolysis, you must depart altogether from the kind *water*. But you are free at any time to rescind your membership in the kind to which you have elected to belong: it is not as if you must, on pain of death, continue to belong to your current natural kind. For this reason, if you currently belong to the kind *dog*, you are free suddenly to present cat-like features; or to respond to being kicked by emitting the hollow ringing of a gong, or by exploding; and you are free to respond to being struck by lightning by forming a pool of molten lava."

The problem, of course, is that nature appears to be populated by entities that take their membership in a particular natural kind quite seriously. Nature's objects and portions of matter appear to treat the duties attendant upon their kind-membership – duties both as to the properties they present and the behavior they perform – as unconditionally binding. They do not respond to impinging influences in ways uncharacteristic of their kind, do not suddenly display properties that lie outside the nature of their kind, and do not appear to be destroyed when exposed to events that members of their kind characteristically survive. If nature *really* is populated by parcels of prime matter, it is an amazing coincidence that each such parcel behaves in the way one would expect of an entity permanently wedded to just one natural kind. To drop the metaphor: the hypothesis that nature is populated by parcels of prime matter is empirically indefensible. There is not a scrap of evidence for supposing it true, and supposing it true requires us to posit a standing and inexplicable coincidence.

Now for Position 4, the position that one and the same object *can* have divergent persistence conditions. What keeps this position from collapsing immediately into absurdity is the contention that no real object ever actually persists – that strictly, the objects of nature are momentary object-*stages*. Sider and Hawley present this contention in a semantic formulation. Phrases employing sortals such as "this banana," "a dog," "that lemon" do typically pick out real objects, Sider and Hawley say, but objects that are extremely short-lived: when we speak of "a dog," for example, we are saying how matters stand with some momentary dog-stage (Hawley 2001, 48–52; Sider 2001, 60, 188–208). For *every* claim that we make about how long familiar objects persist, and what happens to them as they persist, the truth-makers are relations that obtain among momentary stages – those relations, that is, along with the facts as to what each of the related stages is like, in and of itself.[4] Millikan goes half-way toward agreeing with the stage theorists: she rejects endurantism, the view that when it is true that an object lasts across a span of time, one and the same object is present, whole and entire, at each moment within the span. Her view does feature an element largely absent from stage theory, an element to which I will shortly return: "*[s]omething*, we still will want to say, persists *through* time when an object persists," and this "something" is what Millikan calls a "subessence" (LTOBC, 290). But no relation of *identity* links the *object* that exists at t_1 with the object that exists at t_3. If, that is, we set aside the persisting "subessence,"

> ... it appears that the case against genuine identity *over* or *through* time rests.... All there is is the identity of each temporal stage of a thing with that temporal stage, the identity of the whole collection of temporal stages with that collection, and a principle of unity that collects these stages into a unified whole. (LTOBC, 290)

But what is the nature of this "principle of unity?" If we abstract for now from what Millikan says about subessences, it is much the same as the relations that, for stage theorists, bind one stage together with other stages both earlier and later. These are *temporal counterpart* relations, analogous to Lewis's modal counterpart relations. When we speak of

---

[4] With one sort of exception: the comments we make when engaged in "diachronic counting" of bananas and dogs and other objects are *not* comments about individual object-stages, say both Sider and Hawley; but both treat these comments as an isolated and anomalous exception to their general thesis (Sider 1996, 448; Hawley 2001, 63–64).

"this dog," we refer to an object-stage that cannot, in the very nature of the case, itself be present at other moments; but there will be other object-stages (specifically, other dog-stages), located at, say, t_7 and t_8, that "stand in" at those moments for the stage we pick out with "this dog."

If endurantism is false – as both stage theory and Millikan hold – does it follow that we can truly attribute, to objects that really exist in the world, only those properties that can be instantiated in the space of a single moment? Millikan and stage theory both answer no – but they say "no" in different ways. Millikan says "no" in the way that perdurantism does; to put it differently, Millikan is a "worm theorist." That is, she holds that we "divide up" (LTOBC, 293) the world into temporally extended entities, "space-time worms" (OCCI, 70), by "collecting" together object-stages that are located at successive moments. Sider and Hawley avoid positing temporally extended entities[5] and in this respect take a view different from Millikan's. But they still answer no to our question – and indeed they end up endorsing a position that is operationally quite close to Millikan's. They hold that we can truly attribute to real entities in the world "historical properties" (as Sider calls them) – properties such as those picked out by "once grew on a tree," "was once a young boy," "will one day be an old man" (Sider 1996, 437–438, 446–447; Sider 2001, 193–196). We can truly attribute "lingering predicates" (as Hawley calls them), such as "is musing about Vienna" or "is traveling across the room" (Hawley 2001, 53–57). The subject of such true predications is not a temporally extended worm, but instead a single momentary stage. The stage can nevertheless satisfy the predication, by virtue of relations that it bears to suitably propertied temporal counterparts. The relations really obtain, and so the predication is true.

Now for the question on which Millikan diverges sharply from stage theory. Are there limits on which temporally located objects really exist in the world, and on which relations obtain among the real temporally located objects, that *restrict* which true applications we can make of "historical" and "lingering" predicates? For Millikan, this amounts to asking: Is just *any* collection of successive stages, into which we might "divide up" the world, just as ontologically genuine – just as *real* – as any other? For the stage theorists, the same basic question must be asked in rather different terms. For Millikan, a main way that we "collect" successive object-stages is by wielding sortals that pick out particular sorts of space-time "worms," e.g., "tadpole" or "example of *Rana catesbeiana*" (cf. LTOBC, 293). For the stage theorists, our sortals do not pick out worms at all, but rather spotlight particular sorts of temporal counterpart relations obtaining between some individual object-stage and particular other object-stages, located at times both earlier and later. So, for the stage theorists, our question amounts to asking: Does every sortal that we wield spotlight temporal counterpart relations that really obtain, and that can render true the application of a historical or lingering predicate – or do some sortals depict temporal counterpart relations that are spurious, that do not really obtain? The "devil's advocate" in Millikan's discussion answers yes: any collection of successive stages that we might divide out from the world is as real as any other. Millikan herself holds that this answer neglects the role of "subessences," and I will return to this point momentarily. The stage theorists, when

---

[5] Except in the anomalous case of diachronic counting: see note 4.

asked the same question in terms appropriate to their view, likewise answer yes: any sortal that we might wield directs us to heed temporal-counterpart relations that are as real as those made salient by any other sortal. That is why, when we apply first one sortal and then another, to one and the same momentary object – when we say first that the entity is "an adolescent," and then say that it is "a human being," or when we first say that the entity is "a statue," and then say that it is "a portion of gold" – we are in effect attributing different persistence conditions to that one entity, and are *truly* attributing different persistence conditions to it.

Stage theory itself, I now argue, is unfriendly terrain for a realist about objects. But in fact stage theory is a Janus-faced doctrine, and for that reason it is unfriendly terrain for different reasons. Suppose that we concentrate on the "face" of stage theory that says that the only real objects are momentary object-stages, entities that do not persist at all. Then my most basic contention – that realism about the world's objects requires realism about courses of persistence – falls flat. But in that case we have severe problems about which properties can truly be attributed to the real objects of the world. For it then appears to follow that no property can truly be attributed to a real object, the instantiation of which takes longer than a moment. If so, then arguably no velocity can truly be attributed to any real object. Indeed it is arguable no property involving a dispositional component can truly be attributed: no real object can truly be said to have a melting-point or a viscosity or a color; even many of the shapes that we commonly suppose to be present in the world cannot truly be attributed to real objects. The real objects of the world will be little more than sense-data (Elder 2007a). This is an unfriendly position for anyone.

The other "face" of stage theory insists that sentences attributing historical and lingering predicates to real objects in the world can literally be true. If we focus on this aspect, we will judge that there really are truths about how long real objects in the world have been in existence, about what has happened to them in the past, and about what will happen to them in the future; real objects can be credited with velocities, and with properties that involve dispositions. In this case we will judge that, for stage theory, *there is* such a phenomenon as persistence, for temporally located objects. For that very reason, I will be free to reassert my fundamental contention: realism about objects requires *realism* about courses of persistence. But persistence – when we focus on this aspect of stage theory – will be a phenomenon indexed to one conversational context or another. *How long* the real object that we are discussing persists – what that object's *persistence conditions* are – will depend on which sortals are salient in our conversation about that object. The courses of persistence that obtain in the world will be a function of how we are at present talking. And this is the very antithesis of realism about courses of persistence.

What about Millikan's own answer to the question of which temporally extended "worms" really obtain in the world – in other words, about where persisting entities really are to be found? Millikan disagrees with "the devil's advocate." She holds that *not* just any collection of object-stages that we might carve out of the world is as ontologically genuine as any other. The genuine collections are those in which each successive stage is shaped and governed by a "subessence" obtaining at every previous stage. A "subessence" is a principle or nature that ensures – at least given typical circumstances – that properties present in one temporal stage will be reproduced in later temporal stages spatio-temporally continuous with that one stage. (In an analogous way, the nature of the secondary substance *gold* ensures that the properties presented in one *spatially* localized sample of gold will get presented as well in other *spatially* localized samples: LTOBC, 290–291.) The only real worms, in other words, are those for which something causes recurrence of the same properties in stage after stage after stage.

Does this aspect of Millikan's own view mean that she holds a better version of Position 4 than stage theorists do – or at least a more realism-friendly version? No: Millikan does not endorse any version of Position 4 at all, since she holds that there *are* entities that persist – worms that incorporate temporal parts at successive moments, and that thus perdure. Does it mean that, setting Position 4 aside, she holds a view that can dissolve the threat posed by the idea that often, one and the same object belongs to different natural kinds? No again, unfortunately. Typically, if one and the same object were to belong to more than one natural kind, it would be characterized by divergent persistence conditions, and would have different actual courses of persistence. Just that is the case with our human adolescent. At any one moment, he is incorporated in worms of different temporal extension: at the very least, in a worm that ends when he reaches 19 years of age (or so), and in a worm that ends when he dies. Millikan's restriction on which worms are real does not help here. *Something* causally ensures that many properties in the present adolescent-stage recur in subsequent stages of the adolescent, but it is also true that *something* ensures that many of these properties recur in stages of the adult human being who grows from the adolescent. It even is ensured that some properties in the present stage will recur in stages of the corpse that is left after the human being dies: that is why we can identify the victim of an accident by consulting the dental records of an adolescent. "Subessences" link together the stages in three different worms: one picked out by "this adolescent," the second by "this human being," and the third by "this mass of human tissue." As Millikan herself says: "a temporal stage does not by itself determine any particular subessence as *the* subessence that must unify it with other temporal stages into one whole enduring [*sic*][6] thing over time." Consequently, "a great deal of room is left for decision on our part as to how to divide the world into … temporally extended wholes" (LTOBC, 293). This is *anti*-realism about persistence.

This returns us to Position 1 – the position that not all the kinds that Millikan regards as natural kinds really *are* natural kinds, really *do* set up persistence conditions for their members. The idea would be that more is required, for a class to be a natural kind, than what Millikan indicates. Here is an articulation of Position 1 that I have offered (Elder 2007b). At first blush, it might seem that any property that belongs to the very nature of a natural kind must reflect, and be grounded in, other properties encompassed in that nature. But of course this cannot universally be true, as then we would have a vicious regress. Some properties encompassed in the nature of a natural kind must themselves ground, without being grounded by, other properties in that nature. *Having molecular structure H_2O* plays this role in the nature of the natural kind *water*: this microstructural

---

[6] If Millikan were using the terminology from David Lewis that has now become fairly standard, she would say "persisting" rather than "enduring," for she is clearly not an "endurantist." I hasten to add that the Latin word "sic" means only "thus" – it does not by itself say that a given author has made a mistake.

property grounds water's characteristic index of refraction, its specific gravity, its boiling and freezing points, etc. So long as some property grounds many other properties in the nature of a natural kind, and does so all by itself, then it is sufficiently integrated in that nature to count as itself an element of that nature, one might hold. But consider, in contrast, the supposed natural kind *ice*. *Ice*, if a natural kind, would encompass in its nature the property *having a temperature between 32°F and 212°F*. But that property does not grow out of other properties that would lie in ice's nature: if it did, you could keep a chunk of ice frozen, on a hot day, merely by refraining from altering its specific density or its crystalline structure. And while it is true that *having a temperature between 32°F and 212°F* grounds other properties that are distinctive of ice, it is arguable that it does not do so all by itself. Rather it does so by enlisting the powers of *having molecular structure H_2O* – it triggers one way in which that microstructural property can shape other properties. For as realized in steel, *having a temperature between 32°F and 212°F* grounds no special properties beyond itself.

Better articulations of Position 1 may be available. It is hard to believe that *no* version of Position 1 can be defended. But if Position 1 can be defended, then Millikan can be a full-blown realist about courses of persistence. Just beneath the surface of Millikan's writings there lurk powerful "new foundations for realism."

Reply to Elder

I want to thank Tim Elder very much for jumping square into the middle of one of the least well known, most radical, most difficult, and most obviously contentious parts of my work. His questions about the ontology of individuals, of what accounts for their kind of selfsameness, their identity over time, are exciting and I am anxious to discuss them. But first I want to make some clarifications with regard to his exposition of my work.

In his section II, Elder offers some fresh and interesting ways of looking at, explaining, and defending some of my ideas on the semantics of kind terms. Now it is almost always hard to accept someone else's redescription of your views, often because you know from hard experience what kinds of misunderstandings the shortcuts taken, the simplifications, the glossings, may lead to or have led to in the past, even despite careful precautions. Though I think them often helpful, I want to remind the reader that Elder's arguments and statements of results in section II are not my arguments, nor arguments I should be assumed to support. And I want to supply a couple of supplementary footnotes to section II and later sections, to warn against possible misunderstandings. Let me just run these off in a list of short paragraphs.

1. My work on human representation places the emphasis on learning rather than genic evolution, although the basic ways we can learn have, of course, resulted from genic evolution. Especially, it is important that nothing remotely like concept nativism is involved. Instead, I give a detailed discussion of the mechanisms of individual concept development. Prior to that is my own peculiar theory of the nature of empirical concepts, one that does not follow from but is, I believe, compatible with the basic biosemantic theory of representation.
2. My belief is that our personal concepts of commonly recognized real kinds typically do not rest on any descriptions. They rest on perceptual abilities to read diverse proximal signs (Quine would have said diverse sets of "stimulations") that manifest distal members of these kinds to our senses. In my reply to Fumerton, I explain why I think

that *concepts* of properties are not typically involved at all in identifying members of a real kind.

3 I suspect that no creatures other than humans have anything much like our "beliefs and desires" (see VOM, esp. chs. 18–19). Still, all representations, by their very nature, are "compositional" in the sense that they are articulate, having meaningful parts or aspects that appear also in other representations from the same representational system, representations with altered content. In the simplest cases, perhaps only the time, place, or individual concerned is altered, but the effect is still that different information is represented in other tokens that are members of the same representational system. Also, many animals with more primitive forms of inner representation than ours also need to have concepts ("practical" concepts) of real kinds that are important to them. They are able to reidentify things of importance to them in a variety of different ways and to integrate this information.

4 Nowhere do I suggest, I hope anyway, that the identity over time of an individual must be determined by a real kind or kinds it falls under. That is Elder's approach but not mine. So the problems he poses given this approach are not really "[P]roblems that Millikan's position appears to face." It is also well to note why, if it were taken just by itself, this approach would clearly be insufficient. You and I both belong to the natural kind *human*. Something very substantial and important, other than being a member of the same natural kind, must determine that it is still me here, rather than you, tomorrow.

5 The LTOBC quotation beginning "… [i]t appears that the case against genuine identity over or through time rests …" is put by me in the mouth of my "devil's advocate" who holds a position I did not "take quite seriously," but is, instead, the position that I believed to contrast most neatly with my own. (The dialectic of LTOBC chapter 17 was confusing, for which I apologize.)

Now to the real issues. I believe that the quarrel over endurance versus perdurance is largely a verbal matter, but will not defend that here. Instead, without writing LTOBC chapters 16 and 17 over again, I will offer some pointers aimed merely to clarify my position, and then explain why I see no "threat" in the view that "often, one and the same object belongs to different natural kinds."

The keys to understanding LTOBC, chapter 17, "Notes on the Identity of Enduring Objects," are mainly in chapter 16, "Notes on the Identity of Substances and Properties" (which was long and dense; again, I apologize). There I adopted the view that there are real natural necessities, in defiance of Humean and verificationist challenges. Assuming that one could speak of temporal "stages" of an individual (just as one can of a process) without reifying these abstracta, I emphasized that natural causal laws and laws of conservation relate earlier stages of an enduring individual to later ones. (That is an element that would obviously be missing in the bare notion, that remaining within the same real kind determines continuing existence for an individual.)

Real kinds are "substances" in the sense of chapter 16. So are enduring objects. The discussion of the kind of identity that substances have was supposed to apply to both. My claim was that substance and property are relative, not absolute categories. The very identity, the very selfsameness, of a substance is correlative to that of its properties and vice versa (see my reply to Fumerton). What makes a substance a substance is "the fact that it is always the same as itself relative to certain property ranges in accordance with natural necessity – regardless of the perspective from which (for example, the instance through which [and the time at which]) it is observed" (LTOBC, 276). I went on to qualify, admitting "rough" substances, later called "historical kinds" (Millikan 1999a, OCCI, ch. 2), some of whose members may lack some of the properties relative to which the kind is real. In "On Knowing the Meaning" (Millikan 2010), I tell how historical kinds often have rough boundaries, and may fade off through narrow necks into other kinds.

What makes an individual into a selfsame individual over time is the same sort of principle that makes a historical kind into a selfsame kind over space as well as time. The members of a historical kind are the same relative to certain contrary ranges – they have certain properties in common – because of certain kinds of causal connections among them, between them. Roughly, each member has those properties because other members had them, or had what caused them (but see my real discussions of this in LTOBC, ch. 18 and OCCI, ch. 2). A selfsame individual remains the same relative to certain determinable ranges, displaying the same determinate properties at each time because it had them at previous times. The individual remains the same individual because it hangs on to its properties over time in accordance with natural conservation laws or other causal mechanisms. Many of these properties may, of course, be dispositional properties, or general capacities.

Substance identities are relative to limited sets of contrary spaces, so it is not implied that an individual cannot change. Just as cats form a historical real kind relative to rough shape and size, placement of heart and lungs, a disposition to hiss when frightened and so forth, but not with respect to color or subtleties of temperament, Tucker remains Tucker over time with respect to cat properties and to color and temperament as well, but he does not stay the same with respect to posture or mood or emittence of hisses. And there are more parallels. Some historical kinds are nested within others. Siamese are nested within *Felis domesticus*; they form a richer real kind because they are caused to be like one another relative to more contrary spaces. Similarly, Tucker-the-kitten is nested within Tucker-the-member-of-*Felis domesticus*. Real historical kinds often merge into one another. A thinner or thicker neck of related individuals has separated every species from the earlier species from which it evolved. Similarly, there was no sharp separation between Romanesque architecture and Gothic; Romanesque evolved into Gothic. And, of course, Tucker-the-kitten evolved into Tucker the adult cat; human adolescents evolve into human adults.

Which leaves things rather as Elder fears for my position! Speaking in strict technical terms, every individual that would be referred to in common parlance has multiple individuals nested within it, and is probably itself nested within multiple temporally longer individuals. Lively Chipper, the grey squirrel now swinging in our bird feeder, may, sadly, be nested within a continuing body that will lie flattened on the road tomorrow. And his current state is nested within innumerable other individuals as well, most of which it would be hard or impossible precisely to name. There are important considerations, however, that make this claim less bizarre for a realist, I think, than Elder supposes.

These various overlapping individuals are not determined to be other than one another or the same as themselves by any linguistic decisions, nor are "temporal counterparts" in any way involved. Each is perfectly real, objectively the same as itself over time in the natural world before language comes along. Linguistic decisions have to be made only about which of these individuals to give names to. More important, *cognitive* decisions have to be made about which of these individuals to think of, to bother to maintain concepts of. These cognitive and linguistic decisions need to be made in a way that serves purposes for which human cognition and language were designed, the kind of purposes that have accounted for the survival of human cognitive systems, and that account for the survival of names in human languages. They have to be made such that these individuals can be reidentified and knowledge about them accumulated and later employed – never with certainty, of course, but enough of the time – and so that this knowledge can be transmitted. Ironically, however, what that means is that many of these decisions don't have to be made at all! Because language *and thought* both tolerate, indeed positively thrive on, healthy vagueness. Without vagueness they could do no serious work and would not survive. The job of extensional terms is not to cut the world up with sharp-edged cookie cutters, but to locate and anchor to the centers of natural correlations. (For my comments on vagueness, see Millikan 2010, ¶6.3.)

There is a tendency for a historical individual's enduring properties to disappear, when they do, in sudden bursts, many together more or less at once. These steep gradients offer vague demarcation lines to help coagulate substance concepts and fix linguistic expressions. These steep gradients also tend to correspond to episodes in which an outer or encompassing individual leaves – exits from – a real kind, as when a living animal dies. So I think that Elder is right that exiting from a real kind is what marks the end of existence for many of the sorts of individuals that we think and talk about. When a historical individual exits a real kind, the knowledge we had accumulated for dealing with it is suddenly no longer useful, our (always fallible) knowledge of historical individuals depending largely on knowledge either of their kinds, or of idiosyncratic properties of theirs whose permanence depended on the structure of the kind.

Historical kinds, however, are almost always clusters with somewhat vague boundaries. In accord with this, recall how modern medicine and law wrestle with the fact that the line of death has become less and less sharp in recent years. And nature sometimes makes things *really* difficult. Consider the identity over time of geological formations, mountains, rivers, clouds, storms, and so forth.

Am I saying that strictly speaking here in my lap is not just one thing, a cat as it happens, but multitudes of different overlapping things? My colleague John Troyer says "'thing' is not a count noun." Neither, of course, is 'object' or 'individual' or 'entity.' But in the usual way of speaking, of course there is just one thing – his name is 'Tucker' – in my lap. 'Two things,' in this context, would normally require two separate spaces currently occupied by two separate cohesive physical objects, and 'many things' would require many such separate spaces and cohesive objects. Sometimes the ordinary language folks are right. Sometimes apparent philosophical problems do result from entanglement in our language.

9

CRANING THE ULTIMATE SKYHOOK

MILLIKAN ON THE LAW OF NONCONTRADICTION

CHARLES NUSSBAUM

It is also true that if logic is an a priori *science that deals with the relations among concepts, then logical possibility and logical necessity are merely subjective appearances – chimeras – unless our concepts happen to reflect genuine natural necessities, in particular natural identities and natural contrarieties.*

(LTOBC, 273)

Introduction and Conspectus: Naturalizing the Logical Modalities

In *Darwin's Dangerous Idea* (Dennett 1995, 107n1), Daniel Dennett tells of rushing to read a new book entitled *The Possible and the Actual* by the Nobel laureate biologist François Jacob. He was expecting an essay delineating the roles of these modal notions in evolutionary biology. Although he lauds the book, Dennett reports that it did not deliver what he took its title to promise. Like *The Possible and the Actual* according to Dennett, *Darwin's Dangerous Idea* is a fine book with a great title, a title, moreover, that does not fail to deliver on its promise. Still, I experienced a reaction similar to Dennett's when coming upon a section in chapter 5 entitled "Possibility Naturalized." Encountering so resonant a title in this particular book led me to expect a Darwin-inspired naturalized account of the logical modalities, perhaps even a rethinking of the theory of modality as radical as was Quine's rethinking of the theory of knowledge.

Alas, it was not to be. Dennett supplies a fairly standard picture consisting of a series of shells of possibility enclosing a core of actuality, with each enclosing shell enforcing more limited ontological constraints (Dennett 1995, 107). Closest in are historical and biological possibility, furthest out is logical possibility, constrained only by noncontradiction. With logical and physical possibility held "clamped," Dennett's naturalistic innovation is to introduce the notion of possibilities that vary with time: possibilities available 50 million years ago became more remote over time than they once were, and now have virtually been foreclosed by evolutionary developments. Biological possibility in particular concerns the negotiability of a path from here to there in genomic design space by mutation and selection at some historical juncture.

This is all unexceptionable. But if it is a Darwin-inspired naturalized account of the logical modalities we seek, we shall have to look elsewhere. The good news is that we need not search very far, for that is exactly what Millikan supplies in her body of work; or so I shall argue. Yet Dennett will remain very much with us, for I shall be enlisting his potent metaphor of cranes and skyhooks to limn the contours of Millikan's proposal. What is dangerous about Darwin's idea of natural design is its capacity to act as a universal acid that dissolves the skyhooks, or mind-first powers that highly organized phenomena, from life itself to the products of human creative imagination, seem to require. To dissolve a skyhook is to render an apparently intentional design process explanatorily otiose by showing how remarkably sophisticated products that seem to exceed the powers of undirected brute physical causality have been or can be craned, or erected mechanically and incrementally, from the ground up by way of the blind three-stage process of variation, selection, and inheritance.

Although the law of noncontradiction is not *per se* a design principle, it is a principle of conceivability. But it also carries ontological significance. It sets the limits of the logically possible; and what is logically impossible is ontologically impossible. In this, its ontologically legislative or binding capacity,[1] the law of noncontradiction arguably remains the ultimate philosophical skyhook.[2] A *Darwinian* naturalization of the logical modalities would require that this skyhook be transformed into a crane, or a cascade of cranes, and in the course of unpacking this (rather uncontroversial) assertion I shall explain how Millikan's account does just that. In section 2, I compare and contrast Millikan's treatment of modality with that of possible worlds semantics, and in section 3, I explain in detail how she goes about craning the law of noncontradiction. In section 4, I take up the relationship between

---

[1] It has been objected (by Jeremy Byrd) that the principle of noncontradiction need not be seen as legislative or binding. Why not say, the objection continues, that if 'P' correctly describes the world, 'not-P' must be false since it denies that P correctly describes the world? Saying that P and that not-P cannot both be true just notes this fact and does not prevent anything from obtaining. I reply that this is a meta-level piece of discourse, since it makes assertions about what sentences or propositions do and about which ones may or may not be true. The truth predicate applies to sentences at the meta-level. At the object level, however, noncontradiction remains ontologically legislative: it determines that facts whose descriptions are contradictory may not obtain in this or any world.

[2] There is a serious contemporary position going by the name of dialetheism that questions the inviolable status of the law of noncontradiction in limited contexts, notably in the context of the paradoxes concerning self-reference. Dialetheism is, however, a minority view, discussion of which falls outside the scope of this chapter. First, I see no evidence whatsoever that Millikan is sympathetic to it. Second, its aim is to limit the scope of noncontradiction, not to naturalize it. For a good overview see Priest et al. (2004). It is interesting to note, in anticipation of discussion to come, that David Lewis politely declined to contribute a submission to that volume (see his letters included in Priest et al. 2004, 176–177).

natural and metaphysical necessity in Millikan's philosophy, and then conclude in section 5 with some reflections on Millikan and a version of the Myth of the Given.

The Law of Noncontradiction and Possible Worlds

The possible worlds game simply crashes when you dissolve too much real-world glue.
(Millikan 2010, 75)

Millikan is no fan of possible worlds, nor is she particularly keen to discuss them. For one thing, anyone who comments on possible worlds semantics takes on the onus of specifying which of the many versions of the theory he or she has in mind. But whichever version is at issue, possible worlds semantics has established itself as by far the most popular approach to the analysis of modal discourse; and, despite her distaste for the topic, Millikan has allowed herself an occasional comment on it. Not only can possible worlds semantics not be ignored by a modal theorist, it also affords the most direct route into the set of issues that are our present concern.

"According to tradition," Dennett (1995, 104) tells us, logical possibility is "simply a matter of being describable without contradiction." Dennett is of course a naturalist, and he does include in his remark the qualifier concerning tradition; but this formulation should bring any naturalist up short. Why should *describability* without contradiction place modal conditions on material existence? Whence noncontradiction's ontologically legislative or binding authority? This is not to ask the question raised by dialetheism. Nor is it to ask the epistemological question of how we *know* that noncontradiction has the grip on the world it does. The question is this one: What sort of naturalistically acceptable account can be given for the fact that it does? To assert, as have Aristotle and many others, that noncontradiction is an axiom of argumentation because it cannot be proved is one thing. To say that the world must comply with noncontradiction (or any other logical principle) and explain why it must is quite another. It may or may not turn out that any workable general account of modality must take some version of modality to be primitive.[3] Should matters turn out this way, it would be advisable to designate as primitive a version of modality that explains the grip of logic on the world, rather than one that presupposes it. (As we will see, Millikan takes natural necessity to be primitive.) Taking noncontradiction to be a self-evident axiom of *ontology* in the manner of the tradition is to beg this important question of how logic, a formal science of ideal structures, gets its grip on the world. It is to treat noncontradiction as a skyhook, to assume without argument that a condition on what is consistently expressible calls the ontological shots. This suggests a residual Platonism or idealism. The traditional view is very much in evidence in the two most popular versions of possible worlds theory, modal realism and linguistic modal "ersatzism," David Lewis's playfully tendentious sobriquet for antirealism concerning concrete possible worlds.

---

[3] In order to avoid circularity in his analysis of modality, Armstrong (1997, 268) is driven to take seriously the view that the "ultimate metaphysical truths" are neither necessary nor contingent.

The linguistic ersatzist, however, has a response to the charge of Platonism. For her, in contradistinction to the modal realist, possible worlds don't exist as concrete particulars. They are constructed linguistically, much as a novelist constructs a fictional world, except for the requirement that linguistically constructed possible worlds be stipulated to be maximal or complete: for any proposition, it or its contradictory must be true in a possible world. If the linguistic ersatzist is talking semantics and not ontology, there is nothing amiss in holding that there can be no ersatz possible world that defies consistent description. The problem is that the linguistic representation of the actual world is also a possible world, and here description does turn ontologically legislative, for the actual world cannot violate standards of descriptive logical consistency any more than a possible world can. This, once again, begs the question of ontological legislation.

Moreover, by assuming modality in the guise of logical consistency, the modal realist charges, linguistic ersatz theory becomes circular: it analyzes modality by assuming logical necessity (Lewis 1986, 150). But has the modal realist succeeded in avoiding the same circularity? At first, it may seem so. According to the modal realist, the possible worlds are simply all the worlds that exist. (Our own world is distinguished among the existing worlds as "actual" from our standpoint.) If a world does not exist, it is not possible. Necessary propositions are true at all possible worlds; contingent propositions are true at some possible world or worlds, and impossible (which include self-contradictory) propositions are true at no possible world. Naturally or nomologically necessary propositions are true at "closeby" or nomologically similar possible worlds. If a counterfactual is true, there is a possible world (or a state of affairs at some possible world) that makes it true. The modal realist proffers a bloated and incredible ontology, and relies on the slippery notion of degree of nomological similarity between worlds; but if modal realism affords a non-circular analysis of modal discourse, all this may be worth the price. But does it?

The avoidance of circularity is only apparent. "There are so many other worlds, in fact, that absolutely *every* way that a world could possibly be," David Lewis tells us, "is a way that some world *is*" (1986, 2, emphasis original). Notice the modal language, "could possibly be," which already suggests circularity in the definition of a possible world: what worlds there are is a function of the ways a world could possibly be. And again, "whatever way a world might be is a way that some world is" (Lewis 1986, 71). Circularity once more, for "might be" is a modal locution. A dilemma threatens: Do there simply *happen* not to be any worlds whose true descriptions are inconsistent, or is it the case that there *could* not be any such world? If the first, then modal realism's analysis of the logically impossible fails, for it does not correctly describe what Lewis regards as the modal "facts": it does not rule out worlds that cannot be consistently described. If the second, the analysis seems circular.

It might be objected that the disjunctive question in the dilemma is misconceived. For to ask it is in effect to ask whether there are any possible worlds in addition to all the possible worlds there are. And this does seem to be an improper question.[4] But how does Lewis know that he has exhausted the space of possible worlds, if not by way of stipulative

---

4   This is my gloss of a remark made by Charles Hermes.

application of the principle of noncontradiction? This, however, is just the procedure of the linguistic ersatzist. And if Lewis does rely on the principle of noncontradiction in the manner of the linguistic ersatzist, how does he avoid circularity?

The answer is, by evasion and equivocation. First, evasion: Lewis does not tell us why there can be no worlds whose true descriptions are inconsistent, he tells us why he has "no use for" such worlds: "But there is no subject matter, however marvelous, about which you can tell the truth by contradicting yourself" (Lewis 1986, 7n3). Notice, once again, the modal expression "can," just the sort of assumption of modality with which he had charged the linguistic ersatz theorist. The expression "no use for" is vague and evasive, and it does not directly address the circularity problem. Does it mean that the notion of worlds whose true descriptions are inconsistent is *unusable* or simply that is *not needed*? Second, equivocation: "Subject matter" can be interpreted semantically or metaphysically. That is, it can be the propositional content of sentential expressions or it can be their objects of reference. If the former, then Lewis has the same circularity problem with which he had charged the linguistic ersatz theorist. If the latter, then he has not succeeded in avoiding his own circularity problem.[5]

Circularity is not so obviously a problem for another version of possible worlds theory that Lewis rejects, namely pictorial ersatz theory. Here, the ersatz possible worlds are not contents of linguistic representations, they are contents of non-conceptual analog representations, namely pictures or models. Since noncontradiction, strictly speaking, is propositional, it does not apply here:

> I know there are so-called inconsistent pictures, mostly by Escher. But what makes their inconsistency possible is that the representation is not perfectly pictorial. It isn't all done by isomorphism; for instance, there are conventions of perspective to be abused. But the more is done by isomorphism, the less opportunity there is for any inconsistency – can a statue be inconsistent? And our pictorial ersatz worlds are uncompromisingly pictorial. It is *all* done by isomorphism. (Lewis 1986, 168)

Nevertheless, pictorial ersatzism, says Lewis (1986, 168), remains circular because it assumes modality: it assumes that there *could be* a concrete world isomorphic to any one of its coherent pictorial abstract ersatz worlds. Although Lewis is prepared to grant that pictorial ersatzism does not rely on *logical* modality, this may be too hasty. For it is not clear that pictorial ersatzism can do it all by isomorphism. A pictorial ersatz world had better not just be a model, even a maximally detailed model, but a model plus a consistent interpretive commentary or at least an implicit use procedure that can be consistently described. Otherwise, the model loses its non-symmetric representational relationship to the concrete particular world it is modeling and becomes a mere isomorphism, i.e., a structurally similar, if abstract, item. (The simi-

---

[5] Lycan (1988, 46) presses a similar objection against Lewis. "Some sets of sentences describe 'worlds' and some (the inconsistent ones we know them to be) do not; but Lewis cannot make that distinction in any definite way without dragging in some modal primitive or other. Thus, actualists [in this case, linguistic ersatzers] are no worse off in that regard." Thanks to Dan Ryder for this reference.

larity relation is of course symmetric.) The geometry projected by the model, moreover, had better be consistent.

The law of noncontradiction is, then, on display as a skyhook in both versions of possible worlds semantics we have discussed.[6] We can, however, begin to crane noncontradiction the Millikan way by applying a bit of real-world glue. "The law of noncontradiction," she asserts, "is a template of abstract natural-world structure – or it is something that suffices for such a template" (LTOBC, 269). What natural-world structure? It is determinate property exclusion "on the ground of" the substances to which they are ascribed. For Millikan, negation is "an indefinite assertion of contrariety," by which she means predicate contrariety: "to assert that two contraries hold of the same subject can be to contradict oneself. Indeed, it is always to contradict oneself" (LTOBC, 269). Predicate contrariety tracks the mutual exclusion of determinate properties within ranges or categories such as color or shape. Being a determinate color excludes being another determinate color, but has no bearing on shape, a different category; whereas being one determinate shape excludes being another determinate shape. Determinates under a determinable are, as Millikan says (LTOBC, 271), incompatible by "natural necessity," a notion central to her philosophy and one to which we shall return. "If the law of noncontradiction is grasped '*a priori*'," she asserts, "this must be so only in the sense that nature, via evolution, has built this grasp into us as a mirror or a reflection (possibly only a sufficing reflection) of a structural principle in the natural world with which we must deal in order to survive" (LTOBC, 257). Determinate property exclusion is the structural principle of which noncontradiction (or the grasp thereof) is "possibly only a sufficing reflection." In what way "possibly only sufficing"?

"S cannot both be P and not be P" is Millikan's formalization of the principle of predicate contrariety (LTOBC, 273), from which (ignoring the modal locution) a standard version of the law of noncontradiction is easily derived. ('P' in Millikan's formulation stands for a predicate, not a proposition.) But without specifying that 'P' and 'not-P' refer to properties belonging to the same range of determinates, the formal principle of predicate contrariety could turn out false, say if 'P' stands for the property of being teal blue in color and 'not-P' stands for the property of being elliptical in shape. Predicate contrariety, that is to say, is a material logical principle from which formal logic abstracts.[7]

It was, of course, Gilbert Ryle (1949) who introduced the expression "category mistake" into the philosophical lexicon to designate the misapplication of a term based on its categorial membership. Since it involves a misapplication of terms, a category mistake, in my view, is properly regarded as a logical error, but an error of informal or material rather than formal logic. The misapplication of a term is not simply the assertion of a falsehood. "Saturday is in bed" (Ryle's example) commits no formal error, logical or grammatical. Yet it is, I maintain, a logical solecism, a demonstration of conceptual incompetence.

---

[6] Treatment of Lewis's "magical ersatzism" would expand discussion unnecessarily and add nothing new to what is already on the table.

[7] Brandom (1994, 115) puts this as clearly as anyone: "The formal negation of a claim is constructed as its minimal incompatible, the claim that is entailed by each one of the claims incompatible with the claim of which it is the negation."

A description free of formal contradiction will succeed in describing a possible world and a formally contradictory description will be disbarred from doing so only if certain assumptions concerning categorial world structure are in place.[8] Otherwise, the descriptions will allow category mistakes that render them logically problematic, not simply false. We might think of a world whose description is merely formally consistent as a noumenon, a mere thought-object. Whereas contradiction indicates property incompatibility, noncontradiction is a sufficing reflection of property compatibility, but only provided that the properties are members of appropriate categories, something over which it passes in silence:

> Quine was at least close to right about the empirical status of at least one law of logic, the law of noncontradiction applied to empirical judgment. It is an empirical matter that we can carve out concepts of objects along with concepts of properties and their contraries such that the object concepts are suitable to be subject terms for empirical judgment, each consistently taking just one contrary from each of a series of predicate contrary spaces. (OCCI, 106)

Craning Noncontradiction

Negation, Millikan has informed us, is an indefinite assertion of (predicate) contrariety. If we can crane the principle of formal predicate contrariety, the law of noncontradiction will not be far away: "If we read the negative as I have argued (Chapter 14) that it must be read – as operating upon logical predicates of sentences only – this structure is exactly expressed by the law of noncontradiction" (LTOBC, 300). To crane the principle of formal predicate contrariety, which tracks determinate property exclusion, would be to explain how it was constructed by natural selection using more basic designs already in place. For these we must look to representations that are non-sentential. What we seek are those items Millikan in 1984 dubbed "intentional icons"; and the mechanisms whence intentional icons derive their functions are the perceptual systems.

Traditionally, the relationship between determinable properties (e.g., being colored) and the determinate properties that fall under them (e.g., being ochre) has been taken to be fundamentally or even exclusively a conceptual one: we grasp the relationship between determinable and its determinates by understanding the meanings of terms. This is how the relationship was conceived by W. E. Johnson (1921/1964, vol. 1, 177), who revived the venerable idea early in the twentieth century, and by Sir Peter Strawson, as quoted by Millikan (LTOBC, 272). Millikan rejects this view (ibid.).

Mohan Matthen (2005) has argued, controversially but in my opinion convincingly, that the logic of determinable and determinate is not exclusively conceptual but is implicit in the function of human and non-human animal special sensory systems. According to what he terms the Sensory Classification Thesis, formulated independently by him but

---

[8] Cf. Millikan (LTOBC, 311): "Discovery that both a property [more properly, a predicate] and its negation apparently applied to a substance would not indicate a conceptual failure unless the relevant property range *ought* to apply to that substance in accord with one's tentative concepts." Cf. Williams (1973, 69): "... in order to form a contradictory conjunction, predicates must belong to the same category (cf. the difference between 'all prime numbers are green' and 'all prime numbers are divisable [sic] without remainder by two'."

originally postulated by Friedrich von Hayek (1952), these sensory functions are already classificatory and extend beyond mere transduction and input. With color vision the case is particularly compelling. That an object cannot both be red and green all over has been regarded by some as a synthetic proposition that is known *a priori* (cf. BonJour 1998, 41, 100–101). Such aprioricity as this proposition possesses, however, rests on the human red–green opponent-processing system. Yet this is not to deny that the human visual system tracks certain objective property exclusions in the world. They just happen to be ones relevant to the human interests, among others,[9] of locating red berries and other similarly colored fruits against green backgrounds (Matthen 2005, 91–92).[10] Compare Millikan (LTOBC, 270): the "consistent contrariety" of colors "is evidence for their objective reality – evidence that 'red' and 'green' are mapping real variants in the great out-there."

The *component* incompatibility, if not the *feature* incompatibility, of red and green would, however, seem to be an artifact of the human red–green visual opponent processing system. Two features are feature-incompatible if no full determinate that falls under one also falls under the other. Red and blue are feature-incompatible because no shade of red is also a shade of blue. Blue and green, however, are feature-compatible because some determinate shades, like aquamarine, fall under both. Any two *fully determinate* features will be feature-incompatible: no determinates fall under them, so there is consequently no overlap; and, as full determinates, they exclude one another. But this is not to say that they will be component-incompatible. Two features are component-compatible if they can combine to yield a third feature. Red and blue, while feature-incompatible, are component-compatible, since purples are both reddish and bluish. Red and green, on the other hand, are not only feature-incompatible (no determinate shade of red is also a shade of green), they are component-incompatible because no new shade results from the combination of red and green or appears, under normal conditions, both reddish and greenish (cf. Matthen 2005, 103–105).

With human special sensory systems other than vision, the implicit logic of determinate and determinable is a little more difficult to discern. Vision, says Matthen (2005, xxi), presents the world in "object feature terms": we see objects as qualified particulars. The exemplification of a visible property by a particular is a state of affairs. That a particular exemplifies a visible property is a propositional content.[11] Audition and olfaction, on the other hand, are differently organized. We hear sounds and smell smells, but we do not as a rule hear and smell the objects that are their causal sources as qualified particulars. Determinate sounds and smells also seem not to exclude each other in the way determinate colors do: we experience sounds and tastes in combination. But this apparent absence of property exclusion can result

---

[9] See Changizi et al. (2006) for an alternative account of the evolution of human opponent-processing vision. Thanks again to Dan Ryder.

[10] Cf. Quine (1969, 127): "Color is helpful at the food-gathering level."

[11] Matthen actually goes further than this, claiming (Matthen 2005, 78) that visual representations *express* propositional contents and that syntactically structured representations are not required for them to do so. To this one may take exception. It is one thing for a representation that is not syntactically structured to contain *de re* information concerning states of affairs and to represent them non-conceptually, but quite another for a representation to *express the proposition* that this state of affairs obtains.

from a failure to distinguish feature incompatibility from component incompatibility. (For an example of this, see the discussion of determinate properties in Fales 1990, 227ff.)

The pitch of middle C and the pitch of the G a fifth above it in the Western chromatic musical scale are full determinates and therefore feature-incompatible. But they are component-compatible in chords. Although we would not describe an open fifth chord consisting of a C and a G as a G-ish C or a C-ish G, it does, like any musical chord, possess a distinctive, unified resultant sonority. Similar comments apply perhaps more obviously to musical timbre. The qualities of sound of the oboe and of the clarinet are feature-incompatible (no oboe sound is also a clarinet sound) but component-compatible, a fact exploited by Schubert in the first statement of the principal theme of the first movement of his "Unfinished" Symphony. Here we might have recourse to describing the resulting sonority as an oboeish clarinet tone or a clarinetish oboe tone. Sweet and sour (one of Fales' counterexamples to determinate property exclusion) are feature-incompatible but not, as consumers of Chinese cuisine know, component-incompatible: they combine into a sweetish sour or a sourish sweet.

What is important for our purposes is that at least some of the property exclusions and the subordination relations that pertain to the various sensory spaces are *a priori* in Millikan's evolutionary sense of the term. We would have no experiential grasp of determinable qualities, Matthen (2005, 100) claims, if we could not also grasp some of the more determinate property incompatibilities subordinate to them: "One does not grasp *red* unless one can also differentiate some subclasses of *red*, for instance *dark red* and *pink*" (Matthen 2005, 98). Even die-hard empiricists have recognized this: *vide* Hume's granting of *a priori* access to the missing shade of blue and Quine's admission that without prior spacing of qualities in relations of similarity and difference, stimulus generalization and operant conditioning would be impossible: "A standard of similarity is in some sense innate.... Without some such prior spacing of qualities, we could never acquire a habit; all stimuli would be equally alike and equally different" (Quine 1969, 123; cf. Matthen 2005, 40).

Suppose, then, that it is the case that human vision, and perhaps also exploratory touch,[12] present the world in object feature terms and produce intentional icons whose proper function it is, among others, to track property incompatibilities in the environment. Such intentional icons are accurate, or accurate enough, enough of the time, given human interests and given the categorial structure (the substances and the available ranges of applicable properties) of the world in which they were crafted by natural design. The same applies *mutatis mutandis* to the special sensory systems of non-human animals. The logic of predicate contrariety is, then, implicit in the function of these sensory systems and the representations they produce, the non-sentential intentional icons. Predicate negation is implicit in predicate contrariety, and noncontradiction is implicit in predicate negation. If a substance can only have one of a range of categorically appropriate mutually exclusionary determinate properties, it cannot both have that determinate property and not have it, and it cannot be the case that it has it and not be the case that it has it. But this is just noncontradiction. From propositional noncontradiction, $\sim (p \cdot \sim p)$, employing as it does the negation operator and one binary connective, the full complement of propositional

---

[12] Matthen accords this possibility no consideration, which is surprising in light of the interest that Molyneaux's question has stimulated over the centuries.

connectives can be derived. Making these implicit expressions explicit requires cognitive development sufficient to support propositional thought and the production and consumption of the representations that, according to Millikan's early usage in *Language, Thought, and Other Biological Categories*, serve as the vehicles for propositional contents.

It is difficult to exaggerate the significance of negation as a cognitive achievement, for it literally makes logic possible. Negation may be an indefinite assertion of predicate contrariety that tracks property incompatibility in the world. But there is no *negation* in the world itself, only in the representation of the world: there are, I hope you agree, no negative properties, facts, or states of affairs. Yet given the remarkable powers of abstraction and generalization possessed by the human brain, we should not be surprised that our forebears were able to make their way from intentional icons that tracked property incompatibility in specific ranges or categories in the human *Umwelt*, thence to predicate negation, which generalizes to property incompatibility within *any* property ranges in *any* *Umwelt*, and finally to noncontradiction itself, which reflects incompatibilities between states of affairs. An example from J. L. Bermúdez suggests how such capacities for abstraction and generalization might have been craned:

> Imagine a [non-linguistic] creature that has learned that the lion and the gazelle will not be at the watering hole at the same time and, moreover, is in a position to see that the gazelle is drinking happily at the watering hole. This creature can conclude with confidence that the lion is not in the vicinity. (Bermúdez 2003, 141)

Inference and negation are only implicit in this creature's "thinking without words." It simply *perceives* the watering hole as qualified by two exclusionary competing properties: being lioned or being gazelled.[13] Being gazelled affords certain possibilities of action, being lioned affords very different ones. Notice that sensitivity to these specific exclusionary properties and to the range to which they belong is not hard-wired into any of the creature's perceptual systems, but is the result of learning, an advance in cognitive sophistication.

The negation of a proposition, as Millikan states, is not simply an assertion of absence: "absence is itself merely the presence of an incompatibility." "Empty space, zero force, zero acceleration, zero temperature, etc.," she says, "are not nothings but rather somethings that are at the extremes" (LTOBC, 269). Negative sentences map actual states of affairs, she claims, because "the negative operates in the end only upon the logical predicate of a sentence, reversing the sense of this predicate so that it maps a contrary of what it would normally map" (LTOBC, 222). The law of noncontradiction may not legislate for the world, but for Millikan it does reflect, by evolutionary design, its deep ontological structure. The beauty of this tale is that noncontradiction, the skyhook, has been craned: we have a plausible naturalistic account of how it was lifted to its preeminent cognitive position by means of a cascade of cranes, beginning with the function and implicit logic of special perceptual systems.

Like Johnson and Strawson, Robert Brandom, whom we will reencounter in the final section of this chapter, misses or ignores the implicit logic of human and non-human animal

---

[13] Compare Redding's (2007, 213) take on this example and predicate negation.

sensory systems because, like Sellars, he falsely dichotomizes human informational commerce with the environment into entirely non-cognitive, but reliable differential responses (the Realm of Causes), and cognitive and conceptual, inferentially articulated beliefs (the Space of Reasons). For Brandom, animal sensory responses sort or classify only in the sense that chunks of iron "sort" wet and dry environments by reliably rusting in the former but not in the latter (Brandom 2000, 108). "Such responsive classification," he says, "is a primitive kind of taking of something as something. It is in this sense that an animal's eating something can be interpreted as its thereby *taking* what it eats as food" (Brandom 1994, 33–34; emphasis original). But there can be no such taking without the possibility of mis-taking (cf. Dennett 1996, 37). Brandom has failed to distinguish mere "informational sponges" from genuine "epistemic engines," which include systems that wield intentional icons rather than inferentially articulated concepts. The former, things like chunks of iron or expanses of moist clay, merely record and store information. There is nothing intentional about them. The latter include systems that represent, albeit non-conceptually, and are able to learn by modifying their environment-modeling internal states by way of positive and negative reinforcement (Churchland 1979, 143). A system need not be capable of applying concepts and making explicit inferences to perceive that being red falls under the determinable being colored, or to infer implicitly (like Bermúdez's creature at the water hole) that if the area is gazelled, then it is safe to drink.

The argument I have outlined on Millikan's behalf is close kin to the one presented in her "Truth Rules, Hoverflies, and the Kripke–Wittgenstein Paradox"(WQ, 211–239), though the principal aim there is different. Millikan's goal is to provide a straight, as opposed to a skeptical, solution to the Kripke–Wittgenstein paradox concerning rule-following. Rejecting both introspection and dispositionalism, Kripke's Wittgenstein believes he has exhausted the options and concludes that there is no independent fact of the matter concerning what rule or rules determine the present case. Wittgenstein's skeptical solution, according to Kripke, is to invert the standard order of dependence between rule-following and correct computation: if the individual's computational practice (by and large) survives correction by the community, she will be *taken* to be following a rule (cf. Kripke 1982, 93n76). What a community generally agrees is correct and incorrect is a "brute fact" that concerns a form of life.

But this way verificationism lies: rule-governed correctness becomes assertability within a community. Agreeing with Kripke's critique of the dispositionalist account, Millikan attempts to avoid these verificationist consequences by interpreting intending to follow a rule as "purposing" to follow it, thereby avoiding the charge of begging the question and stopping the vicious regress by grounding the chain of antecedent purposes in biological purposes, like the ones that govern the behaviors of the humble hoverfly. A creature need not "purpose" to follow such rules: they are built into its evolutionary design.

By working her way up from hoverflies (Darwinian creatures), which are hard-wired to behave in accordance with rules for intercepting flying females, to rats (Skinnerian creatures), which are designed to learn by association and to behave in accordance with rules ingrained through operant conditioning, to humans (Gregorian creatures), who are designed to internalize public sanctions and external mind tools like arithmetic notation, Millikan succeeds in craning a logical skyhook, namely what she thinks of as correspondence truth rules, the rules that codify the mapping functions of sentential representations, what they are supposed to be doing.[14] Meanwhile, she excoriates the justified assertability rules of Dummett and Putnam as products of "verificationist myopia" (WQ, 236), a tendency to emphasize the proximal rules that govern behaviors concerning stimulations, in this case linguistic responses to linguistic inputs, over the distal rules governing interaction with the genuine intentional objects of language and thought in the great out-there. The motivation for this emphasis is clear: behaviors in accord with proximal rules concerning stimulations lie more fully within the competence of users and enjoy a much higher rate of compliance because proximal rules don't require the environmental cooperation that distal rules require.[15]

I submit that Millikan also applies this style of argumentation to formal logic and the law of noncontradiction, perhaps the preeminent logical rule. Noncontradiction does function as an internal test for consistency of beliefs, but it also functions as a distal principle concerning property incompatibility. But without the real-world glue, without a categorial system, without a specific ontology of substances and ranges of properties that may be ascribed to them, the way in which noncontradiction maintains its grip on a world cannot be specified:

> Language and thought are not a layer of ghostly ectoplasm that can be peeled off the real world and laid overtop any possible world at will. The integrity of extensional language and thought is maintained through deep roots in the actual world. Both language and thought – their meanings – will evaporate if one tries to relocate them in worlds that are too far away. (Millikan 2010, 75)

For Millikan, determinate property exclusion is a principle of natural necessity (LTOBC, 273). But is this all it is? The references to "any possible world at will" and "worlds that are too far away" suggest that there may be more metaphysics here than initially meets the eye.

Natural Necessity and Metaphysical Necessity in Millikan's Philosophy

Because she so often inveighs against confirmation and meaning holism, it is easy to miss the fact that Millikan defends a rather extreme metaphysical holism regarding the identity conditions of properties and individuals:

> Properties are not loners like substances but enter into the world along with contrary properties. A property is itself qua [sic] in naturally necessary opposition to the rest of the property range or contrary range within which it falls, such that whatever substance has that property cannot, in accordance with natural necessity, have any properties from the rest of that range. (LTOBC, 271)

---

[14] The categories of Darwinians, Skinnerians, and Gregorians derive from Dennett (1995, 375ff) and other writings.

[15] We should not allow a misprint in Millikan's statement of the proximal quoverfly rule (the gruelike alternative formulation of the proximal hoverfly rule that accidentally covers all previous hoverfly female interceptions) that made its way into the *White Queen Psychology* reprint (WQ, 221) to obscure her argumentative strategy. The rule should read, "If the vector angular velocity of the target's image is not clockwise and not between 500 and 510 degrees per second, make a turn that equals the (signed) angle of the image minus 1/10 its vector angular velocity, plus or minus 180 degrees; at ease otherwise."

The identity of an individual, on the other hand,

> ... is fixed by its refusal to admit properties that are contrary to one another. In accordance with natural necessity, a selfsame individual takes just one from each of those ranges of contrary properties correlative to which it is a selfsame individual. (LTOBC, 272)

If accepted, such views would have significant repercussions for possible worlds. Many a possible world would have to be a world with an entirely different categorial structure, for the very identities of determinate properties are established by membership in ranges of exclusionary properties instantiated in the actual world. For example, one could not single out the property of being H_2O, change it to being X_YZ, and pretend to keep everything else the same. Being hydrogen hydroxide (H_2O) contrasts with being hydrogen peroxide (H_2O_2). These determinates exclude one another under the determinable property of being a hydrogen–oxygen bonded molecular substance, which, in turn, contrasts with other determinates under a higher determinable. Should someone object that these are natural kind identities and not property ascriptions, I respond that each instance of these natural kinds is an individual, and that in the case of individuals, the substance–property relationship is, for Millikan, who adopts a modernized, but also radicalized[16] version of Aristotle's form–matter distinction, relative.[17] An individual substance can be a property relative to another individual.[18]

This complicates the idea of a closeby possible world. If the metaphysical atomisms of a David Lewis or a David Armstrong (the former of Humean, the latter of Leibnizian provenance)[19] be rejected in favor of a Millikan-style holism, the possible worlds theorist would seem to be faced with a stark choice. All determinate properties derive their identity conditions from their positions in the property ranges that are constitutive of the categorial structure of the natural world, or they are alien properties whose identity conditions cannot be specified, since they belong to different property ranges. Closeby possible worlds will be limited to worlds instantiating properties that are what they are because of their places in these natural property ranges. In light of these considerations, it is hardly

---

[16] Aristotelian primary substances are not predicable of (i.e., cannot qualify) anything.

[17] Cf. Millikan (OCCI, section 2.6), "Ontological Relativity (of a NonQuinean Sort)."

[18] Cf. Millikan (LTOBC, 291, emphases original): "And besides being a substance *with* properties it [an individual enduring object] would also have to *be* a property – a property of each of the temporal stages of the individual that it unifies, as being gold is a property of each of gold's instances. Call this hypothesized substance, for each individual, the individual's 'subessence' ..."

[19] Cf. Armstrong in defense of his Principle of Independence (Armstrong 1997, 157, emphasis original): "Leibniz seemed to have thought it *evident* that simple properties are all compossible. Let us be the last to claim that the intuitions of even the greatest philosophers on such matters are sacrosanct! But they deserve some weight." Contrast Millikan (LTOBC, 268–269): "Leibniz thought that all perfectly simple properties were intrinsically compatible.... What we take to be contrary properties, on Leibniz's view, are of two kinds. Either to assert that both contraries apply to the same thing is simply to contradict oneself, or the contrariety is between properties that are in their own nature – so far as their identity or itselfness is concerned – completely indifferent to one another but prevented by something external to them (God) from ever meeting on common ground. This is the view I am rejecting.... If Aristotle is right, Leibniz is completely mistaken. Contrary properties oppose one another, and their identities or natures are tied up with one another on the most fundamental level there is."

surprising that Millikan regards natural necessity as a primitive (personal communication). If what counts as a closeby possible world depends upon naturally necessary determinate property exclusions within natural property ranges, natural necessity cannot, on her principles, be analyzed using the notion of closeby possible worlds. As a result, it is also not surprising that Millikan believes that causal necessity, as a variety of natural necessity, cannot be analyzed by way of conditional counterfactuals true in closeby possible worlds, or, given her distaste for possible worlds, that she is inclined to collapse metaphysical necessity into natural necessity.[20]

However, two problems remain. First, there is a semantic issue. In earlier times, it seems, Millikan regarded "counterfactual conditionals" as true or false (LTOBC, 263). But what are their truth makers? If conditional causal counterfactuals are denied truth makers in closeby possible worlds, how are we to understand their semantics? Millikan now proposes to regard counterfactuals, causal and otherwise, not as true or false, but merely as reasonably or unreasonably asserted (personal communication). Why is this not a retreat to the "verificationist myopia" she deplores?[21]

Second, it is not clear that Millikan should collapse metaphysical necessity into natural necessity. We may ask why determinate property exclusion within property ranges, first among equals in Millikan's trinity of natural necessity that includes identity and causality, should not be accorded the status of a metaphysically necessary principle, not just a naturally necessary one. To regard worlds with different categorial structures, different property ranges, as metaphysically impossible would be dogmatic. True, the properties at these worlds would be alien. But that need not be a problem for Millikan any more than it is for the modal realist. A world without categorial structure and property exclusion *at all*, however, *is* arguably metaphysically impossible on her principles, for such a world would lack identity conditions for properties and substances.[22] On such a "world" noncontradiction could get no grip. But that is as it should be: a world without identity conditions is no world at all. As Quine taught us, no entity without identity.

The Son, the Daughter, and the Mighty Dead: Debunking the Myth of the Logical Given[23]

"Determinate Negation" is [Hegel's] term for material incompatibility, from which, he takes it, the notion of formal negation is abstracted.

(Brandom 1994, 92)

---

[20] Millikan (LTOBC, 263): "Individual identity ... is in the same metaphysical boat as causality. Causality is a naturally necessary connection between naturally possible occurrences upon which counterfactual conditionals can be founded ... the natural necessity [*of individual identity*] is a *basic* necessity, not merely a natural necessity in situ ... there must be natural laws that are *in* nature – principles that actually account for the patterns in nature rather than merely summing these patterns up."
[21] This is not the place for me to air my own proposal.
[22] It should not be said that a Humean or a Leibnizian world is metaphysically impossible. What should be said is that Hume's and Leibniz's accounts of property identity and incompatibility are wrong if Millikan's is right.
[23] The locution "Myth of the Logical Given" derives from Redding (2007, 59ff).

In 2005 Millikan published a small piece with the intriguing (and revealing) title, "The Son and the Daughter: On Sellars, Brandom, and Millikan." Her second sentence poses a somewhat plaintive question: "How have two siblings, both admirers of the father, come to stand so far apart?" In the course of the article, Millikan attempts to adjudicate differences and emphasize commonalities, but other than briefly invoking the influence on Sellars of the early versus the late Wittgenstein, does not really provide a satisfying answer to her own question. I don't believe that the son and the daughter stand as far apart as Millikan thinks, for they both can be seen to be extending their father's attack on the Myth of the Given to the Myth of the Logical Given, the myth that formal logic and the principle of noncontradiction are philosophically foundational and ontologically legislative. Both siblings, moreover, make fundamental use of the same notion, namely formal abstraction, albeit expressed in different modes: the notion that contradiction is, on the one hand, the formal *expression of material incompatibility* (Brandom, in the logical mode) and, on the other, the formal *representation of property exclusion* (Millikan, in the ontological mode). The fact that the two approaches are to this extent orthogonal should not blind us to their consanguinity. Brandom's pre-Sellarsian patrimony is avowedly Hegelian. What about Millikan's?

Let's start with a few innocent observations about routes of inheritance. Millikan dedicated *Language, Thought, and Other Biological Categories*, her first book, to the semiotician Charles Morris, who was strongly influenced by Peirce, who was strongly influenced by Schelling and Hegel. Her nominal *Doktorvater* was the speculative metaphysician Paul Weiss, an inheritor, via Peirce, of the objective idealist legacy. Modulo translation into his preferred idiom, Hegel would have gotten his mind around Millikan's non-Quinean ontological relativity more easily than might many a philosopher, as he would have her metaphysical holism concerning the identity conditions for individuals and properties and her flirtation with metaphysical monism.[24] "There is not," Millikan explains (OCCI, 27), "one set of ontological 'elements,' one unique way of carving the ontology of the world, but a variety of crisscrossing overlapping equally basic patterns to be discovered there." Each Hegelian logical category (short of the Absolute Idea) is a legitimate, if limited, way of carving the ontology of the world. Aristotle, the principal hero of *Language, Thought, and Other Biological Categories*, on the other hand, would have been puzzled by this (or perhaps any) brand of ontological relativity. For him, there *is* one correct way of carving the ontology of the world, namely at the joints that separate the primary and secondary substances that constitute it. Moreover, as a multiplicity of substances with no essence or substantial form of its own, the world would not qualify as a substance, primary or secondary, at all. Millikan's notion of a historical substance or kind would have been opaque to him.

Nevertheless, Aristotle remains a figure of the greatest significance in Hegel's own patrimony. The Aristotelian form–matter hierarchy profoundly influenced his thinking

---

[24] Given Millikan's version of ontological relativity, the world or cosmic whole is itself an enduring individual historical substance with multiple subessences. After all, the cosmos is an object of scientific study, namely scientific cosmology, that supports non-accidental, causally grounded predictions and retrodictions such as the occurrence of the Big Bang. It therefore satisfies Millikan's epistemological requirements for being a substance. Moreover, it is an individual, indeed, a concrete particular.

regarding the hierarchically organized, recursive categorial structure of the cosmic whole. Lower categories stand as matter, to higher ones as form. Scarcely less significant were the influence on Hegel's conception of determinate negation of the Aristotelian conception of predicate negation of which Millikan makes so much and its descendant, the Kantian "infinite judgment,"[25] which is a particular judgment affirmative in quality but possessing a negative predicate ("Some S is non-P"), the third member of the second triad of judgment forms in the Metaphysical Deduction of the Categories in the *Critique of Pure Reason* (A70/B95). Determinate negation lies at the heart of the Hegelian dialectic, for it designates the movement of the thought of an object from indifferent otherness to unified opposition whereby Spirit, Hegel's term for the cosmic whole, recognizes that what had seemed alien is crucial in determining its own nature and, as such, an inalienable aspect of itself. At this point, the "immediate," in which negation had been only implicit, becomes "mediated" by explicit integration with its opposite.

Hegel, "that great foe of 'immediacy'" as Sellars (1963d, 127) approvingly describes him, had little respect for the formal logic of his day and even less for noncontradiction as an inviolable law. Contradiction, according to the dialectic of reason, was not to be shunned, but embraced as fundamental both to the organization of the cosmos and to its adequate conceptualization culminating in the Absolute Idea. Millikan evidences no sympathy for any such views. Nevertheless, Hegel's treatment of the law of noncontradiction prefigures Millikan's no less than it does Brandom's. Noncontradiction, which Hegel calls the "law of contradiction," both as a result of tradition and because it is for him a prescriptive and not a proscriptive principle, he regards as anything but self-evident ("immediate"). It appears relatively far into both versions of his treatise on logic. It emerges at the beginning of the Doctrine of Essence, the second large section of the work that includes categories that explicitly express dyadic opposition, such as form/matter and appearance/essence, along with the binary laws of identity and excluded middle (Hegel 1811–1816/1976, 439ff; 1830/1991, 185). Both the law of identity and the law of "contradiction," Hegel (1811–1816/1976, 416) claims, are synthetic, not analytic. The law of contradiction (or, as we would have it, noncontradiction) is, then, already a highly mediated result, far from the immediate logical given that most philosophers, including Aristotle, have taken it to be, for Essence includes within itself all the categories of Being, the first large section of Hegel's *Logic*, mediated and *aufgehoben* (superseded but preserved). By replacing the dialectical development of Spirit with natural cosmic and evolutionary history, Millikan arguably has managed to turn Hegel on his head (once again). For she has shown how the law of noncontradiction may be craned, that is, developed historically within a naturalistic setting from less sophisticated cognitive functions that employ intentional icons, a derived proper function of which is to track property incompatibility in the world. Hegel's principle of "contradiction" is similarly rooted in the world, for it is developed dialectically out of the categories of Being. But the fulcrum of its extraction is the great skyhook that goes by the name of Absolute Spirit.

---

[25] "Infinite" because it asserts that a thing is a member of an infinite class: the class that contains everything that is non-P.

Does Millikan's account commit the genetic fallacy by telling a developmental tale that is irrelevant to the logical status of noncontradiction as a self-evident logical given? No, it does not, for it exposes the living roots of the law of noncontradiction in the categorial structure of the natural world. Sever these roots, and noncontradiction *qua* logical given gives little (is merely "possibly sufficing"), for it maintains its grip on the world not by legislation or determination but by reflecting naturally necessary property exclusions within ranges that qualify appropriate substances. Property exclusion within ranges, in turn, is what the ontology of determinate negation comes to outside the setting of Hegel's dialectical logic and objective idealism. Sellars' daughter she may be, but like Brandom the son, Millikan seems to have inherited the recessive Hegelian allele. She may be expected to resist this speculative genealogy, but I believe it important to press it, not only in the interest of sibling reconciliation but also in the interest of theoretical adequacy in the philosophy of mind. If Brandom's Hegel is the author of the *Phenomenology of Spirit*, the story of the self-recognition of mentality, Millikan's is the author of the *Science of Logic*, the complementary story of the emergence of mentality. One story without the other is only half the story. Brandom needs to bridge the gap between the non-normative Realm of Causes and the normative Space of Reasons, but Millikan needs to recognize that explicit norms of rationality enter the natural world only with the social construction of the human extended phenotype.[26]

---

[26] Thanks to Jeremy Byrd, Charles Hermes, Ruth Millikan, Dan Ryder, and Ken Williford for helpful comments and criticisms.

Reply to Nussbaum

Charles Nussbaum's searching essay would require a book as a reply, important parts of which are certainly still missing in my head. I find his initial arguments compelling, for seeking the foundations of logic in the structure of the world rather than supposing the structure of the world to be constrained by a pre-worldly logic. I very much appreciate his addition of this grounding to my musings about identity and contradiction. He has it just right, I believe, that the result is an extension of Sellars' attack on the given to "… the Myth of the Logical Given, the myth that formal logic and the principle of noncontradiction are philosophically foundational and ontologically legislative."

I won't comment on the genealogy of my position, or say very much about its relation to Brandom's, although in my reply to deVries (this volume) I do explain why I remain unconvinced by Sellarsian arguments for the autonomy of reason. The theory of empirical concepts that runs throughout my work is in direct opposition, I believe, to any form of inferentialism. This theory of concepts also leads me to think that the role that noncontradiction plays in human reasoning probably cannot be supported with lower-level cranes in quite the manner Nussbaum suggests. On the other hand, his brief excursion into the neurobiology of contrariety seems to challenge my claim that the law of noncontradiction plays no role prior to human concept development. I will end with a very short and tentative word about the modalities.

I don't see the law of noncontradiction as having been abstracted or derived in any way from experience, perceptual or otherwise (except that its past usefulness undoubtedly accounts for its survival over past generations of genetic and cultural selection). Better to think of it as an implicit hypothesis or theory (Kant might have said "regulative ideal") about a certain kind of skeletal structure to be sought for and that seems, indeed, to have been found ubiquitously in nature. The validity of this theory is evidenced by the fact that we have been able to construct indefinitely many kinds of stable empirical concepts by using it, these concepts reliably providing us with successes, both theoretical and, subsequently, practical. Very likely this regulative ideal grew up with the development of linguistic representation having subject–predicate structure and explicit negation ("with the social construction of the human extended phenotype," as Nussbaum puts it). Learning how not to find oneself in contradiction with others – or, more crucially, with oneself – surely played the central role in this saga, but (*vis-à-vis* Brandom) one that was "normative" only in the sense that it regulated a "practice" that survived because it was successful in practice. It yielded wonderful results. (I have told a just-so story about how the law of noncontradiction along with the development of language enabled the divergence of human theoretical concepts from the merely practical concepts of other higher animals in the last two chapters of VOM.)

That conformity to the law of noncontradiction is, most importantly, an epistemological principle that governs *concept* formation has plausibility only in the context of the theory of empirical concepts developed in LTOBC and OCCI. The claim there is that empirical concepts consist, in the first part, of capacities to identify and reidentify objectively selfsame things (substances, attributes, relations, and so forth) as required for gathering practical skills or theoretical facts. The difficulty is that objectively selfsame substances, attributes, and so forth are generally distal things that have to be detected and correctly reidentified through the medium of a huge variety of different kinds of proximal stimulations. Very different afferent nerve stimulations may indicate the *same* distal things, very similar stimulations often indicate quite *different* distal things. (See my reply to Jesse Prinz, this volume.) For this purpose, props such as the opponent processing system in color vision or Quine's "prior spacing of qualities" (same "quality" here being assimilated to same "sensory stimulation") are not of apparent value. By means of natural selection and learning animals, human and non-human, corroborate and tune their reidentifying abilities first through overt practice, by trying to use, manipulate, or productively navigate among things that they take to be the same so as to achieve the same results again. The laws of identity and noncontradiction are required to step in as judges of conceptual adequacy, of reidentifying proficiency, mainly for developing concepts not honed through immediate practical applications – concepts of things past, of things not in one's practical environment, of things too small, of things that cannot be tracked perceptually in obvious ways. The use of noncontradiction in the development of many scientific concepts is an obvious example here, the locating of real kinds of various sorts together with properties that they consistently possess but that are not observable in straightforward ways. Nussbaum states for us very clearly the necessity of *empirically* discovering "categorial world structure" or "material logic," a grasp of which is developed as different applications of the law of noncontradiction are uncovered.

Nussbaum's excursion into the neurobiology possibly underlying our recognition of certain contrarieties raises a challenging question with regard to this view of mine. Putting the opponent processing theory of color vision to one side,[1] what is hardly merely a

---

[1] There is some evidence that what has appeared to be opponent processing in color vision is itself merely soft-wired inhibition between red and green and between yellow and blue reporting cells (Billock et al. 2001). In an experiment that fatigued retinal cells by prolonged exact exposure to very fine parallel lines of red and green "... when colors were equiluminant, subjects saw reddish greens, bluish yellows, or a multistable spatial color exchange (... entirely novel perceptual phenomena); when the colors were nonequiluminant, subjects saw spurious pattern formation."

"theory" anymore is that connections between neurons helping to constitute the perceptual and cognitive systems are as often inhibitory as they are excitatory. The effect would seem to be to make mandatory for us certain choices between or among certain alternatives during perceptual and cognitive processing. Such inhibitory connections suggest a recognition or an assumption or anyway a hypothesis, on some level or levels, concerning what necessarily excludes what, hence what should not be represented uncensored. The apparent "flipping" back and forth of one's perception of a Necker cube suggests itself as a result of such inhibitory wiring. Likewise the flipping back and forth that occurs when different objects (say a face and a house) are experimentally presented to the two eyes in what is apparently the same place (Hohwy et al. 2008). The waterfall illusion, in which one sees an object as moving rapidly upward while staying in exactly the same place, suggests the absence of appropriate inhibitory wiring of this sort. The implication seems to be, however, that recognition of certain contrarieties occurs in perception well prior to the development of theoretical judgment.

Notice first, however, that opponent processing in early color vision is quickly overridden by the neural mechanisms of color constancy. When you look at the wall of an evenly colored room that has more shadow across one wall than another you do not see the wall in shadow as having a different color. The color you see, in the sense of the color you actually judge to be there, the color you would act on and so forth, is the same color all over, not the different color you would use if you wished to paint rather than to act on the scene before you. (The most compulsive housewife will have no impulse to wash that wall.) Or suppose a small green lamp shines on one part of a white wall, a small red lamp on another, while the whole is also bathed in medium daylight. Despite the red–green inhibition in early neural processing the wall will correctly look to you the same color, white, all over. Red–green incompatibility at the level of opponent processing seems to reflect only incompatibilities in what light mixtures could simultaneously reach the same portion of the retina. On the other hand, in the normal course of things, no wall will ever appear to you to be red and green in the same place, so there seems to be genuinely perceptual inhibition here on some higher level. Sometimes we are aware that we cannot see, in this light, or from here, what color something is – it might be this or it might be that – but we will not see it as two different (feature-incompatible) colors at once.

Still, I think (*contra* Matthen) that what is represented in perception prior to it's occasional use by humans for the making of propositional judgments is not objects having properties. I think it is "things" or "stuff" offering affordances of various kinds, spatially positioned relative to the perceiver. These "things," this "stuff," are not perceived as differentiated between substance and attribute. (Recall the relational nature of the "substance"/"attribute" designation. Nothing is substance or attribute taken just in itself (LTOBC, ch. 16).) I think this because I am a biosemanticist, hence believe that what is represented in perception is a function of what perceptual representations are used for. And I think that they are used, in the first instance anyway (by animals, by infants, and quite continually by human adults as they navigate about their daily business), for representing things only as what is to be dealt with. The significant transformations that help to define our basic perceptual representational systems (see the discussion with Nicholas Shea, this volume) correspond to spatial displacements of affording things or stuff relative to the perceiver, and to replacements of things or stuff with other things having different affordances.

Of course the properties of things are causally involved in a perceiver's abilities to differentiate among affording things, but this does not imply that they are represented in perception *as* attributes of substances, any more, say, than the gradients or edges of early visual perception are represented as such in the final products of visual perception. (See the discussion with Richard Fumerton, this volume.) Colors, for example, are of no practical everyday use at all except on the way to reidentifying useful objects. Recall that there are languages that have few or no words for colors and that children learn color words quite late. Similarly, sounds, odors, and tastes are not of any practical use except as aids in the identification of things and events. We describe them by reference to what they are of (the smell of bacon, a rasping sound, a bell-like sound). When engaged in practical activity (rather than in reporting or reflecting), we do not hear (perceptually represent) sounds or sound qualities but rather doors closing and people walking, or perhaps a something-we-hear-over-there-but-can't-make-out. We do not smell (perceptually represent) odors but rather bacon cooking. We feel and see shapes, not as attributes of things but as guides to handling and manipulating. We see how to move or pick a thing up given its position and shape, how to walk on it if it is rough or slippery, and so forth. (Think how much easier it is to teach a dog to fetch the newspaper than it would be to fetch a black and white thing, whether paper, metal, big, little, rose-smelling, or bacon-smelling.)

I suspect then that the original "contraries" that are recognized in completed perception are contrarily affording things and stuffs, things that would need to be treated or responded to incompatibly. Between such things a perceptual decision has to be made, symptoms on one side pitted against symptoms on the other. The side that loses then needs to be suppressed, prevented from interfering with united action. The cat needs to respond wholeheartedly to the presence of another cat quite differently than to that of a dog. The experimental subject would need to respond wholeheartedly to a real face quite differently than to a real house. The idea might be then that during the process of development of the perceptual systems through evolution or learning, various simpler and quite general principles of distal discrimination that can be combined for use in helping to identify a great variety of different useful things are distilled out in the background, taking their various places in early perception alongside the gradient and edge detectors (distal-color indicators, shape detectors, and so forth).[2] In humans these are later redeployed in processes leading to perceptual judgments about properties of objects as such, coupled in the first instance with linguistic skills that allow communication about objects having as yet no names but that need to be identified to hearers.

Returning to Nussbaum's very early incompatibility recognizers, say, those that refuse to or cannot report that the light currently striking a certain part of the retina is both red and green, it is possible that these very early detectors too are capable of redeployment,

---

[2] I have in the back of my mind here the way Terrence Sejnowski and Charles Rosenberg's classic connectionist program NETtalk, trained whollistically to pronounce English text merely by being shown text as input and matching audio for comparison, programmed itself underneath to sort vowels from consonants and to make many other relevant distinctions within these broader categories.

say, for purposes of realistic painting and also, perhaps, for what is called "phenomenological description" (Millikan forthcoming b).

Nussbaum urges me to talk about the modalities. I did make claims about natural necessity in LTOBC. I hadn't then thought through to my current suspicions about the modalities, nor do I yet have a stable position about them, but here is an offering to be taken for whatever it's worth.

The claim in LTOBC, OCCI, and LBM was that basic linguistic meaning is function ("stabilizing function"), it is not truth-conditions, and that there is good reason to suppose that the functions of linguistic devices such as, for example, "exists," the "is" of identity, and the "means" of Sellars' "translation rubric" are not to produce mental attitudes having truth or satisfaction conditions, but to move the cognitive systems in other ways (LTOBC, ch. 12, LBM, ch. 3). This seems likely to be true of the devices "actual," "necessary," and "possible" as well, the functions of which clearly are not to get us to attribute properties to anything, certainly not to worlds. The result might well be that these modalities are in the first instance "epistemic" in the naturalized sense, their role in "making up one's mind" being basic, their use in the free play of imagination perhaps a mere side effect, perhaps not. A central feature of the biosemantic theory of intentionality was supposed to be that it cleanly washed away all the merely intentional objects. It showed how false and empty representations, rather than representing unrealized possibilities, simply failed to represent.

On the other hand, sentences containing "means" and "exists" and the "is" of identity do have truth-conditions, even though these are not truth-conditions of *mental representations* that it is their job to impart. Their truth-conditions make reference to the functions or lack thereof of adjacent linguistic forms, these not being things they are supposed to produce thoughts about. (See LTOBC, ch. 12, on "proto-reference.") Similarly, I would like to believe, the difference between (naturally) "necessary" and (naturally) "merely actual" is supported in the nature of the world beneath language and thought. It reflects an ontological distinction. I don't think this is true of "possible." The fact that "necessary" and "possible" appear to be interdefinable concerns their epistemic functions only. World affairs are often necessary given other actual world affairs. If they are necessary they are, of course, also actual. It could be that every world affair is necessary, given what else is actual. Necessity concerns relations between affairs, not the affairs themselves, which in themselves are merely actual. Possibility, on the other hand, is nowhere to be found prior to representation. It is a shadow cast by representations that appear to us to be well formed but that fail to represent, or in the epistemic case, that we are uncertain whether they represent. Sentences that express contraries of sentences that are true do not represent possibilities that fail to exist. Strictly speaking, they fail to represent. "Possible" and "impossible" are never more than epistemic directives, the proper results of their uses being the embracings of mental representations or of their contradictories. This view is coherent if one takes negation always to be internal rather than external, true negative sentences having their own ways of mapping onto the *actual* world – a position I argued for in LTOBC, chapter 14.

10

Are Millikan's Concepts Inside-Out?

JESSE PRINZ

Introduction

Ruth Millikan is one of the most original, ambitious, and influential philosophers of our times. Her biological approach to mind and language stands out as one of the most fully and richly developed theories on the market, and it is highly innovative in both broad outline and sumptuous details. But in other respects, Millikan's approach is like many others that were proposed in the 1980s. It echoes themes that emerged during the 1970s revolution in philosophy of language, and can be regarded, for that reason, as conforming to the orthodoxy of that period. Millikan is a realist, an externalist, a naturalist, an opponent of Fregeanism, and a crusader against Meaning Rationalism – her term for a residual Myth of the Given. This familiar constellation of views might be labeled Outerism, because it shifts attention away from inner psychological resources and explains mental content by appeal to features of the mind-independent world. Millikan has been a deft and impassioned defender of the outer. And, like so many, I have long assumed she was on the right side. In this chapter, I want to reconsider that assumption. I think there are ways in which the outer/inner division has been overblown, and other ways in which Outerists have underestimated the contribution of the inner. Ironically, Millikan's work on concepts can be read as reflecting a position that moves beyond these simple dichotomies. She has resources that make mental life more central to meaning than other latter-day heirs to Mill have been willing to recognize. This makes Millikan rather un-Millian, and more of an Innerist than her overt Outerism would suggest. Millikan's theory of concepts might be described as Inside-Outerist. If so, some of the Outerist orthodoxies that pervade her work need to be reexamined. My claim is not that Millikan's views are inconsistent, but rather that her

Millikan and Her Critics, First Edition. Edited by Dan Ryder, Justine Kingsbury, and Kenneth Williford.
© 2013 John Wiley & Sons, Inc. Published 2013 by John Wiley & Sons, Inc.

1980s Outerist terminology may sometimes obscure a key insight of her work – an insight that allows us to move beyond some of the old debates that have divided the field.

Innerism and Outerism

The two perspectives

The Outerist perspective is constituted by a collection of intimately related views, which came into vogue when Kripke and Putnam led their crusade against descriptivism in philosophy of language. These views need not be united (indeed, Kripke and Putnam each rejected at least one of them), but they hang naturally together and were widely accepted by authors in the 1980s and 1990s who were developing accounts of mental content. Notable adherents include Fred Dretske, Jerry Fodor, and, of course, Ruth Millikan.

I will now briefly characterize the theses that constitute Outerism, remarking along the way how each has been endorsed by Millikan. The first is this:

> *Externalism*: mental content is not determined by what's in the head.

This position was most influentially defended, for linguistic content, by Putnam, using Twin Earth thought experiments. Two molecular duplicates may refer to different chemical substances using the word "water" if they reside on planets with different stuff filling the rivers and streams. Likewise for the concept that "water" expresses in these two twins. Millikan has never been a fan of Twin Earth thought experiments and the attendant metaphysics of possible worlds, but she is a committed externalist. For Millikan, the case for externalism hangs on a view about what concepts are selected for. We are evolved to track real substances in the world, even when we don't know about their underlying essences. Thus, referring does not depend on what we explicitly grasp psychologically, but rather on our ability to keep track of things out there, whose unifying features are a matter of empirical discovery.

This brings us to a second Outerist thesis:

> *Realism*: concepts refer to real kinds, i.e., kinds that support inductive inferences in virtue of mind-independent grounds that are responsible for observed correlations and regularities.

Realism of this variety is made explicit in Kripke's appeal to essentialism, which also became a mainstay in the theories of psychologists such as Frank Keil and Susan Gelman. It is also a central assumption in Fodor's psychosemantics, since he insists that concepts get their content in virtue of law-like connections between concepts and the things to which they refer; my HORSE concept refers to all horses rather than, say, prototypical horses, because only the property of being a horse, rather than a typical horse, can figure in such laws. Millikan's notion of real kinds extends beyond those that have Kripkean internal essences; for Millikan, historical factors can also ground real kinds. And her Realism is not motivated by Fodor's nomological considerations. Once again, she thinks it is the evolved function of our representing systems to put us in contact with real things out there in the world.

A third Outerist thesis is:

Millianism: the semantic content of lexical concepts is exhausted by that to which they refer.

Kripke was largely responsible for resuscitating this aspect of Mill's semantics in contemporary philosophy. Putnam is not exactly a Millian, since he thinks concepts have stereotypes as one semantic component, though he does not think stereotypes determine reference. Fodor is a Millian because he has an atomistic theory of concepts, according to which all lexical concepts (concepts expressed by a single word) have no meaningful components, even if they happen to be associated with stereotypes or other bits of knowledge. Dretske's account of digitalization also suggests a Millian picture, insofar as concept acquisition involves a process by which we form representations that carry information about one property without nesting it in information carried about any other. Millikan's Millianism is manifest in her endorsement of P. F. Strawson's "dot" semantics, according to which we have the mental equivalent of a dot corresponding to each referring concept, and the descriptive features associated with each dot are not to be regarded as reference fixing, nor as modes of presentation by which the concept is grasped.

Outerist semantic theories also often come packaged with corresponding claims about epistemic limitations:

Corrigibility: we cannot tell by introspection what we are referring to, or even whether we are referring at all.

This can be seen as a consequence of Externalism. If reference does not depend solely on things in the head, then introspection cannot reveal the determinants of reference. That means we can't know by introspection exactly what we are thinking about. Some Outerists, such as Tyler Burge, have been embarrassed by this, and they have developed theories of self-knowledge that allow them to say that people can, in some respect, know what they are thinking. But Outerists must agree that there are aspects of content that cannot be directly known. Millikan dubs the opposing view, according to which we can introspect the content of thoughts, "Meaning Rationalism." She has been railing against this since the publication of her first book, calling it the last Myth of the Given.

The final thesis that I will include in the Outerist program is somewhat orthogonal to the others. It is:

Naturalism: the determinants of mental content can be fully specified using non-psychological and non-semantic vocabulary.

In principle, one could hold the other theses without being a naturalist. Kripke, for example, does not endorse his Millian Externalism in order to reduce semantics to something else. As a dualist, he is clearly not looking for a reductionist theory of the mental. The commitment to Naturalism is, however, a central component of Dretske's and Fodor's views, and it is equally important to Millikan. Millikan wants to reduce meaning to teleology, and she sees the other Outerist theses as consequences of her effort to do so. Thus, Naturalism plays a central role in motivating her account.

The label "Outerism" draws attention to the fact that the foregoing views seek to explain mental content by appeal to something other than what's in the head. Externalism looks to where we are located; Realism emphasizes mind-independent identity conditions on the things to which we refer; Millianism says that our inner states do not determine content; Corrigibility says introspection does not suffice for semantic knowledge; and Naturalism looks to causation, biology, and other factors that must be empirically discovered to determine mental content. Mental content is relational, according to all these theses, and the relevant relations feature relata that are not psychological in nature.

Each of the Outerist theses contrasts with an Innerist counterpart. In brief, Innerism consists in the following:

Internalism: mental content is determined by what's in the head.
Nominalism: concepts refer to nominal kinds, i.e., kinds whose identity conditions are determined by our minds.
Fregeanism: concepts expressed monomorphemically generally have reference-determining modes of presentation that are part of their semantic content.
Incorrigibility: we can tell by introspection what we are referring to, and whether we are referring (what Millikan calls "Meaning Rationalism").
Non-naturalism: the determinants of mental content cannot be fully specified using non-psychological and non-semantic vocabulary.

The basic idea uniting these theses is that mental content depends very much on what is in the mind, and is therefore discoverable through psychological processes such as introspection and reflection. Descartes seems to have been an Innerist in this sense, as was Frege. In contemporary philosophy, an Innerist orientation can be found in Robert Brandom's Inferentialism, McDowell's second nature semantics, Putnam's post-1970s Internal Realism, and theories developed by some latter-day internalists such as Katalin Farkas. If you are like me, you were raised to think of the defenders of such views as the bad guys. Outerism has been the preferred view since the 1980s, and Millikan's efforts have played no small part in its popularity.

Narrowing the divide

The divide between Innerists and Outerists is among the most polarized in the philosophy of mind. Like the debate between materialists and dualists, it reflects deep ideological divisions in the field. Some authors have mixed views, combining tenets of Innerism and Outerism, but most line up neatly on one side. Millikan is no exception. She seems to be a paradigm case of an Outerist. Throughout the rest of this chapter, however, I want to argue that the division has been exaggerated and that the truth might lie in the middle. I hope to also show that Millikan's Outerism is more qualified than it might initially appear.

I want to start by showing that the basic theses dividing Innerism and Outerism admit of intermediate positions that can help bridge these poles. First consider the contrast between Internalism and Externalism. A textbook version of these views might give the impression that Internalism assigns no role to the environment, and Externalism assigns no role to the mind. But this is not the case. Consider a canonical example of a concept for which Internalism might be true: an explicit description. For example, I might refer to the "last French Monarch." This phrase, and the corresponding concept, refers by satisfaction, and the person who satisfies the description happens to be Louis-Napoléon Bonaparte, also known as Napoleon III. The concept determines the referent. But suppose we lived in a world where the Prussians didn't defeat France, and the French Monarchy had gone on a bit longer. Then the description might refer to Napoleon IV. Thus, the world we are in plays a role in determining the reference of the concept. Likewise, even the least descriptive concepts, such as demonstratives, may refer with the help of psychological states. If I judged "That is delicious!" the concept expressed by "that" will refer to whatever foodstuff is salient in my mind at the time.

The difference between Internalism and Externalism hangs on *how* mental items contribute to content. For the internalist, satisfaction of a description is what does this trick, and for the externalist, descriptions simply serve to put us into the right kind of non-semantic (e.g., causal) contact with the world. This is sometimes expressed by saying that for the internalist, reference is determined by meaning, and not so for the externalist, either because descriptive contents don't count as meanings or because meanings underdetermine reference. Given this way of describing the difference, it should be clear that the views are not polar opposites. There are positions that seem to lie in between the extremes. Consider the incomplete description, "The book on the table." This clearly refers by description, but the table in question is not descriptively specified. It depends on where the speaker is located. If concepts refer in this way, should we say that Externalism is true of them, or Internalism? I would say neither and both. We need an intermediate category:

> *Interactionism*: mental content is partially determined by what's in the head.

Turn next to the debate between nominalists and realists. This, again, looks like a clash of opposites, and, again, there is room for a middle ground. Realists presume that there are joints in nature, and nominalists balk, saying we impose categories on the world. But suppose the situation is like this. There are natural boundaries in the world, but too many of them. For some categories, there are numerous other overlapping categories, whose extensions are similar but not exactly the same. If this were the case, we would be able to say, with realists, that there are real kinds, but reference would depend on a very active process of selection. We'd have to somehow choose which of the many similar real kinds we wanted to refer to, on pain of referring to disjunctive classes of real kinds. The resulting picture could be called:

> *Selectionism*: concepts refer to select real kinds, i.e., kinds whose identity conditions are determined by the world but can be distinguished from other similar ones only via psychological processes that constrain how the kinds expand beyond the individual category members we happen to recognize.

There are also ways to divide the difference between Millian and Fregean theories. Millians want to restrict semantics to reference, but they often admit that there are highly contingent reference-fixing descriptions. They deny that these descriptions are parts of meaning or that they are modes of presentation, because those constructs are presumed to be more stable and sharable within the Fregean tradition. But suppose we relax this requirement and say that reference-fixing descriptions are part of meaning (after all, they contribute to reference determination), and that meanings just turn out to be different than Frege has assumed, because they are contingent and variable. Call this:

> *Conceptionism*: concepts expressed monomorphemically generally have modes of presentation that vary across contexts but play a role in fixing reference (sometimes called "conceptions").

There is also an intermediate position with respect to the epistemology of thought. Defenders of Corrigibility say we don't know what we are referring to, and that we can fail to refer without knowing it. But this is compatible with a very qualified form of Incorrigibility: we might know what we are trying to refer to, or at least what it would take for the world to contain the kind of thing to which we are trying to refer. We might also be able to decide to alter those referential aspirations, should the world fail to comply. On such a picture, we cast out semantic nets with the hope of catching something, but reserve the right to cast differently should we come up empty. In those cases where the world complies, there is a sense in which we refer without complete knowledge of what it is to which we refer – we know only what we were aiming for. In cases where the world doesn't comply, we may know more fully what we are referring to, because, at that point, we may simply make a referential choice. This view, which will become clearer with examples below, can be called:

> *Decisionism*: we can tell by introspection what we aim to refer to, and in some cases this can give us epistemic access to what we are referring to.

The final point of contention, Naturalism, may seem pretty non-negotiable for someone of Millikan's bent. After all, her whole philosophical project is animated by a desire to explain mental and linguistic meaning in biological terms. From this perspective, Non-naturalism may look about as enticing as a Sunday prayer meeting. But there are positions between Naturalism and Non-naturalism that don't give up on materialist ontology or materialist explanation. One can bring this out by comparison to a debate about the mind-body problem. In that debate a distinction is drawn between reductive and non-reductive materialism. Reductionists, such as psychophysical identity theorists, aim for a direct mapping between mental states and states that can be described biologically. Non-reductionists, such as functionalists, define mental states in terms of other mental states, and then postulate physical realizers. Work in psychosemantics has not always drawn such an explicit distinction, and, for that reason, we find authors such as Fodor assuming that there must be some direct explanation of mental content in non-mental terms. The alternative, of course, would be to say that mental states have their content, in part, in virtue of their relationships to each other, and then suppose that the whole apparatus can physically be implemented. Call this:

> *Non-reductive Naturalism*: the determinants of mental content include psychological states, which must be specified as such, though these states are physically realized.

The constellation of views that I have been sketching here is neither wholly Innerist nor wholly Outerist. It might be described as Inside-Outerism, because it embraces the idea that mental content depends on things that are not mental without giving up on the idea that psychological states can exert considerable influence on what we think about. I think there are Inside-Outerists in the literature. I might put Locke, Quine, and Sellars in this camp, though I won't make the case here. I will raise the possibility, however, that Millikan belongs in this camp, notwithstanding her explicit endorsement of Outerism. Before I make this case, I want to motivate Inside-Outerism by suggesting that there are many concepts for which such an intermediate perspective is intuitively plausible.

Are Some Concepts Inside-Out?

Being reared under the influence of 1980s semantic theories, I have long been a committed Outerist, and I continue to see the allure of that tradition. Millikan has always been an intellectual hero of mine, and I've been inspired by her willingness to push the Outerist agenda to its limits – or at least, so it has seemed. But I sometimes wonder whether Outerism would have appeared so attractive if proper names and natural kinds had not been taken as the paradigm cases of concepts in the wake of Kripke and Putnam. And I wonder too whether Outerism even applies, in its purest forms, to these concepts for which it was explicitly designed.

To explore this idea, I want to consider some concepts for which an Inside-Out perspective seems especially attractive. Then I will see whether the case can be generalized. The examples I have in mind might be broadly classified as interest-relative, because they carve up the world in a way that is clearly dependent on our interests. Now, Millikan will say all concepts are in our interest, and it is in our interest, teleologically speaking, to glom onto kinds whose boundaries are given by the mind-independent world. But I want to draw attention, at least at first, to cases where this picture looks less plausible – cases where our interests do not necessarily align with boundaries that exist without those interests.

To be clear, the suggestion I will be exploring is that some, perhaps most concepts, refer in a way that depends on choice. The choices involved are sometimes consciously made by the possessor of a concept. In other cases, the choices may be implicit in the psychological processes used to apply the concept. For example, if one uses a mental image to track a category, the features of that image may place implicit constraints on what instances of the category need to look like. In the case of shared concepts, or linguistic meanings, we might say these choices are distributed across a community, and reference may depend on a collection of choices (compare: an election depends on the choices of the majority). If reference depends on choice for some of our concepts, then the scope of Outerism may be restricted.

To begin with, consider the phenomenon of vagueness. Reading the literature on vagueness, one might think that the most interesting thing about vague predicates (and corresponding vague concepts) is that they admit of borderline cases. This is supposed to embarrass those who would try to model language using classical, bivalent logic. I have come to see such formalist ambitions as a bit dubious (Why bother? What does formalization reveal? Why expect success?), so vagueness strikes me as interesting for other reasons. One thing that is striking about vagueness is that the examples used to illustrate the phenomenon are typically very different from examples used in the theory of reference, where natural kind terms and proper names are the norm. Vagueness theorists focus on concepts such as *bald* or *heap*. Is the boundary between heaps and non-heaps given by nature? Or is this rather a category that we have devised for purposes that don't depend on deference to world-given joints, such as creating a heap of potatoes at the market? Could there be a science of heaps? Balding is clearly a natural phenomenon, but classifying people as bald is not especially concerned with the underlying mechanisms (suppose there are five different sources of balding, which are each biologically different). Rather, it is a largely cosmetic concern. Similar points apply to other notoriously vague predicates, such as empty, tall, or flat. It is up to us, not the world, where we draw the line.

This brings us to another interesting feature of vagueness. When using a vague predicate or concept, boundaries can shift. Is the hotel room empty? Yes, it's unoccupied. Is the rental apartment empty? Yes, it's unfurnished. Is the road flat? Yes, it has no hills. Are you lying flat? Yes, my legs are down. These shifts depend not only on the kind of object we are considering, but on current goals. When laying down pavement, a flat road means something different than when planning a hike. When casting a production of *The King and I*, the threshold of baldness becomes stricter. The boundaries in such cases are given by us. There is still room for error here, because one might fail to realize that a particular object satisfies an operative standard. There is even room for induction: I might discover that most bald actors are self-conscious or that flat hiking paths have few good views. Thus classic examples of vague concepts share much in common with natural kind concepts or concepts of real kinds, even though they are dependent on our classificatory choices and goals.

Similar conclusions follow for concepts that are said to be context sensitive. One trendy example is *knowledge*. Though controversial, some people think that this concept applies differently in different cases depending on operative norms. For example, when stakes are high, standards for knowledge may rise. This would be true, according to defenders of Contextualism, even if knowledge is not vague. Suppose I have always known where the university hospital is located, but then I have a medical emergency and need to make a trip there in a hurry. Under that condition, I may decide to look up the address on the grounds that actually I don't really know, but only have a confident belief. Is there a real kind here, "knowledge," or is it rather that I have flexible norms for possession of information that shift depending on my current needs?

Knowledge is a normative domain, and we can find interest-relativity in other normative domains as well, even ones for which Contextualism may seem less plausible. Consider aesthetic values. If I like certain music or artistic styles, that's as much a fact about me as a fact about them. There is no way to group together the musical forms I like without reference to how they strike me. Likewise for the class of things I find delicious. I think moral concepts, like *right* and *wrong*, also work this way. Some people are moral realists, of course. They assume that moral facts are, in some sense, mind-independent. But I think that Moral Realism is just a false theory that we are willing, when confronted with evidence, to give up.

The story would go something like this. If you think that there is such a thing as objective moral goodness, and I convince you that there is no such thing, you *could* become an error theorist like J. L. Mackie; you could conclude that nothing is morally good. But chances are, you wouldn't. (Compare: if I convince you that there is no objective beauty, you would not conclude that Rita Hayworth is unattractive.) Rather, you'll say that moral facts are mind-dependent. This makes moral concepts a lot like concepts of secondary qualities. When scientists tell us that human color space is not isomorphic with any space of purely physical magnitudes, we do not conclude that bananas are not yellow; we say that yellowness consists in causing a response in us. Yellow really exists, but its boundaries depend in complex ways on how we experience the world. Morality may turn out to be like that. If so, that's because we choose to use our moral concepts that way, even when we give up faith in an objective moral reality.

One might object that there is no need for choice in this picture; if moral concepts track mind-dependent properties, they do from the get go, and it's a matter of discovery, not decree, that they refer to such properties. But this objection will not eliminate the element of choice. Surely there is a difference between trying to track an objective property by use of its appearance, and trying to track an appearance, and surely we can decide in a given case which one we're after. Suppose, in the moral case, we make no such decision, and we are then confronted with a compelling argument against moral objectivism. What, other than choice, would settle whether our moral concepts are empty or subjective at that point? One might say, moral concepts must refer to the subjective properties because their successful application in pursuing goals in the past was evidently not dependent on referring to objective properties. But this reply makes two questionable assumptions. First, it's not clear why success depends on referential success; the belief that killing is objectively wrong might be a very useful fiction. Second, it's not clear how to measure success in the moral case. Suppose I give to charity because I believe that doing so is good, and I want to be good. Did giving to charity help me realize my goal of being good? That depends on whether goodness is an objective or subjective property, which is just what's at issue. I don't see any teleological formula for settling what kind of property we are referring to in the moral case. The issue seems to hang on choice of some kind, a choice that may be implicit in our tendency to continue moralizing even when we lose faith in absolute values.

Choice also enters into concepts of social categories. Consider the concept *gay*. Does it apply to men in ancient Athens, who had relations with boys? Does it apply to men among the Sambia of New Guinea who receive fellatio from male teens? Does it apply to the *hijras* of India who consider themselves a third gender? Millikan sometimes says that such social categories refer to historical kinds, and thus, our concept *gay* would not apply to these historically unrelated groups (and by parity of reasoning, our concept *teacher* would not apply to professional educators in remote cultures). But it seems to me that the choice is up to us. We can use our concept to track something historical or to track something functional. And, on either approach, we can choose which historical links matter (is Athens part of our history?) and which functional roles matter (is being gay a matter of certain kinds of sexual practices or certain desires or certain social roles?). These decisions seem to be up to us.

Millikan could accept this story. Social concepts are acquired and disseminated using language, and linguistic terms refer to that which helps to explain their proliferation within a language community. The choices I've referred to may be understood as factors that determine patterns of proliferation – understood as akin to natural selection – and hence reference. That is an attractive story, but one that incorporates choice rather than denying it. Millikan might respond that choice is playing an indirect role, however: it influences selection, but selection does the real semantic work. But two replies are available. First, whether direct or indirect, the role of choice in reference determination is hard to square with full-blown Outerism. Second, it's not clear why proliferation is even needed once choices are in place. If I choose to use a term in a certain way, then that term seems to have the meaning I give it, provided I am willing to treat it as part of my idiolect. If I want to share the term, I just need to find one language partner willing to use the term the same way. This is just an intuition, of course, but it also has explanatory payoffs; to explain the behavior of a language user, the semantic policies of that user may be more important than the policies of the linguistic community, though communal use can affect choice. On this view, semantics depends directly on choice, and only indirectly on proliferation. Either way, choice is crucial for social concepts.

Similar choices may govern some artifact concepts. Is a junket a boat? What about a Sepik canoe? These sea vessels may lack historical ties to our boats, but they are functionally alike. Are they boats? It seems to be up to us. Suppose two technologies for broadcasting moving images had been developed independently. Would they both be television? When digital televisions were invented, could we have stipulated that they are not televisions because they lack cathode ray tubes? (Recall that we chose not to call CD players "record players.") It seems that our tendency to classify by history or function is a matter of choice. The way we classify observes no consistent pattern, and even when we choose one or the other, we choose which connections and features matter. Note that this point does not depend on lexicalized concepts. Artifacts are inventions, and an inventor can conceptualize what she created before labeling it. Her design intentions can dictate the extension of her concept, regardless of how others would talk about it.

Millikan regards artifacts as real kinds, bound together historically. I am not denying that there are such real kinds. There are artifacts linked together in chains of production, reproduction, imitation, influence, and improvement. My point is just that concept users can exercise some choice in deciding the boundaries of such historical links and even whether those links are necessary for co-classification (recall Sepik boats). The same seems to be true for concepts pertaining to other things that come out as real in Millikan's ontology.

Consider concepts of individuals. As Millikan convincingly argues, we have many different ways of reidentifying the individuals in our lives, and, despite that, we don't seem to possess the modal knowledge necessary for distinguishing those individuals from duplicates on other worlds. Thus, we don't know what is essential to individuals, but we manage to refer to them because of our own histories (we have encountered them in the past), and their histories (they are bound together historically in a way that allows identity over time). But there are always conditions under which we find ourselves compelled to decide whether certain conditions are necessary for identity. Is a person with dissociative identity disorder one person or several? Is a person with advanced dementia the same person she was before the illness? Are decisions made by unconscious brain mechanisms attributable to the person in whose brain they occur? As far as I can tell, there are no real facts that settle such questions. We get to choose. Likewise, we get to choose if my favorite restaurant continues to exist after a change in its location and chef. And we can decide whether basketball teams remain the same as they cycle through members.

Concepts of substances also allow some demarcation by fiat. Is H_2O still water when it has some impurities or is transformed into an isotopic form? Are tears water because they are mostly H_2O? Is steam water? Is salt water the same stuff as lake water? We might defer to experts when considering such questions, but expert decisions are stipulations not discoveries. And, without availing ourselves of any linguistic division of labor, we are entitled to make decisions by fiat as well. Is granola cereal? Is matzo a kind of cracker? Is a bagel a roll? Are rolls bread? Is heavy cream milk? Is fermented cider wine? Is wine juice? Most of our interactions with these substances don't depend on such decisions, and that means that our concepts may be indeterminate. But we can make these decisions if we want to, or if it becomes useful to. People who run supermarkets or organize menus make decisions of this kind when they group things. When such decisions haven't been made, there are indeterminacies in our concepts that reflect the fact that no decisions have been made yet; in this sense, the boundaries of our categories depend very much on our choices.

It might be objected that choice is less important, or at least less directly important, when it comes to publicly used words. We might be able to precisify privately used concepts by fiat, but do we really have the right to dictate the extension of an imprecise public term? In response, I would first note that even if the phenomenon of fixing extension by choice applies only to one's personal concepts, that is already a problem for a strong form of Externalism. It seems that I get to decide (explicitly or implicitly) whether bagels and rolls belong in the same category. My classificatory practices shape the contours of the category. With public language, reference may depend on the practices of numerous language users, so there is still an important role being played by psychological states. It is also possible, as Davidson suggested, that public meanings are renegotiated on the fly in the context of conversation. If a baker puts bagels in the roll section of her bakery, then she may manage to establish a shared concept with her clientele when they visit, even if they use the word "roll" differently elsewhere. I think Millikan might agree with all this, but the point, again, is that a semantic theory should not underestimate the role of psychological processes in reference determination. We don't just point to an example of a category and thereby refer to the whole; we often play an active role in specifying the boundaries. When we fail to specify, we cannot rely on the real kinds to settle the question, because each object belongs to many real kinds, and some kinds have reality only insofar as we group things together (e.g., the category that includes bagels among the rolls but excludes croissants). Absent boundary demarcation, indeterminacy reigns.

Indeterminacy and stipulative demarcation also arise in our concepts of living species. Categories like weed and dog (we exclude wolves) famously show the role of stipulation, but the point is quite general. Scientists can stipulate whether to group species by clade, or genotype, or morphology. Scientists must also settle, in evolutionary time, when one species ends and another begins. In deferring to them, we are, in effect, introducing an element of human decision into the extensions of our concepts. There are also indeterminacies made familiar by philosophers. Is that a rabbit or a collection of rabbit parts? Such indeterminacies don't really matter, because when we discover them, we can decide by stipulation, and before we discover them, they have no impact on us. Either way, the point is that boundaries can depend on what we choose or on our failure to choose. Practice plays a major role in determining extension.

To put the point in a nutshell, most concepts are like jade, at least potentially. They begin life fraught with indeterminacy (did the first uses of "jade" refer to a specific stone or to the whole class of superficially similar stones?), and then, when such indeterminacies become relevant, we make a decision as to how to go on with the concept. When we do so, we may elect to refer to a real kind (unlike the jade case), but which kind we select of many overlapping kinds is up to us. We may arbitrarily choose to exclude kinds that are similar in history and microstructure, and we may include under one umbrella kinds that are similar relative to some human function despite historical and structural differences. Millikan might agree, but to that extent her account is more Innerist than casual readers might realize.

To me, this suggests that the enduring debate between Innerism and Outerism is overly polarized. For the majority of our concepts, the truth seems to lie in the middle. In other words, our concepts are typically Inside-Out. Consider, first, the thesis I called Interactionism, according to which reference is not determined solely by causal connections to the world or by satisfaction, but by descriptive components that, when situated get uniquely satisfied. So I might think of water as the clear liquid that plays the water role around here. That liquid happens to be H_2O, so I have succeeded in picking out a real kind, but I have also ruled out some forms of H_2O, such as tears or steam, because they don't play the water role. If I am right, such stipulations can exert an influence on reference, and that suggests that satisfaction is involved. But descriptive knowledge does not completely demarcate boundaries of our categories; I fail to refer to X_YZ not because I know chemistry, but because there is none around here. Thus, both description and location matter. I hasten to add that what I am calling "descriptions" need not be linguistic or even word-like. A sensory impression might do the trick, or a prototype.

Next consider Selectionism, the idea that I can refer to one of many neighboring real kinds by selective fiat. This is true when, for example, we choose the extension of *gay* or when we defer to experts who choose to circumscribe animal concepts in one of several equally possible ways.

The examples here do not directly support Conceptionism, which is the view that our changing conceptions belong to the semantic content of our concepts, but they offer indirect support. In her Millian moods, Millikan would have us believe that concepts are Strawsonian points whose semantic content is exhausted by what they refer to. Reference, in turn, depends on historical links connecting those points to real kinds in the world. But this picture is less attractive if the methods by which we track things can also have a significant impact on reference. If our conceptions, no matter how ephemeral, alter meaning, then they should be regarded as semantic. This is true in all the examples I've been discussing but perhaps most clear in the cases of vague and context-sensitive terms that change extensions with changing interests.

The examples also favor Decisionism, which is simultaneously a semantic thesis and an epistemic one. I can choose to refer by deferring to real kinds in the world. Indeed, this may be how concepts work by default, so no explicit person-level choice needs to get made. But when I confront the fact that there are multiple real kinds that are historically linked to a concept, or perhaps no real kind (as in the case of moral and other evaluative concepts), then I can make a decision to change my semantic policy in a way that affects reference. This decision need not be conscious. It may just be a function of using my concepts in a particular way. Even so, if my practices of use play an active role in reference determination, then I know a lot about reference by introspection. This is most clear in the case of concepts that I use to refer to response-dependent properties. I can know by introspection that the wines of Bordeaux are more delicious than the wines of Cahors, and I can even know, by introspection, that lying is bad. These judgments are not quite empirical and not quite *a posteriori*. I know them by decision, in some sense, rather than by mere observation or mere analysis.

Finally, consider the question of Naturalism. The examples I have given so far have an important feature in common. Reference of a concept is partially explained by some goal, decision, or policy in the mind of a concept user. For concepts like this, it would be a mistake to provide a naturalistic theory of reference as that term has recently been used. Naturalistic theories usually result in formulas like this: "Concept C refers to referent R if (and only if) C bears such-and-such non-semantic, non-psychological relation to R." For the concepts I have been considering, such a formula might be impossible to apply. Referring to R may depend on a decision, which is a psychological state.

This is a problem for leading forms of semantic Naturalism, such as the causal–historical theory and informational semantics. The decisions that affect reference do not always change the causal history (the decisions might occur after a concept is acquired), and they may not have any bearing on the causal laws linking concepts and world (do tiger clades cause *tiger* tokens or do tiger genotypes?). A causal–historical theory of reference could accommodate such cases only by introducing an elaborate theory of re-dubbing, where changing decisions effectively introduce new concepts on a regular basis; this would delimit the role of causation in the account. An informational theory of reference could accommodate the semantic impact of decisions only by appeal to exotic counterfactuals, which may be more costly to the aims of a reductive naturalist program than the comparatively benign appeal to psychology. In contrast to these other theories, Millikan has two built-in resources accommodating the observation that decisions are semantically important. First, she can say that decisions change the teleological function of a concept, insofar as decisions can introduce functions. But such an appeal to teleology would not be reductive because the function in question would be conferred by a psychological state. Second, Millikan can say that decisions affect the selection processes that propagate linguistic items. But this won't help for idiolects or private concepts. Moreover, the selectional theory conflicts with the intuition that a decision can affect meaning right there and then when the decision occurs, regardless of the effects of propagation. One could deny that decisions have an immediate semantic impact, but that would be costly because they have an immediate behavioral impact, and a theory of meaning is ultimately only as good as the range of behaviors it can explain. Millikan might try to cleave a line between speaker meaning and semantic meaning, but this has become one of the most contested boundaries in contemporary semantics. In any case, drawing that distinction would concede a role for decisions in a theory of speaker meaning, which would imply that naturalistic ambitions in that domain may need to be constrained.

The upshot is that psychological states, such as decisions and choices, may be ineliminable in explaining reference in some cases. No one in the literature denies this. For example, Fodor points out that descriptive concepts or thoughts must be explained by appeal to the concepts that comprise them. The point here is that psychological states may enter into theories of reference for lexical concepts – those that we express with a single word. Since naturalist theories are often introduced to account for those, this would suggest that Naturalism has been a misguided research program. None of this entails, however, that reference is non-natural in any ontological sense, if reference depends on decisions and decisions might ultimately decompose into psychological states that could be characterized functionally or even physically without any semantic or psychological terms. But, with respect to lexical concepts, such reductive ambitions may be overblown. Some of the people who have been trying to naturalize semantics think it would be catastrophic to say that psychological states contribute to the reference of lexical concepts. But that isn't the case. There are standard moves within functionalist philosophy of mind for coping with the threat of circularity. We can reduce semantics to psychology and then reduce the required psychological states to something non-psychological; we get reduction eventually, but in two steps.

In summary, I have been recommending an Inside-Outerist approach to a range of ordinary concepts. I am not saying that Externalism is false, or that Internalism is true, but that both sides have resources we need in our psychosemantic theories. Likewise for the other theses that have keep Innerists and Outerists apart.

Millikan's Concepts

Let's return now to Millikan's views. As noted, Millikan often explicitly endorses Outerist theses. Indeed, she does so with uncommon gusto, raging against incorrigibility claims and crusading for industrial strength Realism. She seems to be an archetypal Outerist – even an activist Outerist, fighting tooth and nail for the cause. But there are also aspects of Millikan's grand theory that bring Innerism to mind. If she would just listen to her Innerist voice, she might conclude that concepts are Inside-Out.

Consider four central themes in Millikan's work. First, she argues that desires are semantically prior to beliefs. The proper function of a desire is to achieve its own fulfillment, and the proper function of a belief is to contribute to desire satisfaction. Thus, roughly, desires represent that which they were evolved to bring about, and beliefs represent that which needs to be the case for them to contribute to desire satisfaction when everything is functioning normally. This aspect of Millikan's theory implies a kind of interest relativity. The content of beliefs depends on goals in a way that suggests a change in goals might influence content.

A second theme in Millikan's work is her emphasis on consumer systems. To count as a representation, a state of an organism must be used in a certain way, and the way it is used helps to determine its meaning. There may be many possible mappings between mental states and states of the world, but consumer systems effectively disambiguate between these, by using mental states in specific ways and thereby determining which of the many mappings gives that state its representational content. It is easy to imagine, on this framework, that a change in use might result in a change in content, provided the change in use affected the consumer functions. This is one way in which the organism contributes to content, and it distinguishes Millikan's theory from theories of people like Fodor, who emphasized production rather than consumption, as well as causal–historical views that afford no semantic role to use.

Third, Millikan explicates concepts in terms of tracking. Tracking is an ability and it depends on the internalization of features that can be used to reidentify something on multiple occasions. Millikan emphasizes that we may use different features to track the same kinds, but concept possession depends on the presence of something inside us that allows us to do this work. It is natural to think of tracking abilities as depending on sets of feature representations, and the features we use can have an impact on what our concepts represent.

Fourth, consider Millikan's theory of substance templates. Concepts of substances are governed by inner templates that tell us what kinds of questions we can intelligibly ask about a given category, and such templates can presumably play a role in determining whether a given concept represents something as an individual, as a natural kind, as a historical kind, and so on. Substance templates constrain inference space in a semantically significant way. Substance templates may themselves have teleological functions, but, as I understand the view, they can be characterized psychologically in a manner similar to how psychologists characterize domain concepts, and this psychological profile is doing real semantic work.

Given all these resources, Millikan is well positioned to be an Inside-Outerist. Desires, consumer systems, tracking mechanisms, and substance templates can contribute to content. I think Millikan should be an Inside-Outerist and that she probably is, despite her association with the Outerist orthodoxy. First, I think she is probably an Interactionist. Suppose you encounter a dog, and a little light goes off in your head (perhaps it's a perceptual state caused by seeing a Rottweiler, which then gets used as a concept). Does that perceptual state represent this particular dog, or Rottweilers, or dogs, or canids, or animals, or prototypical dogs? For Millikan, that will depend on our tracking abilities, which depend, in turn, on what features we use to reidentify this dog on future occasions.

I think Millikan is also a Selectionist (in the sense defined above, as well as the Darwinian sense). Suppose you see a gray squirrel for the first time, and form a concept.

Millikan's Realism entails that your concept will also refer to black squirrels because they belong to the same sub-species, even if you don't recognize them as such (assuming it's random whether a squirrel ends up gray or black). But what about red squirrels, a different sub-species? This, I think we should say, depends on how we extend the concept. It's a matter of selection.

What about Conceptionism? Millikan insists there are no modes of presentation, but she is thinking of the fixed, sharable Fregean variety. She admits that tracking abilities involve representations of features, which are therefore (unlike Fodor) partially constitutive of concept possession, and (unlike Putnam's stereotypes) implicated in reference determination. There are no specific features any concept possessor must use, but each of us uses some features or others (whether consciously or unconsciously), and these contribute to our conceptualization of the things we track on any give occasion. These conceptualizations presumably contribute to what we track, and hence to the semantics of our concepts.

Millikan is not explicitly committed to Decisionism and she has written reams, since her first book, arguing that we lack privileged access to what our concepts represent. But it seems to me that she should allow for access at least in cases of evaluative concepts, which, I claim, refer to response-dependent properties; for example, we know what we find delicious, and that knowledge coincides with the extension of the deliciousness concept that each of us possesses. Millikan should also allow that we can decide what kind of substance template to use when classifying something for the first time, and that gives us access, by fiat, to what kind of thing it is (though, of course, we may end up failing to refer as a result).

Millikan should also be a Non-reductive Naturalist. Consider her account of beliefs, which makes explicit reference to desires. Or consider her account of substance templates as constraints on conceptual content. In these places, she implies that reference should not be explained by directly relating mental states to, say, biological functions, but by relating mental to mental, and then integrating the whole apparatus with biology in a way that leaves no ontological remainders. This departs from Fodor-style reductionism, which tries to directly reduce meaning to causal laws. Millikan's story may be amenable to a two-step reduction, but the semantic theory itself need not be stated in reductionist terms.

In other words, Millikan can handle the kinds of examples that I've been considering, precisely because she places more emphasis than her rivals on things inside the head. Millikan seems to be an Inside-Outerist. Why, then, does she so often express deep faith in Outerism? If she is an Outerist, and not an Inside-Outerist, then the cases I've presented can be interpreted as problems for her theory. That would leave her with three possible lines of response. She could say:

1 The kinds of cases I offered are rare.
2 The kinds of cases I offered are compatible with Outerism.
3 I am mistaken about the content of the kinds of cases I offered, because the goals and decisions do not alter content.

I invite Millikan to tell us which, if any, of these options she prefers. Let me just indicate, before concluding, why I don't think any of them are especially attractive.

Are the cases I've been considering rare? I think the answer is obviously not. Concepts such as heap and bald and other vague concepts are used all the time. (Other examples: "My keys are over *there*," "I won't be home *late*," "*That* was a big meal.") Evaluative concepts are also frequent; we reside in a world of pretty, nice, yummy, and noble things, as well as their opposites. Concepts of individuals are generally bounded by views about personal identity (e.g., observant Catholics might use the person concept to cover embryos while Jews might not). Concepts of natural substances, animals, and artifacts are also frequently used, and whether we realize it or not, the ones we have words for (i.e., most of them) have extensions that have been circumscribed by the fiat of experts. In short, I suspect these cases are common.

Are these cases compatible with Outerism? To answer this, I need to underscore that I am not here defending Innerism. My claim is not that Frege was right about sense, or that people always know what they are thinking. I want very much to preserve aspects of the Outerist picture. But the Inside-Out view also preserves aspects of Innerism, and this is important. To illustrate, consider the contrast between Realism, Nominalism, and Selectionism. Realists like Fodor and Dretske often let ontology pick up the slack in their theories. They say we represent real kinds because we can detect *some* of their instances: we detect A, B, and C, and since these belong to the same natural category as D, E, and F, we manage to refer to the whole category. Nominalists say that we must explicitly represent the principle by which we extend the category: we represent A, B, and C, and similarity criterion S, which extends the category to D, E, and F. Selectionists try to occupy an intermediate position. There are many natural ways to extend the category A, B, and C, so we can restrict the extension by specifying principles of exclusion. Strictly speaking such a view is realist about ontology, but it limits the role that Realism can play in a theory of reference by saying that any simple deference to the joints of nature will result in concepts that are overly polysemous for some purposes. The kinds of cases I've offered may conform to the letter of some tenets of Outerism, but not to the spirit. Trumpeting your Realist horn is misleading if semantic theory must also import aspects of Nominalism. Likewise for the other contested theses.

Option (3) tries to defuse the cases under consideration by saying that my analyses of them are wrong. Suppose one decides that steam is not water. Perhaps this is just a mistake. Perhaps the natural kind determines reference in a way that is indifferent to such decisions. Suppose Moral Realism is false. Then it may follow that my moral concepts are confused, even if I elect to treat morality as if it were a response-dependent kind. These moves strike me as rather desperate. I'm not just appealing to intuition. My aim here is not to provide counterexamples to any semantic theory that makes counterintuitive claims about what our concepts refer to. Rather, I am trying to point out that decisions of various kinds seem to be a pervasive part of conceptual practice. I think naturalistic semantic theories have failed to acknowledge the extent to which concepts undergo shifts and refinements that reflect our goals, needs, whims, and theories. To stipulate that these things have no impact would suggest dogmatic allegiance to Outerism.

In conclusion, I have been trying to suggest that the deepest fault line in contemporary psychosemantics – the division between Innerism and Outerism – is shallower than we have sometimes appreciated. Innerists underestimate the role of the world in determining content, and Outerists underestimate the contributions of the organism. The most satisfying approach may be Inside-Out. I think Millikan has the resources to provide such an account. In fact, her theory posits much more on the inside than her main competitors', and this may be one of her most important contributions to the field. Despite much emphasis on Outerist themes, Millikan assigns a central place to the inner. Her work might best be seen as bridging this unfortunate gap.[1]

---

[1] It is customary to write acknowledgments at the end of an essay, and I can say that my sense of gratitude has never been more heartfelt. Dan Ryder provided an enormous amount of patient, helpful feedback; when I articulate an objection in the essay, it's usually just a rephrasing of a worry that he put to me in his comments. Ruth Millikan also provided invaluable feedback, on a talk version, and convinced me, along with Dan, that there is nothing here that poses a serious threat to her philosophical views. But Ruth also deserves thanks of another kind, for years of inspiration, which helped draw me into this subfield of philosophy.

Reply to Prinz

Jesse Prinz offers an interesting discussion of what he believes are excesses connected with current mainline meaning externalism and kindred doctrines that originated, he says, from the revolutions in philosophy of language of the 1970s and 1980s. But I must begin by fussing a little that the fuzziness of his general claim that "psychological processes" or "what's inside the head" "determines" or "contributes to" or "affects" meaning makes it very hard to know how much he is jousting with windmills.[1] Have any externalists really held that we could have concepts without any "role being played by psychological states?" Surely even the most radical "outerists" don't hold that the fact that there is a world out there, say, with natural kinds in it, *constitutes* our having concepts of those kinds without our brains or our minds, or whatever, having to do something about it. Obviously some kind of psychological/neural mechanisms must be involved in determining which of the things out there we are thinking of when. But which kinds of things that our brains or minds might do to make these connections, whether with real kinds or not, would be "outerist" and which "innerist?" I find Prinz unclear on this matter. A sensible move might be to exclude from "innerism" reference to mechanisms that realize *abilities* to deal with prior structures of the world. For these mechanisms clearly cannot be constructed arbitrarily; they must be framed within limits imposed from without by those world structures themselves. Prinz's examples of innerism, however, do not respect this boundary.

I must also fuss that it was not part of the classical meaning-externalist program, nor has it been part of my externalist program, to discuss the roles of all kinds of terms in natural language. Kripke talked about proper names, Putnam talked about natural kind terms of a certain limited sort, I have talked about terms for "substances" and for "properties," carefully defining these (quite separate) notions. What is going on with various evaluative terms (Millikan 1996a, and LBM, ch. 9) and with functional categories (see my

---

[1] I feel I must warn that a number of his references to my own position seem to me to be misleading or misreadable.

reply to Braddon-Mitchell in this volume), for example, have been considered additional issues by pretty much everyone, I think. Nor, as Prinz does note, does anyone suppose that there are no lexical items with analytical meanings. My own statement in Millikan (2010), for example, was only that "[i]t is my belief that all empirical terms either fail to have any handed-down intensions, or that what is handed down are criteria the basic elements of which themselves have no defining intensions" (para. 6.2). (In LTOBC, ch. 15, there is also a discussion of the difference between "analytical concepts" and "synthetical concepts," though I don't wish to mix words with concepts in mentioning this.)

Prinz says his impulses toward externalism derived from the literature of the 1970s and 1980s. Mine were developed during the 1960s (Millikan 1969, *Empirical Identity*) and they depart from much of the later externalist tradition and also from Prinz's own assumptions in two ways that are important here. One concerns the relation between words and concepts. The other concerns the nature and importance of the distinction between identifying and classifying. I am very grateful to Prinz for the opportunity to try to make these divergences plainer, for both are fundamental, nor have I brought out or emphasized them clearly enough at the right points in previous writings.

Perhaps the most prevalent obfuscating proclivity of contemporary philosophy of mind and language is constant vacillation between talking of words and talking of concepts. Words cannot be used alone to label or pick out concepts: "the concept *dog*," "the concept *purple*." Focusing here just on those words and concepts that have natural-world extensions, an extensional term in a public language corresponds to at least as many concepts as there are people who know it. This is because, putting it simply, a concept is a particular, not a universal. It exists in a particular person over a certain time (explained a bit more clearly, finally, in Millikan 2011). Moreover, various different concepts may be expressed by the same person using the same word on different occasions.

Even when it is typical for a word to be used to express only one concept in each user's idiolect, what is common to the various concepts expressed is likely to be only extension. And there are, of course, ways of *classifying* concepts, the most obvious and useful being, exactly, by extension. Given a word, such as "London," understood as meant with a certain referent/extension, you can often use it uniquely to describe a certain person's concept, say, "*Pierre's concept of London.*" But should Pierre turn out to have two concepts of London (say, he expresses one using the word "London," the other using "Londres") this will not distinguish between them. Alternatively, you could mean by "*Pierre's concept of London*" the one he himself expresses with/understands by the word "London." But this designation could turn out to be equivocal in another way, for perhaps Pierre thinks there are two cities called "*London*," people saying "*London*" sometimes talking about one and sometimes the other, depending on context. I have argued that the same word expressing concepts having exactly the same extensions in two people's idiolects will usually express concepts governed by very different "conceptions" or ways of identifying that extension, including that one person's concept may be analytical while the other's is synthetical. (See, for example, Millikan 2010, n19 on "vixen.") Easy cases are where the term itself has a conventional intension (it is analytical) but where many users who reidentify the extension in part via recurrence of the term itself do not grasp this intension (Millikan 2010, para. 8).

The above bears on many points in Prinz's discussion, most obviously perhaps on his concern with cases in which we must make decisions about how to use words. I appreciate that the examples Prinz gives us are not mere possible-worlds cases but real ones, cases where there actually are different things in the world that might usefully be thought of and talked about. For example, he wants to know (concerning some of the things I would call "substances") whether a person with dissociative identity disorder is "one person" or "several," whether a person with advanced dementia is still "the same person," whether his favorite restaurant continues to exist after a change in its location and chef, whether steam is "water," whether a bagel is a "roll." In each of these cases it is clear that the particular portion of the whole substance being wondered about is neither imaginary nor arbitrary, but a portion that has a good number of interdependent properties that diverge from those generally characterizing the whole. In each case these portions can be considered to be rough substances (in my sense) in their own right or to be what I called "humps" or "bumps" on the "clots" constituted by the larger wholes of which they are portions (Millikan 2010). Thus Prinz is respecting the outerist insight that we use common (substance) words not to cover arbitrary classes but rather empirically discovered clumps, recognition of which can support inductions. But if one can wonder whether to count steam as "water" or bagels as "rolls" and so forth, obviously one must have a concept of steam and a concept of bagels as well as a concept of water-the-chemical-compound (a concept with that extension) and a concept that covers other rolls as well as bagels. So the problem must not concern concepts but only words. Nor would the decision one makes on what word to use on this occasion or that (sometimes I include ice when I say "water," sometimes not) have any effect on these concepts. Indeed, whenever one knows exactly what the situation is, when the thought is clear, as when wondering about Theseus' ship(s) or about tele-transport or whether to apply the (philosophical) term (of art) "person," how to apply concepts cannot be what's at stake. Everybody understands what happened, the question is only what to say. (Does one ever *decide* what concepts to have? One simply has/acquires them or one doesn't, I think.)[2]

Causing trouble here, I believe, is the assumption that the job of a concept, and also of a word, is to *classify*. A classification system fails if it does not draw determinate boundaries between all the actual cases, putting each either within or without each category. (Prinz's outerist sensitivities save him from the more common and radical view that these boundaries must be drawn for all *possible* cases.) A classification system operates over a domain of individual items to be classified, disposing these according to various specified properties that each item must either possess or not. Its job is to determine for each of these individual items (perhaps themselves properties) whether or not it has each of the specified properties, hence whether it falls inside or outside of each category. There are two central difficulties with the idea that all extensional concepts are classifiers. The first is a failure to recognize that the purposes of our concepts, and of the words with which we express them, seldom require them to have determinate or sharp boundaries. A great deal

---

[2] I have argued that being able in practice to recognize a word in context as the same word bearing the same meaning when one encounters it again is already to have a concept of its referent (OCCI, ch. 6, VOM, ch. 9, LBM, ch. 10).

of vagueness, even equivocation, in both concepts and words is compatible with their ordinary, powerful, and successful usage. The second is that any act of classifying presupposes that one possesses a prior concept both of the item to be classified and of the properties relevant to its classification. Obviously then, classification cannot take us all the way down. At some level concepts/thoughts must identify before they can classify. I will clarify these matters in turn.

Words and concepts are not like cake knives or cookie cutters. Rather, they are *anchored* to various striking features of the world such as substance "clots" and property extremities or "peaks," elastically tied to these features, constrained but also allowed much latitude. Paragraph 6.3 in Millikan (2010) is a discussion of the way many of our words for substances, properties, and ways (adverbs) are anchored to natural "peaks" such as pure water, perfect roundness, flatness, emptiness, or baldness. What gets called "water" is whatever is close enough to pure water for the communication needs of the moment, what gets called "round" close enough to pure circularity, and so forth. Similarly, where words are anchored to historical kinds, how close to a paradigm of the kind something must be and in what respects to be appropriately referred to by that word depends on the needs of the moment. If it is unclear whether "water" or "round" or "cat" will be understood within the right range in a certain context, we use more words or different words. We explain. Only the law tries to force the issue of boundaries.

The same is true for what is *thought* of as being water or as being round or a cat. Depending on what the thought is to be used for, something can be grasped as being close enough to the extreme, or close enough to various paradigms in enough ways for one's current practical or theoretical concerns, without a grasp of exactly *how* close or *what* ways. Everyday thinking is pervasively and deeply inductive; indeterminacies/uncertainties fuzz all its edges. Nor, although this model is often useful, should we analogize concepts too closely to discrete, countable words. The paradigm of a pet fish, as Fodor and Lepore (1996) observe, is neither a paradigm pet nor a paradigm fish. And *contra* Fodor and Lepore, most concepts people would express with "pet fish" would not merely be compositional either. They would be of a certain bump on the clot that makes up fish, a bump that pulls in many traits of *pet* with high probability as well. Substance concepts may sometimes lie roughly on a continuum, following the course of a substance clot that changes many of its dimensions over property space in a continuous, coordinated way. (Think of overlapping connectionist nets conceptually following the track of babies turning into little boys, teens, young adults, old men.) Property concepts may also lie on a continuum or overlap one another.

Finally, if a thing is to be classified, first it must be identified, and each of its relevant properties must be identified as well. Identifying is moving from proximal sensory inputs to recognition of distal properties, objects or kinds, and so forth, of which these inputs are signs. This movement may involve elaborate processing including sophisticated explicit inference, or it may involve only very quick and hidden neural processing. The difficulty to be surmounted is that the same distal property, object, or kind may have any of numberless different proximal impacts on the sensory systems, depending on which senses are involved, on distance, direction, mediating circumstances, disturbances or interferences, occlusions, all in uncountable variety. For starters, consider the work that has gone into trying to figure out how color constancy, phoneme constancy, shape constancy, and object constancy are achieved. Consider how many different ways you might recognize the presence of a piece of the dog clump – by a bark or whine, by the feel of any of many different body parts, by looks in different lights, at different distances, from different angles, in different postures, in different breeds. Consider how many different kinds of stimulations impinging on your senses may result in your recognizing an individual member of your family, from the front, from the back, from different distances at different angles, sitting, standing, walking in the distance, in daylight, at dusk, by lamplight, in spotted sunlight, by their voice nearby or over the phone or across a crowded noisy room, by their clothes, by any of thousands of descriptions, by their handwriting, by what they are doing (in context), by visual or auditory tracking as they move or as you move, and so forth.

Contrast Quine (1960, 83) : "The denizens of what we have been calling the child's quality space are, we can as well say, the stimulations; what needs to have been peculiarly 'within' the child is just the spacing of them." And contrast Prinz: "reference is [sometimes] ... determined ... by descriptive components ... A sensory impression might do the trick...." Identifying is not classifying sensations; it is reaching through a radical diversity of sensory impressions to find the same distal object again, and sorting through similar or identical sensory impressions to find the diverse distal objects behind them. To understand how evolution and learning could discover how to do this – to understand how empirical concept formation works – requires, I have argued, an entirely separate and new kind of natural epistemology. (See my reply to Braddon-Mitchell, with references, this volume.) Certainly you can decide what to use your words to mean, but you can't decide what there is to mean by your words. You have to find the world before you can cut it up with categories.

(Prinz is also concerned that the externalist position implies that "we lack privileged access to what our concepts represent." On knowing what our concepts represent, see, for example, Millikan 1993c, 2011, and my reply to Cynthia and Graham Macdonald in this volume.)

11

THE EPISTEMOLOGY OF MEANING[1]

CYNTHIA MACDONALD AND GRAHAM MACDONALD

The externalist is obliged to accompany claims about the ontology of meaning with a plausible epistemology of adequacy for empirical concepts. She must construct an epistemology of meaning to support her claims in the philosophy of mind.

(LBM, 72)

Introduction

Ever since Hilary Putnam proclaimed that "meanings ain't in the head," philosophers have worried about how it could be, if semantic externalism were true, that we know what it is that we mean by our words when we speak. Traditionally, it has been assumed that linguistic meaning must be transparent to speakers of a language, so that, whatever one's words mean, one must, in the normal case, know what they mean.[2] This claim is even more attractive, and seems more obviously true, when applied to thoughts: whatever the contents of one's attitudes are, one must, in the normal case, know what they are. Both semantic and psychological content externalism pose a challenge to these claims. If linguistic meaning and psychological content are individuation-dependent on empirical factors beyond one's mind about which one may know nothing, then a conclusion appears inevitable: one may not, in the normal case – the case in which one's sayings and thinkings are world-involving – know what one is saying or thinking, since one may be ignorant of the empirically discoverable external factors that determine the contents of one's sayings and thinkings to be what they are.

---

[1] This chapter continues a conversation on this topic with Ruth Millikan which began after she delivered "On Knowing the Meaning" (Millikan 2010) at Queen's University Belfast in 2007. We are grateful to Ruth for many such conversations, and for her friendship over the years.
[2] For a recent defense of this view see Jackson (2006).

Some regard the traditional claim as so incontestable, and the externalist conclusion so ineluctable, that they are driven to reject externalism. Others find the arguments for externalism to be so persuasive that they reject the traditional claim. Ruth Millikan has, in our view, presented one of the most powerful cases for the rejection of the traditional position and for the acceptance of a radical externalism. In a number of groundbreaking works she has articulated a detailed vision of how mind and language work. One aspect of her view is a rejection of neo-Fregean orthodoxy in the philosophy of language, a rejection of anything like a Fregean sense, *Sinn*, in her account of the meaning of linguistic items. The radical externalism Millikan advocates requires, she thinks, that for the most part,

> ... the public meaning of a simple referential term typically includes only its stabilizing function and its reference, since the stabilizing function depends almost entirely on sentential context, the public meaning is essentially *just* reference. (LBM, 66)

A corollary of this is the downplaying of the importance to meaning of the contents of speakers' or hearers' psychological states. Here Millikan stands in stark contrast to two important traditions in the philosophy of language, the Gricean approach to meaning, and Davidsonian truth-conditional semantics. Both of these ways of theorizing about meaning make essential reference to the contents of psychological states of the users of language, though they do this in very different ways.

Millikan has discussed at length the Gricean program in several of her works (e.g., LBM, ch. 10). She has spent less time discussing the Davidsonian one. In this chapter we want to compare her anti-psychologistic theory of meaning with an externalist program in the Davidsonian tradition, that of John McDowell. There are two reasons why we think that this comparison is worth pursuing. First, the programs of both Millikan and McDowell are similar in being radically externalist with respect to meaning. But second, they are markedly dissimilar in the way in which an appeal is made, or not made, to the intentional contents of the psychological states of users of the language. Of particular interest to us here is Millikan's view that her externalism is incompatible with the postulation of Fregean senses.

In what follows, we first adumbrate Millikan's approach, highlighting what we consider to be the essential difference between the Millikanian and Fregean before going on to identify the source of dispute (section 1). We then argue that McDowell's notion of a *de re* sense can both serve the Fregean purpose of rationalizing speech and satisfy Millikan's externalist requirements, one that is capable of reconciling Fregean sense with teleosemantics (section 2). A consequence will be that users of a language can, consistently with a radical externalism, be said to know what they mean by their words. We conclude by mentioning some of the theoretical advantages of the appeal to Fregean senses.

Section 1

Millikan thinks that the public meaning of a referential term can be specified in terms of its reference alone; there is no need, on her account, to postulate a realm of senses. Frege, she says, "... made a mistake in positing something common beyond *Bedeutung* that is grasped by the mind of every competent speaker using the same unambiguous linguistic form" (LBM, 66). Note, however, that the rejection of Fregean-style senses does not involve a behavioristic rejection of *any*thing in the mind accompanying talking and listening: what is at stake is whether there is necessarily anything in common in the minds of the speaker and hearer, beyond grasp of a term's reference, such a common element bearing some relation to Fregean senses (a "grasping" of the sense).[3] In order to appreciate how Millikan deals with the interaction between meaning and mind, we need briefly to identify those features of her overall theory of meaning that are essential to understanding her epistemology of meaning.

Millikan organizes the domain of linguistic meaning into three types:

1. Conventional linguistic cooperative functions (stabilizing functions);
2. Conventional semantic-mapping functions (in the mathematical sense of 'function') which determine truth and other satisfaction conditions; and (crucially)
3. Conceptions and conceptual components: "… methods of identifying … that govern individual speakers' grasp of referents and of truth and satisfaction conditions, hence help to determine their dispositions to use and understand various conventional language forms" (LBM, 54).

Before elaborating on what is involved in (1) and (2), it is worth pointing out that the conceptions of Millikan's theory mentioned in (3) can in at least some respects play the role of concepts in the neo-Fregean tradition: the disavowal of a concept *common* to all users of a name (say) is compatible with their being *different* conceptions available to individual speakers. Given that there is no need for these conceptions to be common across a linguistic community, there is no requirement that the sense I associate with 'Mark Twain' be the same as the one you associate with that same term. Equally, the conception I have that accompanies my use of 'Mark Twain' may be different from that accompanying *my* use of 'Samuel Clemens.' In this case the difference in conceptions may be sufficient to explain my otherwise peculiar linguistic behavior. If, for example, I know that 'Mark Twain wrote *Huckleberry Finn*' is true but do not know that 'Samuel Clemens wrote *Huckleberry Finn*' is true, then I am liable to ask questions about the identity of Samuel Clemens, ones that would be difficult to make sense of if I knew the relevant identity (such as, for example, 'Was Samuel Clemens a writer?'). What explains this ignorance? The Fregean answer is that because the names have different meanings, the speaker does not know that they co-refer. To put it as neutrally as possible, the psychological profile of the speaker must include different 'markers' for the names 'Mark Twain' and 'Samuel Clemens.' Crucially, the Fregean finds it theoretically fruitful to treat these markers as meanings, or senses. The Millikanian appeals to conceptions to explain this ignorance.

Turn, then, to (1). If the coordination brought about by a convention is important and obvious, it will proliferate without the need for speakers/hearers to think about each others' thoughts. The conventional functions of a linguistic form will remain stable only

---

[3] The cautious way of putting this is due to Frege's rejection of psychologized senses.

if it continues to serve the interests of speaker and hearer. The function has to be performed only enough times to avoid extinction.

For the purposes of semantics two tokens are of the same linguistic type "... only if they have been copied from the same pool of tokens reproducing in the same language community. They must be segments from the same historical lineage" (LBM, 61). New meanings can be created using the same forms when a new use is introduced which generates a new coordinating function, and a new stabilizing function.

Turn now to (2). Communicative forms work in part by mapping; "They correspond to states of affairs in accordance with semantic-mapping functions that have been determined by convention" (LBM, 63). Descriptive communicative forms are designed to produce in hearers true beliefs, "... but a true belief will be formed by normal mechanisms only if the sentence corresponds to a world affair in accordance with its conventional mapping function" (LBM, 63). The semantic mapping function of a sentence determines the satisfaction condition, but not vice versa. ('It is raining' differs from 'There is rain here now' in its mapping functions, but not its satisfaction condition.) As we have noted, Millikan takes the consequence of (1) and (2) to be that public, or shared, meaning amounts to reference alone. As a result, the conceptions mentioned in (3) cannot occupy the role of Fregean senses.

So the picture is this. We may, and often do, have different ways of identifying the referents of our terms, these different recognitional pathways being our conceptions. Conceptions bestow on us a recognitional ability, and provided that this ability is reasonably robust (i.e., that our identification is not a matter of chance), we have a concept. But the concept is just the ability, so we share a concept insofar as we have the ability to identify the same referent, even though these abilities may be grounded in different psychological capacities. Our psychologies are backgrounded, and as far as knowledge of public meaning is concerned (knowledge of the stabilizing function of the terms), we need not have any such knowledge.

The contrast between Millikanian conceptions and Fregean senses seems stark. As we have depicted the contrast between the two, the Fregean is committed to the idea that we have knowledge of the meanings – senses – of our terms, and that only such knowledge can explain how it is that we understand one another when we communicate linguistically. Millikan denies that that we have knowledge of Fregean meanings, and insists that they do no useful work in explaining either how language functions or how it is that we understand one another when we speak and listen. Instead, all that is needed to explain how language functions is an account of linguistic convention that is tied to (sufficiently) successful public performance; and all that is needed to explain what we are doing when we speak and listen is a notion of idiosyncratic conception, not a common Fregean sense. If Millikan is right, there is no useful work for Fregean senses to do, given that linguistic actions can be explained without them. What can the Fregean say in response to this?

The Fregean starts with an account of what it is that we understand a person to be doing when they utter a sentence. Given that the uttering is an action, it has an intentional description, and the claim is that this description will invoke semantic properties of the sentence: In uttering s, Mary said that p, where p provides the content of the sentence uttered and it is assumed that Mary intended to say that *p*. A hearer will, in the successful case, understand that that is what Mary said, and will assume that Mary intended to say that. The kind of understanding that is at stake here is *semantic* understanding, the kind of understanding that is concerned with truth. On this account, any rational explanation of Mary's uttering *s* will advert to Mary's intention to say that *p* and to her belief that in uttering *s* that is what she will be doing.

Now, Millikan is in agreement with us that this is what the Fregean notion of sense is meant to capture. However, she argues that Frege's senses cannot capture what is involved in semantic understanding and rationality. Let us see why.

According to Millikan, Frege's senses are intentional contents that have a number of fundamental characteristics. First, and perhaps most importantly for her purposes, they are "intermediaries" between mind and world that are transparent to, because directly apprehended by, the mind:

> Frege's senses ... are his "intermediaries," given our gloss, for beliefs about the world. Graspings of senses of the kind Frege calls "thoughts" are what stand between mind and world, making errors in thought possible when harnessed by mental acts of assertion. (OCCI, 129)
>
> The fact of sameness or difference in content can be read off the sameness or difference of thoughts and vice versa. Thus for the rational thinker no misidentification of thought content should ever occur. Contradictions show up right on the surface of thought so that no inconsistencies should occur either. The relation between thought and its content is perfectly transparent, indeed, it entirely disappears. There is no vehicle moving the mind but the very content itself. (OCCI, 131)

In Millikan's terminology, Frege not only externalizes sames (effectively, what the thesis that sameness of sense determines sameness of reference amounts to), but also externalizes same*ness*, with the consequence that "... if senses are the same, then the corresponding referents are necessarily *conceptually visaged* as same or necessarily available to the rational mind *as* same" (OCCI, 130; our emphasis). This, says Millikan, is why, for Frege, a subject cannot take contradictory attitudes toward objects under the same mode of presentation.

A second important, and related, characteristic of senses is that they are what constitutes "having a determinate object in mind." A third, as the second quotation above states, is that they are what "moves the mind" (OCCI, 129). Finally, sameness of sense determines sameness of reference (what Millikan calls "externalizing sames"), although difference in sense does not determine difference in reference. Millikan takes the first and fourth of these characteristics of Frege's senses – specifically, that senses are transparent to the mind and that sameness of sense determines sameness of reference – to lead immediately to semantic internalism about thought contents, this being the view that such contents are autonomous with regard to (i.e., are not individuation-dependent on) factors in the world beyond the mind. And it is this perceived commitment to semantic internalism, it seems, that leads Millikan to conclude that Frege's senses cannot capture what is involved in semantic understanding and rationality.

Fregeans such as McDowell and Sainsbury have objected to a number of Millikan's claims about Frege's senses, mainly in the context of her discussion of "meaning rationalism" (McDowell 2009, Sainsbury 1997, 2002). Both remark, for example, that Frege's views about senses first and foremost concern knowledge of ordinary objects in the world rather than knowledge of the semantic features of senses or of what they constitute – thought contents – themselves. That is, Frege's senses are invoked in order to explain, not knowledge of the contents of one's own mental states, but knowledge of ordinary objects in the world. To this extent, Sainsbury claims, the doctrine of "meaning rationalism" that Millikan ascribes to Frege (WQ, ch. 14), concerning as it does claims about subjects' knowledge of the semantic features of their thoughts or language, is not one that he himself finds any basis for in Frege's work. In a similar vein, while McDowell concedes that Frege did hold a version of "meaning rationalism," he claims that it is much weaker than Millikan's version, importantly in being restricted to transparency of sameness and difference in content elements present to the mind *at the same time*. Both McDowell and Sainsbury insist that Frege's view that senses are to be individuated in accordance with the requirement that the rational mind cannot take contradictory attitudes to a single object under the same mode of presentation specifically concerns attitudes taken toward that object *at the same time* and that the view has no clear implications for attitudes taken toward that object at *different* times. As a result, they argue, Frege is not guilty of supposing that one cannot make errors in thoughts about a single object over time, or have confused thoughts about that object, even to the extent of taking contradictory attitudes toward it under the same mode of presentation at different times.[4]

Our purpose in raising these objections is not to engage in a debate about Frege's claims about senses or what he himself took the commitment to senses to entail, nor is it to provide a justification of Fregean senses, since, like Sainsbury, we think that this project could not be carried out within the space of a single essay. Rather, we are interested in considering how a Fregean – someone who appeals to a notion of sense as what is grasped by rational creatures when they speak and understand language and what explains agreement in linguistic communication – might respond to Millikan's reasons for claiming that they can serve no useful explanatory role in helping to make intelligible the rational speaker's speech and understanding of language.

Millikan's reason for thinking that Fregean senses cannot carry out the role that they are designed to carry out seems to be motivated largely if not exclusively by the view that Frege's senses commit him to semantic (and content) internalism. As we see it, however,

---

[4] In this connection it is perhaps worth mentioning that Sainsbury explicitly takes senses to be properties of utterances, and accordingly of expression tokens, thereby allowing utterances of intuitively non-synonymous expression types to have the same sense, so that, for example, when Graham utters today 'The bank is closed today' and I utter tomorrow 'Graham said that the bank was closed yesterday,' Graham's token of 'The bank is closed today' and my token of 'The bank was closed yesterday' can have the same sense. If sense is a property of token expressions, it is not hard to see how it might not be manifest to a subject that two of her utterances at different times about the same object, even when they present that object under the same type of mode of presentation, express different tokens of the *same* sense type. Similarly, McDowell's *de re* senses are first and foremost properties of token expressions, constancy of linguistic meaning among different tokens of the same expression type being determined by these.

commitment to Fregean senses is not incompatible with semantic externalism, since there is a way of viewing Fregean senses that stems from the work of McDowell that can help illuminate how a Fregean who is a semantic externalist can supply a useful explanatory role for the notion of sense. Allowing for that role to be played by Fregean senses (thus understood) need not be viewed as incompatible with the central core of a Millikan-type view of language use and understanding.

McDowell, like us, discerns in Millikan's arguments the assumption that Frege's senses require a commitment to semantic internalism – or at least a commitment to the rejection of semantic externalism. According to McDowell, semantic understanding and rationality, as Millikan rightly observes, while being something that Frege wants the notion of sense to help explain, is not something that one's "intact mind" (WQ, ch. 14) – the head that is in good mechanical working order, when nothing is wrong or broken – can do, since that mind cannot even think. As he sees it, however, Millikan argues that since this is not where semantic understanding and rationality can be located, and yet this is where Fregean sense is located, Fregean sense cannot do the kind of work it is meant to do, namely, explain what a rational thinker understands when they understand a sentence. The only alternative that she can see, the alternative that brings the world into the picture and with it semantic understanding, must be a system that goes beyond the "intact mind," a "head-world" system.

In supposing that Frege's senses are something that the intact mind – effectively the machinery within the head, devoid of semantic connections to factors in the world beyond it – must have access to, Millikan interprets commitment to Frege's senses as entailing commitment to semantic internalism. It is not that Millikan supposes that the machinery of a mind in good working order does not require the presence of factors beyond the head – after all, brains, like kidneys, to be the kind of biological organs that they are, must be individuated in terms of causal-etiological factors in the world beyond the bodies in which they carry out their biological functions when in good working order, functioning biologically normally. So it is not that Millikan is assuming that the "internal mechanics of having one's mind on objects" (McDowell 2009, 272) can occur in a world in which there exists nothing beyond the mind of a person. Millikan herself is an externalist – and a semantic externalist at that. What drives her rejection of Fregean senses, we suggest, is the radical transparency of sense that she attributes to the Fregean: she assumes that Frege's senses are items to which the mind – viewed in terms of its internal workings alone – must have infallible access, and so must be items that really do "stand between" mind and world. This makes the relation between senses and the items that they present seem to be contingent and merely causal, not essentially semantic.

There is, we believe, another concern at work in Millikan's rejection of Fregean senses. Millikan conceives of such senses as necessarily very finely individuated, so finely that they could not be what is shared by, grasped by, speakers when they communicate linguistically. Fregean senses impose a requirement on speakers that they very often do not and cannot meet, namely, that they share a common "pool" of information associated with an expression, when very often speakers know virtually nothing about its referent yet still communicate effectively with it. Further, speakers can differ wildly in the information they associate with the terms they use, and their information can be, and often is, idiosyncratic, peculiar to their uses of those terms. Fregean senses cannot serve the role of either type of situation. On the one hand, speakers who know virtually nothing at all about the referent of a term communicate perfectly effectively with the use of it, and so do not need Fregean "modes of presentation" to do so. On the other hand, speakers who have idiosyncratic information that they associate with the referent of a term, as when my use of 'mama' connects me with information about my mother, and your use of 'mama' connects you with information about your mother, manage to communicate with the use of the term 'mama' even though their information, their "conceptions," to use Millikan's term for such information, is so different as to have virtually nothing in common. Fregean senses, construed here as idiosyncratic information, cannot do any explanatory work here either, since communication succeeds despite the lack of shared information.

Section 2

In Millikan's defense it must be admitted that there are those who adopt a Fregean internalism. Their thought might be: it is only under a mode of presentation that a *Bedeutung* is "present to the mind," and if a Fregean sense is transparent to, and directly apprehended by, the mind, then one cannot, in grasping a sense, be having a *Bedeutung* in mind, since senses are distinct from referents, and referents are (in the case of world-directed utterances and thoughts) in the world beyond the mind. Senses must therefore stand between mind and world.

The problematic assumption in the above reasoning is that Fregean senses are the direct objects of thought and cognition and ordinary objects in the world are *not*, senses being only contingently and causally connected with such objects. But there is no reason to think that the Fregean is committed to this assumption. For one thing, doing so makes it virtually impossible to understand why a Fregean would introduce the notion of sense to explain what the rational subject grasps when she speaks and understands language, a semantic achievement. What we need here is another way of understanding the notion of a Fregean sense, one that does not assume that, in grasping senses, minds are thereby somehow blocked from grasping – indeed, grasping directly – objects in the world. And this is something that the notion of a Fregean *de re* sense, properly understood, can provide.

McDowell's views about how Fregean senses and the thought contents that they constitute relate to the objects they present, objects in the world beyond the mind, are essentially connected with his use of the notion of a *de re*, or object-dependent, sense. For McDowell, there could be no thought contents were there not *de re* senses. Such senses present objects to subjects in certain ways and are such that, were there no such objects to present in these ways, there would be no such ways of presenting them. So the view that senses – *de re* or any other – act as mental intermediaries between objects presented and thinking subjects is rejected; they are not "in" the mind, in contrast with being "in" the world. We do not grasp objects and properties by first grasping senses – entities in the mental domain – and then trying to connect them to objects and properties. That is to say, senses, and in particular *de re* senses, are not themselves objects, the grasping of which enables us to somehow reach beyond them to grasp things in the world.[5] They are ways *through* which, in thought, we are brought into direct contact with things in the world. The notion of "association" that McDowell has in mind when he says that "... for an object to figure in a thought is for it to be the *Bedeutung* associated with a *Sinn*" (McDowell 1999, 94–95) is no mere contingent connection, and this comes out clearly when considering the case of a *de re* sense.

In a response to Charles Travis, he characterizes the division between the realm of sense and the realm of reference in terms of "Frege's line" – the line that "... separates non-conceptual items, on the left, from conceptual items, on the right" (McDowell 2009, 259). Here McDowell clearly makes the distinction between the realm of sense (on the right-hand side of Frege's line) and the realm of reference (left-hand side). The distinction must, however, be understood against the background various other claims that McDowell makes about the contents of true thoughts. Chief among these are (1) that the content of a true thought is a fact, and (2) that facts are elements of the world. McDowell invites us to recognize that there are two ways of conceiving the world – as the realm of facts/sense, and as the realm of objects/reference – but that these are not two distinct realms, but rather two alternative conceptions of the world. Taking the constituents of propositions to be Fregean senses, then, of the following four claims:

1 In thinking truly we grasp facts.
2 Facts are propositional in structure.
3 Propositions are what we think.
4 Facts are what we think about.

(1)–(3) are endorsed, but (4) is rejected.

In distinguishing between the realm of sense and the realm of reference, McDowell does not endorse any two-world view of facts, on the one hand, and senses, on the other. On the contrary, he goes on to explicitly state that the kinds of expressions one uses to denote facts, "... expressions of the form 'the meat's being ... on the rug', [and] 'the meat's being ... underdone' ... characterize right-hand side items [or alternatively] belong in the realm of *sense* and hence can be *thought*" (McDowell 2009, 261).

Of course, this insistence that there are two alternative conceptions of the world in play raises the question of how they are connected. McDowell (1999, 94) says that objects – elements from the realm of reference/left-hand side of Frege's line – *figure in* facts – elements from the realm of sense/right-hand side of Frege's line. To understand the connection between the two realms, then, we need to know what "figuring in" amounts to. McDowell insists that we must distinguish the assertion that an object figures in content from the assertion that an object is a constituent of content (McDowell, 1986, 237). The reason for this is clear: we need to distinguish between *what* one thinks, a Fregean Thought or thinkable, and what one thinks *about*, when one thinks what one thinks. And once we understand this, we are explicitly prohibited from reading the claim that objects "figure

---

[5] "To construe knowledge of the sense of an expression ... as, at some different level, knowledge of (perhaps acquaintance with) an entity (the sense of the expression) seems ... gratuitous" (McDowell 1977, 175).

in" thoughts in the Russellian way on which objects are literally the constituents of thoughts, because we don't think objects, we think about them. The relationship between facts – elements from the realm of sense – and objects – elements from the realm of reference – is that of sense to referent. This is not a reductive account of the key "figuring in" relation; the relationship of "figuring in" is simply the relationship of being the referent of a sense.

It is for this reason that a Fregean *de re* sense neither "stands between" mind and world nor reduces to the realm of reference; while it presents objects in the world to the mind, it is neither "inside" the mind nor "outside" the mind, nor are the objects it presents "in" the mind since they do not constitute it. And so, we need a different analogy altogether from that of a mental intermediary in order to understand the way in which *de re* senses function in a Fregean account of what the rational subject grasps when she speaks and understands language.

How are we to understand the claim that Fregean *de re* senses are ways *through* which objects in the world are presented to the mind? Well, we might begin to do so by means of an analogy with a certain understanding of what the senses (visual, auditory, etc.) are. A common way of characterizing a sense modality is as an informational channel or avenue. Keeley, for example, does so as follows:

> Modality is an "avenue into" an organism. Question: What travels on an avenue? Answer: information about the physical state of the world exterior to the central nervous system (CNS). What constitutes an "avenue"? An evolutionary dedicated sense organ that converts energy into nerve impulses and conveys those impulses to the CNS. This captures the original sense of the term: the different senses are different "modes" of perceptual interaction with the world. (Keeley 2002, 6)

Keeley speaks of a sense modality as an "avenue *on* which information travels," and this might seem not to be quite what we are looking for, but the metaphor of an avenue is easily combined with the thought that avenues can take us through cities, towns, and suburbs, from one city to another, and so on; and a sense organ seems more appropriately viewed as something through which (or through the exercise of which) energy is converted, rather than as something on which energy is converted. Also talk of an avenue as an "evolutionary dedicated sense organ" might seem not to be quite what we are looking for, since *de re* senses, whatever else they are, are not credibly thought of as organs of any kind. Indeed, there may be sense modalities for which there is no clear organ of sensation (proprioception may be one such modality, and Keeley himself notes that medical students are typically taught that there is no sense organ associated with the vestibular sense (though his view is that there is such an organ, and that in general having an evolutionary dedicated sense organ which processes information about the physical state of the world external to the CNS is a requirement on having a sense modality)). In view of the controversial nature of the requirement that a sense modality have a dedicated sense organ through which it is exercised, it seems more appropriate to speak of a sense modality as a channel – or a conduit, or medium – through which information about the world travels, which may or may not be opened up or exercised by a dedicated sense organ.

So, a sense modality is an informational channel through which information about the world beyond the sentient organism travels. This characterization is one that we have reason to think Millikan herself is sympathetic to, since she describes perception – and language itself – as something through which one gathers information about the world, and speaks of it as a medium.[6] It is a characterization that in no way encourages one to think of a sense modality, or perception, or language, as something that acts as an intermediary, standing between the sentient organism, perceiver, or speaker, and objects in the world with which she interacts. Importantly, it is compatible with the view, endorsed by Millikan and McDowell alike, that language and its semantic features can be as direct a way of being put in contact with objects in the world beyond the mind as perception itself; and it is compatible, too, with the view that the medium by which one is put into such contact is not something the organism need be aware of in any way in the process of grasping objects in the world beyond the mind. Just as a drinking straw serves as a medium through which one's tongue is put into direct contact with liquid, a sense modality, and – if Millikan and McDowell are right – the vehicles by which the rational subject speaks and understands language, are mediums that put organisms into direct contact with objects in the world beyond them. In the case of language use and understanding, Fregean *de re* senses have a claim to be just those vehicles; they are a conduit, a medium, through which the rational subject is put into direct contact with objects in the world around her. Just as it would be wrong to view a sensory channel as a thing which mediates between the sentient organism and the world, the awareness of which enables the organism to be aware of objects in that world, it would be wrong to view a *de re* sense as a thing which mediates between the rational subject and objects in the world around her, the awareness of which enables her to be aware of those objects.

We are not suggesting that there are no important differences between a sensory channel and a *de re* sense, any more than Millikan is suggesting that there are no important differences between perception of the world through sight and perception of the world through language (as she puts it). But just as she takes these differences to be irrelevant to her claim that perception of the world through language is no less direct than perception of the world through sight, we take the differences between a sensory channel and a *de re* sense to be irrelevant to our claim that awareness of objects in the world through a *de re* sense is no less direct than awareness of objects in the world through a sensory channel. As a result, we do not see that a commitment to Fregean *de re* senses is incompatible with two core commitments of a Millikan-type account of language use and understanding: the commitment to semantic externalism, and the commitment to the direct awareness of the world beyond the mind through language use and understanding.

What of Millikan's second concern, mentioned in the concluding paragraph of section 1, that senses are too finely individuated to be shared by speakers when they communicate linguistically? We agree with Millikan that many uses of language are ones in which speakers

---

[6] So, when arguing for the view that language is as direct a way as perception to acquire information about the world, she says: "Interpreting what you hear through the medium of speech sounds is in relevant ways just like interpreting what you see through the medium of the structured light that strikes your eyes. Understanding speech is a form of perception of the world, as direct as seeing" (Millikan 2009b).

possess virtually no information about the referents of their expressions, yet they do manage to communicate effectively. And one cannot but agree with her that, whatever explanatory work Fregean senses can do, they cannot require speakers to be in possession of information about the referents of their expressions that speakers typically do not, and need not, have in order to communicate effectively. Where we disagree with her is in the claim that Fregean senses cannot respect this condition.

Fregean senses must indeed meet the constraint that they have sufficient "fineness of grain" to explain how a speaker could take contradictory attitudes to the same object at the same time, since that is a constraint on rational language use and understanding. However, this constraint can be met by a very austere theory of sense. One such theory might take the form of a truth-conditional approach to the theory of meaning, such as that articulated by Davidson, which requires assigning denotations to the singular terms of the language under study and satisfaction conditions to the predicates. A condition on the adequacy of these assignments is that, for every indicative sentence of the language, the theory must assign to it truth-conditions in such a way that so-called "T-sentences," theorems derived from the assignments plus a syntax for the language, come out true. However, because the logic used is standardly extensional, one can be faced with unsatisfactory but true T-sentences: for example, "'Grass is green' is true iff snow is white." And if the right-hand side of the bi-conditional T-sentence is meant to specify the truth-condition giving the meaning of the sentence mentioned on the left-hand side, this consequence is clearly unpalatable.

It could be that most of these ill-matching T-sentences will be eliminated as inadequate because theories yielding them as theorems will also yield false-T-sentences, such as "'Snow melts at 10° Centigrade' is true iff grass melts at 10° Centigrade." But it is unlikely that one could eliminate all mismatches of meaning, not without the use of further machinery. For example, one could not be confident that an assignment of denotations to 'Samuel Clemens' and to 'Mark Twain' would eliminate the (true) T-sentence "'Samuel Clemens wrote *Huckleberry Finn*' is true iff Mark Twain wrote *Huckleberry Finn*." And now the crucial question arises: Should a theory of meaning be judged as inadequate if it had this as a consequence? Note that the consequence for the truth-conditional semanticist is this: 'Samuel Clemens wrote *Huckleberry Finn*' means that Mark Twain wrote *Huckleberry Finn*. So a constraint on a putative theory of meaning is that the contents the theory ascribes to assertions must form part of the overall understanding the speaker has of their world, and this requires that those contents be provided using concepts grasped by the speaker. Such a theory can be expected to have, for each name in the language for which it is a theory, an axiom giving its denotation (e.g., for the expression 'Hesperus' the axiom governing its denotation would take the form "'Hesperus" denotes Hesperus'), which gives the sense of the expression 'Hesperus.' Similarly, the sense of the expression 'Phosphorus' might be specified by an axiom such as "'Phosphorus" denotes Phosphorus.' A speaker would be credited with possession of the sense of a name, say, the sense of 'Hesperus' if and only if they knew the truth expressed by that axiom (and not merely if they knew that the axiom is true) – that 'Hesperus' denotes Hesperus.

The Fregean constraint that prohibits attributions of synchronous contradictory attitudes toward the same object would prohibit attributing attitudes involving the sense of

'Phosphorus' to a speaker who asserts 'Hesperus is the morning star' at the time that that assertion is made. Why? Because although a speaker who knows that 'Hesperus' denotes Hesperus, and so knows the sense of 'Hesperus,' might know also that 'Phosphorus' denotes Phosphorus, and so know the sense of 'Phosphorus,' she may not know that Hesperus is Phosphorus, and so cannot be credited with knowing that 'Hesperus' denotes Phosphorus.[7] Since she cannot be credited with knowing this, even if she does know the truths expressed by the two axioms governing the denotation of, respectively, 'Hesperus' and 'Phosphorus,' she cannot be credited with any attitudes that involve the use of the name 'Hesperus' and the sense of 'Phosphorus.' She cannot help to make rational sense of her linguistic behavior in a situation in which she asserts both 'Hesperus is the morning star' and 'Phosphorus is not the morning star,' a situation that is likely to occur in cases where she simply does not know that Hesperus is Phosphorous.

We have been discussing Millikan's rejection of Fregean sense on the assumption that her own preferred way of thinking about meaning is to elide it altogether in favor of reference: meaning just is reference. This is what Millikan sometimes says, but she also suggests that meaning is "stabilizing function."[8] She applies this idea to linguistic forms (for example, the indicative mood), but it is also essential to her idea of a Reproductively Established Family (REF). An example from *Language, Thought, and Other Biological Categories* illustrates the view vividly.

Consider the fabulous tribes, the Hubots and Rumans. These two tribes inhabit the same environment, but are constituted differently, the Hubots needing (for their survival) some mineral supplement they find only in gold, the Rumans requiring (for their survival) a mineral supplement they find only in copper. Fortunately, both lumps of gold and lumps of copper are in plentiful supply in their territory, the only "problem" being that neither Hubots nor Rumans can tell them apart. Furthermore, they both have the (phonetic and orthographic) same term 'Golper' for the lumps they find around them. Question: Does H(ubot)-'Golper' *mean the same as* R(uman)-'Golper?' Well, an internalist notion of meaning would deliver the verdict that the terms in both languages mean the same: the users of the terms are in the same epistemological predicament, not being able to tell the substances apart, so the internalist would be inclined to make the extension of H-'Golper' be the same as the extension of R-'Golper,' the extension including lumps of both gold and copper. Given the epistemological situation (with consequent non-discriminating use of 'Golper'), this internalist will judge that the terms mean the same for Rumans and Hubots.

The teleosemanticist, however, asks the question: What makes the two terms proliferate and remain in the respective languages of the two communities? What purposes are served by the use of these terms? Given the different dietary needs, one can surmise that

---

[7] Put in the terminology of *de re* senses as informational channels, the sense of 'Hesperus' is one way *through* which that planet is grasped, the sense of 'Phosphorus' another, distinct, way. These are different informational channels, even if what travels through them is information about the same planet.

[8] Thus, she says, "Looked at this way, the function – I call it a 'stabilizing function' – of a conventional language form is roughly its survival value. It is an effect it has had that encouraged speakers to keep reproducing it and hearers to keep responding to it in a roughly uniform way, each relying on the settled dispositions of the others" (Millikan 2010, 53).

the Hubots use 'Golper' to communicate about lumps of gold, the Rumans use 'Golper' to talk about lumps of copper, and so the extensions of the terms in the two languages differ accordingly. What stabilizes the use of 'Golper' for Hubots is that it enables communication about (a sufficient number of) lumps of gold, such uses serving their dietary purposes, with the stabilizing function for Rumans being that use of the term picks out (a sufficient number of) lumps of copper. H-'Golper' and R-'Golper' tokens belong to different REFs, the source of their respective proliferations and stabilizations being different.

This fable is (and was originally intended to be) merely suggestive, and was never intended to provide a picture of meaning for human languages. It does, however, provide a glimpse of the strength of the teleosemanticist's approach to the determination of extensions: that determination proceeds from outside the heads of the users of the language.[9] In the fable as presented there is no need to postulate any knowledge of anything, let alone meanings, but we can embellish it as we think fit, so that the tribes have fully fledged intentional attitudes, a complex language, the full variety of human "forms of living," but are saddled with the original predicament: somehow they need gold and copper for different purposes, use the same (orthographic and phonetic) term for both, and cannot discriminate between the two. Now, we surmise that Millikan would say that we still have no need to postulate any knowledge of the meaning of 'Golper' in this scenario; what holds the various uses of the term together are the different grounds that hold copper and gold together as substances. That is, the explanation for the unity of use is determined by whatever it is in the world that makes the gold-properties (on the one hand) hang together, and the copper-properties (on the other hand) hang together. Given that there are different grounds for this coherence of properties in the two cases, this is sufficient to generate the verdict that H-'Golper' and R-'Golper' have different extensions, and the same difference in the grounds for the coherence of properties is sufficient to explain how it is that the terms are used in the respective communities. As Millikan puts it in recent work:

> Moreover, as a practical matter, their own agreement in judgment with others is often the only thing actually discerned by language learners and users as a check on their usage, hence the only factor (of this kind) controlling proliferation of an extensional term's tokens. So it cannot be part of such a term's stabilizing function to implant intentional attitudes towards its extension or members of its extension ... *as recognized in any particular way*, or *as thought of under any particular description*. Our basic extensional terms do not have handed-down conventional intensions. (Millikan 2010, 57)

We have argued that a Fregean *de re* sense need not be envisaged as requiring any particular way of describing, or method of identifying, the object to which it refers. The manner in which an object, or substance, is picked out by various users of a referential term does not figure in the slimmed down version of a Fregean sense envisaged by the austere Fregean account. And that version of what constitutes the sense of such expressions is

---

[9] "At the bottom level, then, what determines its extension is not the knowing of anything by its users" (Millikan 2010, 69).

surely radical enough to satisfy the most ardent externalist, given that the object itself contributes to the individuation of the sense. So our suggestion is that Millikan can afford to revise her rejection of Fregean senses, and permit users of a language to possess knowledge of meaning in a more full-blooded way than she appears to allow.

Conclusion

If sense is to play the role we have suggested, it is critical that one does not require a speaker's behavior to display a *theoretical* knowledge of meanings, which nearly all speakers could not be expected to possess.[10] Millikan is surely right that the myriad ways in which speakers identify the objects to which they refer cannot be serious candidates for the sense of the referential expressions used: if the folk do have knowledge of sense, it must be a common sense. It is right, too, to say that speakers cannot be attributed knowledge of what it is that makes the various properties used in such various ways of identifying a substance cohere together over time. But the informational channels that we have identified with *de re* senses need not require speakers to have such knowledge, and the intentional attitudes we seek to instill in our hearers by our use of a referential term might not have anything to do with our way of identifying the referent of the term.

Clearly the plausibility of any cognitive account rests on the idea that our linguistic actions are intentional under semantic descriptions. The underlying rationale is: no understanding of meaning, then no intention, so no rationality. In short, in assigning truth-conditions to assertions, one views the speaker as being engaged in a rational activity, and as a rational agent the speaker will be required to have a rational psychological profile. The assigned content of an assertion will need to fit into this profile as a content believed by the speaker, the "best fit" yielding the relevant truth-condition, where "best fit" is determined by it being the "most rational" way to augment this psychological profile given this action. Of course, any such procedure will need to make allowance for imperfect rationality, as well as deceit, irony, and other complicating factors. For present purposes, however, these are worries to be put aside.

One essential part of making sense of an action requires seeing the world from the agent's point of view, thus making it intelligible why it is that the agent is intervening in the world in that particular, perhaps peculiar, way. For example, the attribution to the agent of a false belief can help to explain why the agent acted in a way guaranteed to make them fail in their purpose. But this possibility, explaining failure as being the consequence of a defective understanding of the world, requires us to render that defective understanding as being the agent's defective understanding, one we comprehend as being an understanding of the world *as* appreciated *by* the agent. And this seems to require that the

---

[10] Exactly how the behavior that manifests knowledge of meaning is to be described is the subject of much disputation, most of it focusing on whether the behavior can be described in terms attributing a content to the behavior that only a language-user could possess (for a recent round of the argument, see McDowell 2007 and Dummett 2007).

content of an assertion be given in terms appreciable by the agent, on pain of not describing accurately what the agent was doing in uttering that sentence then.

Our concern in this chapter has been to consider how a Fregean – someone who appeals to a notion of sense as what is grasped by rational creatures when they speak and understand language and what explains agreement in linguistic communication – might respond to Millikan's reasons for claiming that they can serve no useful explanatory role in helping to make intelligible the rational speaker's speech and understanding of language. An important part of this project is to understand the role that sense might play in describing indirect discourse, because making intelligible the phenomenon of indirect discourse requires one to take into consideration the point of view of the rational subject, a subject's perspective on her actions in relation to those of others and the world around her.

It may be, as we have said, that the intentional attitudes that we set out to instill in others have nothing to do with our way of identifying the referent of a term. It would, however, be strange if our use of such terms were not intended to bring the referent to our hearer's attention. It is, as we have argued, the connection of sense with the rational lives of individuals that makes this a plausible claim, and one which we suggest Millikan can accept without jettisoning any of her substantive complaints against more ambitious uses of the Fregean framework. After all, we think that accepting such a Fregean sense requires no more than this: "For a descriptive extensional term to survive without change in how people are using it there has to be enough internal and external agreement on its application, *enough agreement on its extension*" (Millikan 2010, 56; our emphasis).

Reply to Macdonalds

Cynthia and Graham Macdonald urge me to accept what they term "a slimmed down version of Fregean sense," arguing that this addition would be pretty much compatible with the rest of my views. But I am uncertain what lacks they see in my current views that would call for this. I always assumed that the reason for introducing a Fregean notion of sense was to solve well-known problems that arise on a purely referential theory of meaning that also collapses linguistic meaning and speaker–hearer meaning into one. There is the problem (1) about the function of assertions of identity, (2) about the function of sentences asserting existence, (3) about empty names, (4) the problem that terms with the same referents or extensions cannot always be substituted for one another without a change in truth value, and – some would make this a separate problem – (5) the problem that the description of a psychological attitude into which a different term with the same referent has been substituted may fail to explain a behavior that the original description did explain. If these problems could be solved without introducing any sort of Fregean sense, I figured I was, as an anti-Fregean, home free.

So I described "conceptions" (OCCI, LBM, ch. 3) or "intensions" (my earlier term for conceptions in LTOBC) as ways, including both basic perceptual ways and ways involving inference (e.g., from prior knowledge of an identifying description) – ways of identifying incoming information about a thing (individual, property, relation, real kind and so forth) through a variety of different manifesting media. Conceptions, often a whole lot of them, I have said, are what support any individual empirical concept had by an individual person. But people who have concepts of the same thing generally have somewhat different conceptions of it; not all the ways that one person knows how to identify the thing will be known to the other and vice versa. (Indeed, maybe almost none. Consider Helen Keller.) Add to this that recognizing a word in context is often all that's needed for recognizing incoming information about that word's referent (allow properties and real kinds, etc. to be "referents" too). Thus it is possible that the only part of their conceptions of a thing that two people using the same referential word may have in common is their ability to recognize that thing through that word. (In LTOBC I called this special kind of ability to recognize a "language bound intension.") Language bound intensions/conceptions are the only conceptions of its referent that two competent users of the same term necessarily have in common – the only possible candidate for a required mutual "sense" in many cases.

Having laid out my theory of empirical concepts as involving the above claims (there is considerably more to the theory – my replies to Braddon-Mitchell and to Jesse Prinz, this volume, fill in a few more details), I proceeded, in various places, to address the Fregean problems (1) through (5) above. (1) and (2) were addressed perhaps most carefully in LTOBC, ch. 12; (3) was addressed in LTOBC and especially in OCCI, and again in Millikan (2006); (4) was addressed perhaps best in VOM, ch. 7, and (5) was addressed in OCCI, §12.6 and more carefully in Millikan (2006). Perhaps most relevant to the Macdonalds' concerns, the position I have taken on (5), on explaining behavior, is that knowing merely what a person's intentional attitudes are does not predict behavior. One would also have to know *at least* in what ways they are able to identify the objects with which they would interact – something of their conceptions of these (Millikan 2006).

Assuming, then, that the notion of conceptions can be used to solve all the classic Frege problems (the Macdonalds have not argued that it fails), and assuming that my conceptions cannot reasonably be taken to be Fregean senses ("Millikan is surely right that the myriad ways in which speakers identify the objects to which they refer cannot be serious candidates for the sense of the referential expressions used"), what then have I left out?

I think it is the following concerns that I am perceived as failing to address:

- … How it could be, if semantic externalism were true, that we know what it is that we mean by our words when we speak?
- … Whatever the contents of one's attitudes are, one must, in the normal case, know what they are….
- … This seems to require that the content of an assertion be given in terms appreciable by the agent.
- It would, however, be strange if our use of such terms [referential terms used when speaking] were not intended to bring the referent to our hearer's attention.

Why do the Macdonalds think that these concerns are not addressed on my view as it stands?

I think because they hold the prior view that the only way to answer questions of this sort is with a view of the relation between thought and world of the sort that John McDowell holds. On McDowell's view, in perception we have, as they put it, a "direct awareness of the world," and "the vehicles by which the rational subject speaks and understands language, are mediums that put organisms into direct contact with objects in the world beyond them." They take it that I could easily adopt such a view, indeed, that I already have adopted a "commitment to the direct awareness of the world," in saying, for example, that "[u]nderstanding speech is a form of perception of the world, as direct as seeing" and in rejecting Fregean senses understood as "intermediaries," or in a way such that in "grasping senses, minds are thereby somehow blocked from grasping – indeed, grasping directly – objects in the world." I need then to explain better my position on the relation of mind to world, and I am very grateful to the Macdonalds for offering me the opportunity to do so in such friendly circumstances.

I think that perception and thought are neither "direct" nor "indirect," as these terms have traditionally been understood. Turning to Kant for terminology, the objects-in-themselves referred to in perception are neither in the mind nor directly before the mind, nor do I hold that thought or language is a medium "that put[s] organisms into direct contact with objects in the world beyond them." The senses can indeed be considered as information channels, but information is carried by signals or signs that do not carry their meanings on their sleeves. These signs need to be interpreted. Never do we get a direct view of the thing-in-itself. In this respect, perception is never direct,[1] nor is thought.

But perception is not indirect either. It is not that something other than the object of perception is before the mind, some intermediary, so that an inference has to be made to its cause or so that it needs to be interpreted as a picture must be. On a representational theory of mind – a teleosemanticist is, of course, a representationalist – the representations that are perceptions and thoughts help *constitute* the thinking mind; they are not in front of it. To be used/understood by the mind as representations they need, of course, to be interpreted. But what this interpretation consists in is the appropriate guiding of actions, making of decisions and inferences and so forth, not some kind of reflective awareness. (Phenomenal objects are the objects of a false theory. It is as though philosophers looking through a window thought that what they were seeing was painted or projected on the inside of the window pane and then proceeded to "describe" what they understood to be the two-dimensional patterns and colors on that inside surface (see Millikan forthcoming b.)

How then do we "… know what it is that we mean by our words when we speak?" A help here is, first, to keep questions about the meanings of words in a public language separate from questions about what a speaker means with her words, for these can come apart. Conventionally, stabilizing functions and so forth are aspects only of public language meaning.[2] There is, of course, the question of how a public language is learned by new speakers so that they come, for the most part, to mean what their words mean, that is, how the purposes of new speakers in speaking come to accord with the stabilizing functions of their sentences. (Language learning is addressed in Millikan 2010.) But the basic question worrying the Macdonalds seems to be how speakers who mean anything, either by their words or *just in their own thoughts* when these involve external objects or properties, can know what they mean.

Two questions are rolled into one here. There is a question for philosophy of mind: What would constitute correctly understanding what one means by a word or a thought? And there is a question for epistemology: On the basis of what kind of evidence does one achieve an understanding of what one means by a word or a thought? Further, the epistemological question has two readings. It can be taken as a request for an ultimate justification for coming to a certain understanding, a demand for a proof of the soundness

---

[1] In saying "Understanding speech is a form of perception of the world, as direct as seeing," I intended only that it involves no more in the way of neural processing than ordinary perception. The point was intended to be anti-Gricean (Millikan 2010, 71n).

[2] I didn't intend the story of the Hubots and Rumans as it is used by the Macdonalds. The points there were separate from the points about agreement in judgments in Millikan (2010). Hubots and Rumans were hard-wired; their language was not conventional.

of an understanding of what one means. Or it can be taken as a request only for a natural explanation of how such a correct understanding is normally reached, a demand for the empirical evidence that one knows what one means.

First, on a representational theory of mind, understanding what one means cannot be some kind of seeing through a transparent channel to the essence of the thing thought of, to what the thing is in itself. The various Frege problems acknowledged in the first paragraph of this essay are good enough evidence that the representations we have of objects, properties, and so forth are not transparent in this way. Rather, knowing what one is thinking of is having the capacity correctly to identify that thing with what one may have encountered before and with what one may encounter later, through a variety of information channels; the greater the variety and the greater one's accuracy the better one knows what one is thinking of. Knowing what is the same as what – *that again* – *oh, that once more* – is the firmest hold one can have on the identity of things thought of given a representational theory of mind (OCCI, ch. 13, Millikan 2011). Indeed, my proposal has been that what a valid empirical concept in part is, is exactly a capacity to reidentify its object, typically using a variety of means, a variety of conceptions (see my reply to Braddon-Mitchell, this volume). A valid, adequate, or clear empirical concept is, exactly, a knowing of what one is thinking of.

This leaves the epistemological question. How do we know when our concepts are non-empty, adequate, and clear? How do we know that we are reidentifying correctly? Here enters my opposition to Meaning Rationalism. Trying to ward off radical skepticism is a fool's errand. There is no such thing as Cartesian certainty that our concepts are clear. There is no *a priori* proof that they are adequate. But there are excellent empirical tests for conceptual clarity, both practical and theoretical. These are briefly reviewed, with references, in my reply to Braddon-Mitchell.

12

Weasels and the *A Priori*

DAVID BRADDON-MITCHELL

There's a philosophical conundrum that seems to arise for many of us rather often. You hear a paper, you nod along with it, you admire many insights, and just before it's time to applaud and go for a drink the speaker announces that therefore a whole range of doctrines you favor must be false. Or perhaps they announced this at the beginning, but in the nodding and applauding, it's slipped your mind. You think "But that can't be right! What's going on here?" Often this is very fruitful. Maybe you discover that you shouldn't have been nodding. Maybe you discover your beloved doctrines are indeed false. Maybe you discover that fewer people than you hoped understood what your doctrines amounted to. Maybe it turns out that the words used to express your doctrines are more ambiguous that you thought, and there are indeed some false doctrines that you could reasonably express with those words, but, phew, the ones you intended were fine.

Something like this has been how I have responded to Millikan's (2010) most recent work on knowing meanings. After admiring the idea of a lumpy, clotty natural world whose topography binds together the extensions of empirical terms, after nodding along with the account of the functions that explain the use and persistence of words and concepts, before she concludes I am reminded that at the outset she tells us that the upshot will be that:

> … purely *a priori* or "armchair" analysis is not the right tool for examining meanings of basic empirical terms. This is because their meanings are not determined by methods of application that are necessarily common to all competent users. And because non-basic empirical terms always rest ultimately on empirical concepts that are basic, these being held together and in place by empirical-world bonds, the meanings of empirical terms can never ultimately be modeled merely as functions from possible worlds to extensions. All empirical meaning is, in the end, immutably embedded in the actual world. (Millikan 2010, 45)

How then is one to reconcile that agreement with a commitment to there being analytic truths about the content of concepts, with the meanings of words somehow supervening on these? With two-dimensional semantics and its championing of the idea of a sort of meaning which is a function from centered possible worlds understood as actual (roughly, a way things could be, combined with a context within that way) to functions from world to truth values?

All of the above responses will be right to some degree, I think (except of course that bit where it turns out that all my preferred doctrines are false). Examining how and in what degree will be helpful indeed for anyone who cares about a priority, analyticity, and how to reconcile them with a naturalistic picture of the mind.

The strategy in this chapter will be twofold. On the one hand I'll examine the negative aspects of Millikan's remarks, and show that they don't impact on the plausibility of two-dimensional semantics with armchair analyticities, at least as I understand them. Some of her remarks aren't explicitly negative moves against the *a priori*; rather they are a kind of job description for a semantics. In those cases treat my comments as ways in which I think that job can be filled by two-dimensional semantics. Then I'll move on to her positive story and say how I think it is largely consistent with the story I prefer.

The Proliferation of Handles

According to two-dimensional semantics, reference is often determined indirectly. Associated with terms in a language, or with concepts, there is a description which is analytic and which may only serve to pick out samples. The reference in the actual world is given in these cases by some kind of similarity relation between the sample so picked out, and other things. The intension across possible worlds is picked out by the same similarity relation. So, for example, associated with 'water' might be 'that actual substance which is drinkable and which forms the dominant fluid found in the lakes and rivers on Earth.' The reference of the term is then everything that is similar in the same-substance way with what is picked out by that description. And across possible worlds the water is whatever is similar in that way to that sample; all and only the H_2O, to simplify the chemistry somewhat. The view says that that description is analytic. If we were to consider as actual a world in which the substance answering to that description was different, the reference of 'water' would be different from what it is, but nonetheless in one sense of 'meaning' agents would mean the same thing – they would possess the same concept. The word would have the same "A-intension" or "primary intension." Similarly in the actual world, there could be languages that possess words which share the same A-intension yet differ in reference. The speakers of twin-English, for example, have a word 'water' which has the same A-intension, but whose reference is X_YZ. Thus the term agrees with ours along one dimension of meaning, disagrees along another.

It might seem that it is crucial to this story that there is a univocal, introspectable, shared description that functions as the A-intension. And this is what Millikan challenges.

> Basic empirical terms are learned in a way that does not pass on to the learner any particular manner of thinking of or identifying their extensions, but only some kind of grip or other, perhaps completely idiosyncratic, on a naturally bonded extension having multiple handles. Individual users may each use a variety of alternative handles, none of which are criterial or definitional either within the public language or within the user's idiolect. (Millikan 2010, 44)

To take our example, there seem to be two ways in which there just is no univocal descriptive handle associated with water: no community-wide agreed description, and no intrapersonal agreed description over time.

What do some Inuit people use as their reference-fixing description for 'Imiq?' Probably something to do with solids; snow, ice. Or take an Australian word for water, 'abma.' What were the descriptions associated with this before European settlement? Something to do with the sap of certain plants? What is found when digging deep? In contemporary California perhaps the key reference-fixing descriptions might be 'that which is found in blue bottles.' It's absolutely right that the exemplary descriptions often used in discussion of two-dimensional semantics ('the actual substance which falls from the sky, fills the rivers and lakes, etc.') is something explicitly believed by few, and plausibly not even tacitly believed by many who seem able to talk about water.

A discussion later in Millikan's chapter makes the same point about learning. To make the point in a way relevant to the example at hand, over our lives the kind of reference-fixing description we use will vary a lot. When first learned how did we identify samples of water? Perhaps that they came from a tap. Later we might use clarity and potability. Perhaps one grip some young children have on it is that it is the tasteless thirst satisfier. In any case it's very varied, both between individuals and over the life span of individuals. What role could be played by such unstable identifying descriptions?

Thinking back to the various responses to disagreement I mentioned in the opening paragraphs, the first one I'll go with is to agree that all this is true, and it shows that at the level of concepts, we really do have massively varied concepts in two-dimensional semantics, and neo-descriptivism is right. The actual substance that is hard and icy, the actual substance that is the majority of what is in the sea, the substance that satisfies thirst that is left when you take all the rich goodness out of milk: all of these descriptions might give the content of very different concepts. As what world is taken to be actual is varied, these different concepts will correspond to different extensions. In a world where the oceans are filled with nitrogen, the rocks are water, and the icy stuff that can melt is methane, concepts like these will pick out quite different extensions. Perhaps we need to appeal to weird worlds that seem superficially like ours to make the point perfectly, but in any case all that is granted. A two-dimensionalist has to admit that the concepts we use to think about water in the actual world are very different indeed. Thus, in one dimension of content, the contents of our thoughts vary dramatically.

How much does that matter? Not much: in fact, it just seems like the obvious truth. Of course in the actual world all these concepts pick out the same extension. And what's more they pick out the same extension across possible worlds: the Inuit and I, insofar as we are appealing to natural kind terms, rigidify on the same substance and thus are thinking about the same substance in every possible world. It is only if we vary the world considered to be actual (or compare me to a counterpart Inuit in the actual world who lives somewhere superficially very like this world but where the ice isn't water-ice) that the contents of our thoughts (in one sense of 'content') come apart. So far this is all about concepts: I'll come to language in a moment.

This, then, would be to bite the bullet firmly. But in fact there's no need to bite it *that* firmly. The idea is that the associated descriptions that govern A-intensions are not ones that you would get by interrogating people, or by simple armchair reflection. They are descriptions that are tacitly believed. Tacit beliefs are those that are manifested in behavioral dispositions of various sorts. Now the most obvious such dispositions are ones that might result in directly identifying objects or substances. More subtle, though, are the dispositions to coordinate with others and defer to others not in the strong sense of expertise in what the true nature of the (in this case) substance is, but even in the matter of identifying descriptions. The relevant identifying descriptions are those that might be agreed on after a degree of equilibration, and this will reduce the extent by which the idiolects vary.

This is not quite the same as a Putnamesque claim that as a matter of analysis there is or ought to be a linguistic or conceptual division of labor; rather it is that as a matter of fact most humans are disposed to defer to either expertise on the one hand, or to the results of equilibration with others in their community on the other. Thus there could be no such division of labor in organisms that were not so disposed. There could also be no such division of labor among individual humans who are not so disposed. Such humans – call them *ornery* as a technical term – would not be disposed to care what alleged experts think, nor to equilibrate with their fellows. On this kind of account the contents of *their* thoughts would not be standard in the community. Of course, what they *say* may have the standard content. It may be that they find it very hard to say what they mean – for the meaning of words in a language is a rather different affair. The meanings of words in languages supervene on what concepts the words are typically used to communicate, and if most speakers use 'water' to communicate that something is the same substance as the majority of the potable stuff which actually fills the rivers etc. (for example), then that's what the word means, whatever the content of the concept that the speaker is trying to use it to mean.

This leads nicely to another negative comment that Millikan makes. She points out that as we learn natural language, we do it so quickly that we could not possibly be learning anything like the A-intensions of words. If we are learning many words a day, the relevant information wouldn't even be available, let alone be able to be absorbed.

But I agree with this. The bullet (one of a selection she offers) is that, as we learn language, we often don't know what *words* mean. Indeed in adulthood most of us don't know what many words mean. That's because we use the words to express concepts that we possess. A child may have a concept of something that satisfies thirst without being rich and opaque, and uses the word 'water.' In the environment in which she grows up, nothing goes wrong doing this. You get the water when you ask. The adult has a concept of water whose activity is triggered by the child's use of the word; it too might be a little idiosyncratic and not defer entirely to the community average. But it will likely, in the current environment, overlap enough in extension with the child's concept that the word serves its purpose in bringing water to the lips of the child.

Perhaps the community average that we would mostly defer to after equilibration is that the extension of the word 'water' is one that is given by the same-substance relation and a wide set of identifying descriptions. So water is the same substance as that which (insert list of identifiers descended from the ones that most of us use), and that gives its actual extension as well as telling us what to judge as water under various possible circumstances.

Here is where I start to nod at Millikan's positive story. In the next sections I'll first show how useful I think that story is and how it fits in with two-dimensional semantics, and then move on to ask where, therefore, the difference is between us.

Why the Two-Dimensionalist Needs Millikan's Positive Story

The positive story that Millikan tells is, in part, about the lumpy shape of the natural world and the ways in which its contours explain and underpin language and thought.

Here's the idea in the rough. The world is not continuous. Take a particular organism: suppose it is a weasel (*Mustela nivalis*). It's not as though there is a smooth gradient of similarity between it and other existing organisms, so that they get progressively less similar in an incremental way until you reach starfish. Instead, there are more and more dissimilar organisms across a range of dimensions, and then you reach a chasm, a void of similarity where the differences are very large until you reach, say, other members of the *Mustela* genus,[1] or even *Neovision*. The same story applies to substance terms. Take a sample of water. There is no continuous chain of substances less and less similar to that sample which takes you to pure liquid lead, or lumps of rock. Instead you get a chain of slightly dissimilar things – different impurities, different states of matter, and so forth, but eventually you reach a chasm at the other side of which are very unlike things. These groups of objectively similar entities or substances are like lumps in the natural fabric of the world. They are what make nature non-continuous and structured.

This lumpiness, Millikan holds, underwrites the extensions of our terms. We often have no idea what the logical geography of these lumps is. A child who learns to use the word 'water' is in no position to have any opinion about how many substances there are that can be reached by, as it were, chaining together substances by a relation of very small dissimilarity. Nor is she in a position to have any idea what organisms constitute the island of similarity that corresponds to *Mustela nivalis* (or indeed *Mustela*). Despite this, our terms (or at least our empirical terms) pick out these groups. They do this because, roughly, it is the fact that the world is lumpy in this way that underwrites the possibility and utility of semantics. 'Water' refers to water, if it does, because nature is lumpy in that way. The non-continuous structure of what we might call substance space is what makes general terms useful. It's also what makes them possible. It explains the very existence of the words and concepts, and governs their persistence as well. But it is because the world is lumpy in the water way, among others, that theorists can be confident that whatever else happens to your concepts and language, there will likely always be a concept WATER and a word 'water.'

And so 'water' refers to the substance that forms that island in substance space, and it does so whatever idiosyncratic handle we may have on it: that which we use to build igloos, that which is non-milky and slakes thirst, or simply recognitional cues.

---

[1] This may even be true over time if speciation is a relatively rapid thing, though it is less clear that this is the case. For the purposes of practical semantics this rarely matters, as the extension of our terms in the past is rarely practically relevant. It might be theoretically important, though.

Nodding Along to the Positive Story

This is only a rough outline of the story, of course, and it can be fleshed out in many ways. No doubt one way is a fully-fledged teleosemantics. But at this level of abstraction I find it very helpful, and not inconsistent with the aspirations of two-dimensional semantics, which insists that there are A-intensions.

So the nodding first. A two-dimensional armchair analysis of a concept like WATER has a major placeholder. What has to be placed into that placeholder is an account of 'same substance' or 'same kind.' So consider the concept of 'the substance which is the same as the actual watery stuff,' where 'watery stuff' picks out, in the linguistic case, an averaged equilibrated list of identifying descriptions, or in the conceptual case, a particular, idiosyncratic identifying description. Identifying the extension, and indeed the intension,[2] requires an account of same substance. Or, talking not in the epistemic mode, what the extension and intension actually are depends on the nature of substancehood and kindhood.

Saying what that is is a difficult task, and not one often undertaken by anyone in the two-dimensionalism literature. The usual attempt is something along the lines of deference to the ideal scientific view of naturalness.

This is supposed to deal with something like the following worry. Suppose we are in a society before the advent of chemistry. Suppose someone has a concept of 'the same substance as whatever constitutes most of the fluid in the Thames.' What is supposed to govern the meaning of 'same substance' in that description? At first blush we are not entitled to simply substitute whatever we take to be the correct account of what it means. But that pre-chemical Londoner may have no explicit views on what the same substance relation is, and the underlying folk conception[3] may massively underdetermine the extension. This leads to odd consequences – a contemporary speaker of English who is chemically uninformed might have a concept whose extension is H_2O in virtue of their deference to current science, whereas our imagined fifteenth-century Londoner might have a concept whose extension is very odd (any non-alcoholic substance that is potable) or, worse, a concept whose extension is *almost empty* if it turns out that they defer to alchemy or some theory which has an account of same substancehood according to which (unbeknownst to the alchemists) in fact nothing stands in the 'same substance' relation to anything else.

My solution has been to say that there is something like deference to equilibration over improved knowledge. It's not what our Londoner takes the same substance relation to be, nor what her contemporary shamans take it to be that matters, it's her disposition to go with what would be decided as knowledge improves. In short, it's something like deference to ideal science.

Millikan's account in terms of the lumpy nature of the world gives us a story about why ideal science matters, and what ideal science is a science of. It's the right theory of where the lumps are, and it matters because it's the non-continuous lumpy nature of nature that causally impacts on us. What makes the various different samples of water behave similarly is that underlying similarity. What makes the pre-chemical Londoner defer to ideal science is that

---

[2] Understood as the extension at each world.
[3] See, for example, Medin and Atran (1999) on folk biology.

ideal science will answer to why they are using the term, for it will explain the very patterns of similarities which cause having a term like 'water' or a concept like WATER to be useful.

So the two-dimensionalist could use what Millikan says and substitute it into the placeholder so we get two-dimensional analyses of terms like 'water' which proceed like 'whatever is in the same objective structural lump of similarity space as the actual substance which is picked out by [substitute idiosyncratic handle in the case of concepts, substitute equilibrated description in the case of words].'

This is, though, no departure from what Jackson, Chalmers, and I[4] say – it is just substituting in a particular account. It is a very good particular account, I think, for it does what the two-dimensionalist wants. It's both (as far as it goes) probably a correct account of natural kindhood, and an account which explains why we would expect it to be one to which ultimately speakers might be expected to tacitly defer (insofar as they might be disposed to tacitly defer to the best account of what makes concepts or words useful to them, and what promotes and regulates their continued use).[5] Thus empirical meaning, insofar as this means extension or even intension, is indeed "… immutably embedded in the actual world" (Millikan 2010, 45) as Millikan puts it. The extension is indeed fixed by the lumpy structure of the actual world, together with the actual narrow structure of the agents' dispositions.

So What is There to Disagree About?

So now it's time to see where the agreement ends. Surely there can't be that much agreement. After all, the view is supposed to be inconsistent with two-dimensional semantics. So in this section I will explore where those disagreements might lie. Some matters that look like disagreements may turn out to be issues of mutual incomprehension, but others will be real.

I'll start by setting aside really fundamental conceptual topics (I don't think basic perceptual concepts are what is at issue here; rather, it's empirical kind terms – dog, cat, weasel, and so forth). I will focus on three related issues:

1 When are terms in fact natural kind terms? Or at the level of concepts, when are they natural kind concepts? The irenic responses I have given above are ones that make sense for natural kind terms, for in those cases we can plug in Millikan's story to the two-dimensional story to determine extension.
2 Where terms are natural kind terms, does the descriptive handle play any role, or ought one just use the Millikan-style analysis directly (in other words, does it matter that the agent in some way tacitly authorizes rigidification on the exemplified lump in nature's fabric)?
3 What commonality is there in semantics between agents in narrowly the same state, but whose environments are different in where the bumps are – for example in the twin-earth case between Oscar and Twin Oscar, or between agents in worlds where different organisms are picked out by the same psychological handles and A-intensions?

---

[4] Jackson (1998), Braddon-Mitchell (2004), and Chalmers (2006).
[5] Not that these can't come apart, and when they do, that is I think grist to the two-dimensionalist mill.

When is a Term a Natural Kind Term?

It's common now to think that Putnam and Kripke went too far in their discussion of the semantics of terms like 'water.' Perhaps overall the semantics of 'water' in this community is as of a natural kind term that picks out the actual watery stuff. But we do at least have the *concept* of 'clear drinkable liquid, whatever its nature,' and the practitioners of empirical philosophy tell us that at least some of us think that's what 'water' means. Among philosophers, David Lewis,[6] for example, took the view that 'water' is ambiguous in English between a natural kind term that rigidly picks out H_2O and a functional kind term that picks out all and only reasonably clear potable liquids. Lots of our terms for animals don't track natural kinds – 'bug,' 'pest,' 'vermin,' 'predator' (though perhaps they track ecological kinds). Someone could use the word 'dog' to refer to things phenetically like dogs, and perhaps would in a society with less deference to good theories of the underlying natures of things.

So in virtue of what is it the case that someone does use a concept that is not a natural kind concept, whose extension is not governed by the real grain in nature?

Here's one story: it's when, and only when, there is a persistent deliberate conscious stipulative definition, which in addition reflects dispositions to identify substances in contexts.

Perhaps Millikan and the two-dimensionalists can agree on this as a sufficient condition. If an individual, or speech community, explicitly claimed that their word 'water' meant 'clear potable fluid found in rivers and lakes locally,' and systematically had no disposition to change this in response to the fact that the underlying nature is little changed when it becomes ice (ice is not water, they would claim), and thought it ceased to be water when poison was added and so forth, then this would just be the empirical linguistic discovery that their term 'water,' while it has substantial overlap in local extension between our term 'water,' means something very different. It's not a natural kind term, or a substance term; rather, it's a functional kind term that does not apply to all water (vapor or ice) and applies counterfactually to substances other than water (X_YZ).

But we are looking for disagreement. Perhaps the disagreement is about whether this is a necessary condition. Put another way, perhaps the disagreement is as to what the default content of concepts and words is.

If it's not a necessary condition, we might find that there is a community (I take it there were such communities) whose behavioral dispositions are such that they are inclined to take to be water anything they come across that is functionally like the paradigms that their idiosyncratic handles pick out. That's how the term is used, even if there is no explicit definition in anyone's mind, and even if the idiosyncratic handles vary a lot from individual to individual.

There are two ways to take this case. One is to say it is a case in which the term is a functional, non-natural kind term, due to the lack of an explicit definition in terms of natural same-substance relations (or perhaps lack of dispositions to be influenced by the discovery of such relations etc.); the other is to say that in this case it's clearly a natural kind term, because ultimately it's water that explains the term's persistence etc., and

---

[6] Lewis (1994).

there is no explicit or near-explicit overriding of that presupposition by a stipulative definition in terms of functional kinds.

The two-dimensionalist will say that here we have a non-natural kind term; the community as a whole may be using different handles that lead them to the samples, but the similarity they care about is the functional similarity in terms of potability (to any speaker, say, not necessarily to them – thus they see that someone rightly calls something water even if they, the original speaker, would not find it potable). It's only if there are dispositions to take discoveries about the fundamental nature of the local watery stuff to be relevant to the extension of 'water' that it will become a natural kind term.

I'm guessing, and I'll be corrected if wrong, that for Millikan the default goes the other way. The handles people use, the behavioral dispositions they in fact have, are not what governs the extension. The default supposition is that it is the nature of the world that supports the use of the term that gives the extension. In virtue of the fact that the term connects to water, and so a massive island in the ocean of real structure is landed on by thought, the term is about water, and its extension is all and only the water. This presupposition only gets cancelled by explicit definition, and perhaps by behavioral dispositions consonant with that definition.

So for the two-dimensionalist, a natural kind term and a natural kind concept is a relatively sophisticated thing. It requires concepts of intrinsic nature, or underlying kind, or scientific investigation, or deference to the explanation of why words and concepts are being used. Natural kind terms and concepts arise when the idea of naturalness, of the grain of nature, of seeking to explain the variety of nature with which we come into contact, is able to affect our dispositions to identify kinds and substances, and start to affect our behaviors and judgments.

On the other view thoughts just are about the naturally bonded extensions that they latch on to by varied means. This has to be *cancelled* by something sophisticated. It's the functional terms and concepts, the ones that aren't about the natural groupings of things, but are rather about our explicit interests, which are the recherché exceptions, born of the conceptual sophistication of humans and perhaps some other organisms.

So on this account we'll get a very different take on animal content for the two views. Views consonant with the two-dimensionalist will be on the very proximal end of the proximal-distal debate for animal content. Since there is no concept of natural kind in at least most animals, the content to attribute to them is more or less phenomenal; it's something like whatever answers to the recognitional capacities. They won't have a concept of water, but will have a concept of anything that satisfies thirst and smells a certain way and can be recognized a certain way. On the other, natural kind default view, it'll turn out that animals do have thoughts about water, whose extension is all and only the water.

What to make of this disagreement? I'll come to that later. Now, the next disagreement.

What Role Does the Deference to Naturalness Play in Natural Kind Terms and Concepts?

Now let's turn our attention to cases where both sides agree that something is a natural kind term or concept, picking out all and only a group in nature whose exact delineation may be both unknown and vague.

Suppose that Fred and the speakers of Fred's language are inclined to say that they mean by 'weasel' the animals they have contact with by [insert the recognitional criteria they have in mind] and whatever forms a natural group with them. Suppose also that he and his community are inclined to defer not just to current best opinion about that, but also to modify their judgments as more is learned (both about the natural group, and how to identify the experts).

We'll now have a case where both sides agree that the extension of 'weasel is *M nivalis* (in British English; in American English, the extension of 'weasel' includes other members of the genus *Mustela*). These sorts of dispositions are likely widespread enough that there will rarely be a difference of opinion between the sides of this debate on the extension of terms in contemporary English.

But where we will have a disagreement, I think, is in the role that those dispositions play. The two-dimensionalist will say that it is they, together with the nature of the world, that make a term a natural kind term and govern its extension. Millikan (perhaps) will say that they are idle and unnecessary theory – per what I said in the previous section, the default presupposition will do the work, uncancelled by any explicit definition extending the term to all organisms that share a certain appearance or role or some such.

The Commonality between Narrowly Similar Agents

Above I said that there is agreement that the extension, and indeed intension (in one sense) of a term is immutably governed by the actual world. But of course according to two-dimensionalism there is more to semantics (broadly understood) than extension and intension. There is a kind of similarity between us in the actual world, and other agents either differently located in the actual world, or in other possible states of affairs, who are narrowly similar but in very different environments.

The extension and intension of 'weasel' are governed by the particular grain of lumpiness of the actual world, granted. But what to say of agents in a possible world very much like ours, but in which evolution has gone slightly differently for the Mustelidae? Such agents could be narrowly *exactly* like me, since I'm pretty ignorant of the evolution of the Mustelidae. Indeed if we turn the clock back a bit, every actual person might be narrowly like the agents in this other world.

Agreement again: we agree that the extension of our concept/word is *M nivalis*. Let's stipulate that the lumpy grouping that the same handles pick out in this other world pulls together a quite different group (though with some historical and intrinsic similarities) – let's call it *Exomustela*.

Our extension is immutably grounded in the actual world, and so is our intension. Their extension is immutably grounded in their world, and so is their intension. So is there no similarity of content between us and them? Millikan, I think, has to say no. We and they have similar handles that glom on to different organisms, but that's a superficial matter that says nothing about meaning, in any sense of that word that does any theoretical work.

Two-dimensionalists, (or at least some of them)[7] admit into the realm of semantics something else – A-intensions. And this is how we are semantically similar to our counterparts in the *Exomustela* world. Our term 'weasel' and theirs share the same A-intension, and it is that together with the structure of our respective worlds that fixes the different extensions and different intensions of our respective worlds. It's sameness of A-intension that explains why they would behave like us in our world and vice versa. It's grasp, however feeble, of A-intension that gives us some sense that we are competent with the term 'weasel' when we are ignorant of Mustelid studies. It's comparison at that level which allows us to tell whether we in some sense or other mean the same thing or whether we and another are using homonyms. (I don't need to be an expert in Weasels to know that my friend and I mean differently when he accuses someone of being a weasel – "Oh I see," I say, "you don't mean by 'weasel' 'whatever organisms are naturally like the samples called "weasel" by expert zoologists' – you mean instead 'someone of dishonest and subtle disposition'" – this is something I am competent to judge, while not having a clue about the extension or intension of 'weasel.')

Now of course one possibility is that this thing we do have a defeasible grip on plays no role in settling the extension. There might be *something* that I have a grip on which someone who uses the term in the other way also has a grip on. There might be *something* that is in common between me and my counterpart in the *Exomustella* world. But perhaps that something plays no role in settling the extension of the term. If that were so, however, it would be hard to see how access to it played any part in determining whether this was a case of homonymy or not. The right conclusion would instead be that it's not just that we can't be sure *a priori*[8] that the terms aren't co-extensional (which is often right, except perhaps where the A-intensions are inconsistent) but that we can't tell whether the terms are homonyms (or worse, can't tell that they obviously are *not* homonyms).

Some Arguments and Some Diagnoses

Having identified three areas of disagreement, what should we think? What I want to suggest is a kind of possible explanatory pluralism that leads to very different but compatible accounts of meaning. I say *possible* explanatory pluralism because all I will be doing here is giving a job description for a couple of explanatory projects, not arguing in any detail that these descriptions can be filled.

I'll start with an example that shows an area of disagreement, and draw my distinction in explanatory methods from it. The example is that of error theories of reasonably large domains. We generally agree there are not, and never have been, any witches. Many philosophers are tempted by the idea that positive ethical discourse is largely false. Other philosophers are tempted by error theories of belief, desire, aesthetic goodness, and much else.[9]

---

[7] Stalnaker (2004).
[8] Careful: 'analytically' is perhaps a better way of saying it.
[9] For example, Joyce (2001).

Error theorists are (usually) error theorists because they think that there are things that are analytically true[10] of the relevant discourse that they take to be impossible, or at least not actually instantiated. In the case of ethics, for example, a usual route to an error theory is the thought that in order for there to be a property of rightness there would need to be something which inconsistently behaved like a belief and a desire; or else something which in an un-Humean way could rationalize desire in a non-instrumental way just by judging that something is true.

What leads someone to be tempted to think that the judgments that we make about such cases tell us enough about the content of the relevant concepts that we can be error theorists? I take it that it is a tacit bet on what would happen when agents come to realize that these properties are not instantiated. The error theorist about witches is betting that, if it were discovered that no one ever slept with the devil or rode on broomsticks, that most witch allegations would be withdrawn. The error theorist about ethics is betting that if it is discovered that making a cognitive judgment can never rationalize a non-instrumental desire,[11] most folk would (ideally) accept that rightness was something of an illusion. The error theorist about phlogiston[12] would have been betting that on discovery that nothing is emitted on combustion that held things together, we would accept there was no phlogiston, and the concept would lie idle and the term be ignored.

Notice that the interest here is predictive. It is about what will happen in circumstances that have not yet arisen.

Even error theorists, however, will in many cases not deny that there is something that agents are tracking in their use of words and concepts of which the error theory is correct. There's some kind of agreement in many communities about what is right, what is beautiful, and who the witches are. There's an important explanatory project to be done to figure out why this is so. What properties, possibly complex social properties, explain and support the use of these terms? No doubt there are some. It may turn out that single women living alone of bedraggled appearance and inexplicable skills were being tracked by people using the witch discourse, and maybe an important psychological or social role was being played by stigmatizing such people. No doubt something is being tracked when we engage in ethical discourse; some say it is guesses at social consensus on desire, or guesses as to what our own ideas might more ideally be. Perhaps oxygen was being tracked in phlogiston talk.

Why not say that the outcome of the investigation into the properties that support and explain these practices is the content of the relevant concepts and words? Actually I'm happy to be pluralist about this. By all means do. Note that in doing so we are showing an interest in past-directed explanation: we are concerned about why these practices are there, and how they function currently, rather than how they will function in different environments.

---

[10] Actually many would abjure this claim, but it is nonetheless true of them.
[11] This is a species of internalism which some take to be true of our ethical concepts.
[12] Caveat to all of these: they are actually betting not on outcomes, but on the current presence of dispositions that would potentially lead to these outcomes.

Two Projects

That is, then, the bare bones of a suggestion. We need some kind of reason to make content attributions to ourselves and other organisms, and to language, rather than just being quietist about it. But the reasons might be plural. If we are interested in explaining the existence and persistence of mental or linguistic phenomena, we'll be pursuing a project that will give one set of answers in the cases where two-dimensionalism and Millikan's new externalism disagree. If we are interested instead in the dispositions we have to respond to new information and new environments, we'll be giving answers in line with the two-dimensionalist account of A-intensions given by semantic dispositions.

The projects might have more or less interest in different cases. I can see an argument that the two-dimensionalist concept is not very fruitful for animals. Few if any have a cognitive apparatus that has semantic dispositions lurking to change use on making various sorts of discoveries, and any dispositions they do have to behave differently in strange circumstances are plausibly uninteresting: just facts about how they would behave in environments in which they didn't evolve.

In the case of people, though, we seem to be very interested in how we would behave in different circumstances, in what we would regard the extension of a term if things were different, how we will react to various possible discoveries, about how to regard beings who are like us narrowly but are situated very differently. It's an important part of our social and imaginative ability to be able to place ourselves in others' shoes, and make comparisons between us as we are and us in those shoes.

None of this is to say that either project is vindicated of course. It's just to set the agenda for vindicating them. Ruth Millikan has done a lifetime of fascinating work vindicating one of those projects. Perhaps it is time that we believers in analytic semantics did some work defending not so much the coherence of the other project, but its explanatory utility.

So, finally, which of the possible reactions at the beginning am I advocating? Well, a tiny bit of each, but largely another. The literature on content has been at cross-purposes because there are meanings for 'meaning' that serve different purposes. The job now is to get those purposes clear, and evaluate their utility.

Reply to Braddon-Mitchell

David Braddon-Mitchell ends his essay by suggesting that it may be that "there are meanings for 'meaning' that serve different purposes." He and I seem to agree that the kind of meaning we would like to discuss is the kind that determines reference or extension. So his suggestion seems to imply that there may be more than one interesting kind of reference/extension to discuss. (If so, that would be a radically new suggestion.) Braddon-Mitchell has opened up the issue in a way that I very much appreciate, not by asking, as Putnam did, how we would apply the words "meaning" and "reference" in strange counterfactual situations, thus implicitly assuming that the references/extensions of the words "meaning" and "reference" are determined by intensions "in people's heads," but by asking what different kinds of interesting phenomena there might be in connection with language use that might be studied productively. What I find myself questioning, however, is prior to whether it is fruitful to study A-intensions and the dispositions people have to apply words given new information or in new circumstances. I question whether there exist such things either as A-intensions or as stable dispositions, of the kind Braddon-Mitchell suggests, to react with words in new circumstances. So I have to be recalcitrant and disagreeable and reject Braddon-Mitchell's proposed compromise. I am extremely grateful, however, for the opportunity he has given me to clarify my views, both in my own head and on paper.

Before turning directly to the above issue, I need to explain a bit about my position on empirical concepts. For although the "weasels" paper on which Braddon-Mitchell comments was almost entirely about words, in my own thinking it is inextricably entwined with previous work that I had done on concepts (LTOBC; OCCI; VOM, chs.14, 18, 19; LBM, ch. 3; Millikan 2009a, 2011). Explaining some of the connections will help both in answering Braddon-Mitchell's very helpful questions about functional kinds and in explaining what I believe governs changes over time in what I have termed people's "conceptions," these being the things most akin, in my view, to Braddon-Mitchell's A-intensions.

Empirical concepts are best understood as, first, abilities to reidentify objectively self-same things, such as properties, relations, individuals, or real kinds (the lumps Braddon-Mitchell refers to), paradigmatically in a great variety of different ways. To reidentify them, for example, by sight at various angles and distances, under various lighting conditions, by shape, color, movement, by various identifying sounds, by feel in the hand held this way or that, by taste or smell, by any of a very great number of descriptions, by traces or tracks or other manifestations of various kinds, and so forth. Consider, for example, the enormous variety of alterative proximal stimulations that, impinging on your sensory surfaces, might tell you of present, past, or future locations of your spouse or of a dog or of water or of a city, a great many of these tellings involving pure recognitional capacities rather than any inference (see Fumerton's "Properties Over Substance" and my reply, this volume). I have called the various ways a person knows how (fallibly) to reidentify a thing "conceptions," or taken collectively, that person's "conception" of that thing (OCCI). (In LTOBC I called these "intensions" – a bad choice of words.)

Possessing an empirical concept requires, second, a use or uses to which this capacity to reidentify is tailored to be put. Empirical concepts are developed and used for various kinds of learning and for applying what one has learned. Their criteria of adequacy, hence the ways they are readjusted and improved over time, are determined by the specific learning purposes to which they are put. These uses divide very roughly into two non-exclusive kinds that I have called "practical" and "theoretical."

Practical concepts are used for reidentifying things to be responded to or used directly in the guidance of various practical activities. They help to identify affordances. A single practical concept may be used to identify many affordances; for example, many animals are able not only to reidentify water in many different ways, they also know how to react to it or use it in a variety of different ways – to drink it, to avoid getting wet in it, perhaps how to hide in it, perhaps to bathe or cool off in it, to find minnows in it, and so forth. For these animals, learning to recognize water and learning how to use it or deal with it successfully are two sides of a coin; their recognitional skills are honed and perfected using practical success as the measure. Nor have humans grown out of practical concepts. Although practical concepts need not be expressed with words, words that designate functional kinds – "can opener," "cup," "piano," "house," "clothes," "food" – surely originated as merely practical concepts, concepts of what is for opening cans with, for drinking out of in a certain way (by holding the handle), and so forth. Later in human thought and in child development, practical concepts may move over into the theoretical realm, "being for" this or that turning into "intended or designed for" this or that or, more weakly, "capable of being used" for this or for that.

Theoretical concepts are tailored for gathering theoretical knowledge, paradigmatically, knowledge of what things have what properties, relations, and so forth. They are honed and corrected by the measure that is avoiding contradiction or, put positively, maintaining stability in judgment. Subjects of judgment should be things you can gain stable knowledge about. That is, having judged a thing to have certain properties and relations by using *these* methods of identification, *these* conceptions (telling, say, from the front in dim light), one should not be disposed, using *those* methods of identification, using other conceptions (telling, say, from the side in speckled light) to judge it to have contrary properties or relations. That my methods of reidentifying are successful is evidenced, in solid cases, by multiple triangulations upon the same subject matters. The ideal is, no matter how I go about checking, the law of identity should hold: *if P, then P.* This result supports confidence in each of the empirical concepts – those embedded in the subject and in the predicate terms – used in judging that P. It helps to confirm that each of these would-be concepts really does consist in the ability to recognize one and the same objective thing over time in a variety of different ways (hence is a real concept).

It follows that theoretical subject concepts have to be developed along with predicate concepts and along with concepts of the contraries of these predicates, for all are tested together. The concept of any subject of judgment has to consist in part of a "substance template" (OCCI), a grasp of some, at least, of the contrary spaces within which its properties/relations are to be found (LTOBC, esp. chs. 16, 19; OCCI, esp. ch. 5). Individual physical objects, for example, have weight, shape, and size; chemical substances do not. A theoretical predicate concept has to consist in part of a grasp of – a capacity to recognize – what is contrary to it, and some ability to recognize the kinds of subjects that tend to choose just one property from the contrary space in which it lies. Theoretical concepts are adjusted then not holistically but in a variety of different combinations, subject concepts tested along with a variety of predicate concepts, predicate concepts with a variety of subject concepts. Successful readjustments of the conceptions supporting an empirical concept bring it to focus more clearly on just one real subject, real kind, real property, or whatever.

Thus neither practical nor theoretical concepts are merely dispositions to track some particular thing. Rather, they are *abilities*. They have purposes, purposes for which the identifications they support are used, and which determine what can count as "the same again," hence as a correct reidentification. Routinely failing to accomplish these purposes they would not be abilities. For example, even if the idea of witches managed to track some particular thing, it surely did not track anything that actually had the kinds of imaginary properties that witches were judged to have. The idea of witches may have had some social function, but that does not imply that it was a concept of anything real, indeed, that it was, speaking strictly, a concept at all. Merely would-be abilities are not, strictly speaking, abilities at all.

Because Braddon-Mitchell focuses pretty exclusively on concepts of real kinds (he calls them "natural kinds"), I will do the same in what follows. Real kinds – Braddon-Mitchell's lumps – are clumps, or clumps within clumps, or clumps within clumps within clumps (say, Dachshunds within dogs within mammals) or humps on clumps (say, Gothic churches within churches), most of the members of which have most of some (typically very large) set of properties in common *for a reason*. They do not resemble one another accidentally. A concept of such a clump includes a way, typically multiple ways – recognitional capacities, descriptions – to identify something as a part of the clump. These ways, capacities, are conceptions that support the concept of that real kind. These conceptions differ from traditional intensions and, I believe, from the two dimensionalist's A-intensions, in a number of important ways.

Conceptions do not attach to words but to concepts, concepts being personal possessions, components of individual people's psychologies. We each have our own concepts, each one supported by our own idiosyncratic collection of conceptions.

Ways of identifying a clump are not fully unpacked by descriptions, either definite or indefinite. Thinking of water as "that actual substance which is drinkable and which forms the dominant fluid found in the lakes and rivers on Earth" would be of use only if one could already recognize Earth, lakes, rivers, drinkableness, fluidness, dominance (for this kind of case), and when something is "the same substance" as the above ... (Like it in what respects; like it for what reason?).

The purpose of one's conceptions of a real kind is not primarily to determine the boundary between members and non-members of the kind. Conceptions are not "reference fixing." Their job is not to define a class, but merely to help reidentify a clump – "oh, part of that clump again." Because the typical interesting clump does not have completely determinate boundaries nor completely determinate properties, the typical concept of a real kind is somewhat indeterminate as well, its degree of determinacy being an empirical matter.

One's conceptions of a real kind are always in principle fallible. It is not their job to determine extension but only to serve as practical methods of recognizing more of the same clump again when it turns up. They are empirically discovered rather than *a priori* known. Methods routinely used to identify a clump may fail in unusual cases.

In the usual case there is every reason to multiply one's ways of identifying a real kind, so long as each way proves relatively reliable. Paring them down, relying on one or two central ways only, or trying to use just the same ways as other people to identify membership in a clump, would be useful only if we needed to use our words and concepts in other possible worlds, or if we needed to cut neatly between all the real cases, but we don't. Empirical concepts are not classifiers. They are practical tools for building inductive knowledge *in this world*.

How then do concepts get together with words? In the case of basic empirical concepts, only their referents (the individuals, the properties, the clumps) need to get together with words. One among other ways of identifying an object or property, as it occurs in some configuration in the world, is through its manifestations via the thought and then configured speech of other people – through a word used for it in one's language community. This is a fallible way of identifying, of course, but all ways of identifying are fallible. Being able, in context, to recognize a word for a thing constitutes one sort of conception of it (see LTOBC on "intensions, language bound"; OCCI, ch. 6; VOM, ch. 9; LBM, ch. 10, Millikan ms).

Now if all this were correct – an "if" that may, of course, cast a long shadow – then it would seem that we do need an argument for why current dispositions to identify, say, what one thinks of as "water," or current dispositions to change these dispositions in this or other possible worlds, should hold an interest for us. Since people don't generally tell others how they go about identifying ordinary real kinds, nor indeed need they know themselves, not only is there no reason, in ordinary cases, to drop one's own methods in deference to the methods of others, but also no obvious way to discover just what other people's methods are. Agreement in judgments, not methods, is the goal, and agreement in this world, not in possible worlds. What makes basic empirical terms synonymous is coinciding reference or extension in this world, not methods of identifying, not conceptions. (Surely Braddon-Mitchell's 'aqua' and 'Imiq' and 'abma' were/are all synonymous with 'water.')

It is true, and important, of course, that knowing what sort of glue holds a real-kind clump together can often give guidance in deciding whether queer cases really belong to the clump rather than being similar to the clump's members by accident (XYZ, twin-earth weasels), and it is true that this kind of knowledge is often acquired from others. It is not necessarily the job of scientists, however, to discover this glue. What holds McDonald's restaurants together is not a matter for science, nor is what holds various renditions of *The Star Spangled Banner* together or various Gothic churches.

A last observation – about our dispositions to use words in new circumstances. Many words seem to change their meanings over time not because we change our minds about how to use them, but because the world changes or because we become acquainted with new portions of it. For example, sometimes the actual clumps in the world slowly extend themselves over time into new areas of logical space, or slowly shift their positions in it, or they may separate into subclumps, or subclumps may merge together. Consider, for example, why the word "computer" has shifted over the last fifty years, or what has happened to the dog clump due to breeding over the last two thousand years. Would the Romans have called a poodle or a Pekinese "canis" had they encountered one? If not, would they have been mistaken, would they have failed to identify what really was in the extension of their own term 'canis?' What would determine this? The clumping that is being used to carry a word forward with understanding by peoples located in a certain place during a certain time may be a local clumping, one that merges seamlessly into a larger and perhaps more interesting clump when other places or times are considered. The boundaries of the word, given a broader perspective, are then very seriously underdetermined by its current usage.

Add to this that when one needs to communicate one must use the language at hand with the people at hand. So the question *what would you call it?* (perhaps while also giving some further explanation) and the question *what would you think it was?* tend to come apart.

The following problems then seem to arise for thought experiments that describe merely possible things and then ask, *would this be an X or not?* Most words have been designed to apply to clumps in this world, not in other possible worlds where the clumps may have come apart or where there may be no such clumps. The clumps that a word names may have vague edges or be, in some dimensions, completely indeterminate (were poodles examples of canae?). Even if the description given fits nicely with a person's current ways of identifying Xs, it does not follow that the thing would be an X even in this person's idiolect. *I would take this to be an X* does not imply *this is an X*. *Would you take it really to be an X?* can come apart, in various degrees, from *would you call it an X (perhaps with qualification or further explanation)?*

That useful information might be acquired by forcing questions of this kind seems to me to require an argument that has not been given. The factors that figure in actual language change are, of course, well worth studying, but actual language change takes place in response to actual and detailed changing circumstances, not to single anomalous cases. Moreover it is unclear how studying the paths it will or might take would throw any light on current usage.

For these various reasons I find myself sadly unwilling to compromise with Braddon-Mitchell. Conceptual analysis in all of its forms – from two-dimensionalism to projects within experimental philosophy, among others – seems to me to arise from mistaken assumptions about both language and thought.

13

All in the Family*

WILLEM A. DEVRIES

In "The Father, The Son, and The Daughter: Sellars, Brandom, and Millikan," Ruth Millikan considers how two scions of Wilfrid Sellars, both of whom see their own work as a further development of Sellarsian ideas, could come to be as philosophically distant as she and Brandom. Though Millikan contemplates the notion that (at least) one of Sellars' spawn is apostate, she favors the idea that there is a crack in Sellars' own position. She and Brandom stand on opposite sides of that crack and have wedged it still further apart. Discord among the children is traced back to the parent's inner conflict. Family dynamics are almost always interesting, but I have a personal stake in this story. Academically, Sellars was not only my teacher, but my *Doktorvater*, and I feel no Freudian urge to kill him off.

Sellars tries to sketch a coherent view of humanity-in-the-world that portrays us as perceptive, thinking, willing agents and accommodates naturalism and scientific realism without slighting the "autonomy of reason." In several places Sellars uses a distinction between the causal or *real* order and the conceptual or *rational* order to set up the problems that must be resolved.[1] Persons have access to and, indeed, sit at the intersection of

---

*In this chapter all references to Sellars' works will employ the now standard abbreviations. This note correlates these abbreviations with their titles and bibliographical entries in this volume. "Some Reflections on Language Games" (SRLG, Sellars 1954). "Counterfactuals, Dispositions, and the Causal Modalities" (CDCM, Sellars 1957). "Being and Being Known" (BBK, Sellars 1960). "Philosophy and the Scientific Image of Man" (PSIM, Sellars 1962a). "Truth and Correspondence" (TC, Sellars 1962b). "Abstract Entities" (AE, Sellars 1963a). *Science, Perception and Reality* (SPR, Sellars 1963b). *Philosophical Perspectives* (PP, Sellars 1967a). *Science and Metaphysics: Variations on Kantian Themes* (SM, Sellars 1967b). "Language as Thought and as Communication" (LTC, Sellars 1969). "Meaning as Functional Classification (A Perspective on the Relation of Syntax to Semantics)" (MFC, Sellars 1974). *Naturalism and Ontology* (NAO, Sellars 1979). "On Reasoning About Values" (ORAV, Sellars 1980). "Mental Events" (MEV, Sellars 1981). *In the Space of Reasons: Selected Essays of Wilfrid Sellars* (ISR, Sellars 2007).

[1] He doesn't use a consistent vocabulary. In BBK, for instance, the distinction faces the 'real order' off against the 'logical order' or 'intentional order' (BBK, ¶¶31–33; in SPR, 50; in ISR, 218–219), or the 'order of signification' (BBK, ¶¶15, 46, 52–53; in SPR, 45, 55, 57; in ISR, 213, 223, 225–226). See also AE, in PP, 234, 267; in ISR, 168, 203. In MEV he talks about the order of being and the order of knowing (MEV, ¶5, 326; in ISR, 283). It is important not to confuse this distinction with his famous distinction between the manifest and scientific images.

Millikan and Her Critics, First Edition. Edited by Dan Ryder, Justine Kingsbury, and Kenneth Williford.
© 2013 John Wiley & Sons, Inc. Published 2013 by John Wiley & Sons, Inc.

these two "orders"; trying to understand their relations is the central crux in Sellarsian philosophy. Neither Millikan nor Brandom seems to think that Sellars was finally successful in giving us a unified image. Each chooses to elaborate and extend one side of Sellars' distinction in an attempt to tell a more unified story. Brandom emphasizes the rational, intentional order but says disappointingly little about its relation to the causal or natural order.

Millikan, in contrast, emphasizes the natural and causal order, and her articulation of a more thorough naturalism is a major contribution to contemporary philosophy. However, Millikan is not as clear as she could be about how she disagrees with Brandom or just what crack she finds in the Sellarsian edifice, so there is a clarificatory task to be performed. I will argue that she neglects certain aspects of the rational, intentional order, in particular what Kantians call "the autonomy of reason." There are important hints in Sellars for constructing a more adequate naturalistic treatment of reason.

Locating the Fault Line: Rules and Roles, Norms and Causes

Millikan quickly dismisses the notion that the crack in Sellars' edifice derives from his attempt to join themes from the early and late Wittgenstein, picturing and language games; "… there is … no obvious crack in the bridge Sellars built between the *Tractatus* and the *Investigations*" (Millikan 2005c, 60). Rather, Millikan worries about Sellars' "… treatment of the nature of linguistic rules and the relation of these to conceptual roles and thus to intentionality" (Millikan 2005c, 60). It isn't clear, however, exactly what the conflict is that Millikan finds within Sellarsian philosophy, so let me explore this issue with some care.

Sellars treats language as a rule-constituted set of practices that are embodied in certain complex patterns of behavior.

> The key to the concept of a linguistic rule is its complex relation to pattern-governed linguistic behavior. The general concept of pattern-governed behavior is a familiar one. Roughly it is the concept of behavior which exhibits a pattern, not because it is brought about by the intention that it exhibit this pattern, but because the propensity to emit behavior of the pattern has been selectively reinforced, and the propensity to emit behavior which does not conform to this pattern selectively extinguished. (MFC, 423)

Pattern-governed behavior is present wherever there is learning. *Linguistic* patterns include patterns of responding to the world with words and responding to words with more words. Calling a certain pattern of response *linguistic*, however, is not a purely *descriptive* characterization of it. Linguistic events and objects occur in a context of rules and occur in part *because of the rules*. Crucially, the appropriate rules are not themselves descriptive generalizations about *de facto* patterns found in the world, but generalizations with *prescriptive* or *normative* force. Both linguistic behaviors and the rules that govern them are, as Sellars liked to say, "fraught with 'ought'" (TC, 212). And the point of an 'ought' is to motivate behavior:

> Learning the use of normative expressions involves ... acquiring the tendency to make the transition from 'I ought now to do A' to the doing of A ... it could not be true of a word that "it means *ought*" unless this word had motivating force in the language to which it belongs. (SRLG, 350)

Sellars distinguishes two different 'oughts' with which behaviors can be fraught. There are ought-to-dos, rules of action. The basic form of such a rule is a conditional imperative: "If in circumstances C, do A!" This is the kind of 'ought' involved in what I will call paradigmatic rule-obeying. Paradigmatic rule-obedience requires complex cognitive and conative capacities on the part of the agent: knowledge of the rule, recognition of the circumstances as appropriate to the application of the rule, and conative structures that motivate one to apply the rule and act on it. Sellars absolutely rejects the idea that linguistic behaviors all occur *because of the rules* as cases of paradigmatic rule-obedience.

Sellars' other 'ought' is the ought-to-be, rules of criticism. For example, it ought to be the case that dogs come when their masters call. Such a rule speaks to no agent in particular, and it is certainly not a rule that dogs *obey* in the paradigmatic sense. It simply endorses a particular state of affairs without regard for any mode of achieving it. Still, dogs can exhibit a pattern of behavior that accords with the rule, and they can do so *because* of the rule, if their masters train them to come when called because the masters have reasoned along the following lines:

> It ought to be the case that dogs come when their masters call.
> Therefore, it ought to be the case that my dog comes when I call.
> My dog will come when called only if I train it to do so.
> Therefore, I ought to train my dog to come when called.

This reasoning moves from an ought-to-be to a relevant ought-to-do and comes to full fruition not in a belief about one's obligations, but in a set of actions that result in one's dog learning to come when called. Ought-to-be's imply ought-to-do's.

Linguistic rules are primarily rules of criticism, ought-to-be's, especially the rules that define the structures constitutive of the language. Language is possible only because our linguistic behavior exhibits numerous interconnected regular patterns shared by most of the community because of the early learning that made those patterns second nature to community members. We can think about this set of patterns in two different ways, however, and this is significant for understanding the relations among Sellars, Millikan, and Brandom.

On one view, an anthropological and causal story dominates. It concerns the evolution and proliferation of complex patterns of behaviors and dispositions that provide for enhanced communication and coordination in the community. On this first view, the rule-governedness of linguistic behavioral patterns recedes into the background; linguistic behaviors take the shape they do, not *because of the rules*, but because the functionality of their historical ancestry explains their current appearance. At best, a rule summarizes a historical pattern. Indeed, the question is whether, given this view, the rule-governedness of linguistic behavior doesn't simply evaporate altogether.

On the other view, the sanctioning of the patterns and dispositions in the community, particularly the passing on of these patterns to new members, is taken to embody an *endorsement* of that set of patterns by the community, and the actual behaviors in the community are taken to realize (more or less well) a set of ought-to-be's operative in the community. On this view the base-level linguistic behavioral patterns are still imbued with normativity – as Brandom says, there are norms "all the way down" (Brandom 1994, 44, 625, 638) – but one has to ask: Communities of all kinds share patterns of behavior (isn't that what it means to be a "community?"), so what distinguishes a *de facto* from an *endorsed* and thus *de jure* behavioral pattern? When and how does a behavioral pattern acquire normative force?

We can recognize Millikan and Brandom in these two different views of the base-level patterns of linguistic behavior. Apparently, they do not believe these two views can be made fully compatible; the tension that Millikan finds lurking in Sellars is just this. On the one hand, because rules are normative or prescriptive, they cannot be translated or reduced into non-normative, purely descriptive terms, yet, on the other hand, "… on Sellars' view the presence of normative rules in the natural world appears in the end as just one more level of fact in that world" (Millikan 2005c, 62). Brandom seems to agree that the two views are not compatible, but he disagrees with Millikan about which view is dominant.

Some Sellarsian Geophysics

Rules, rational agents, and the institution of norms

Let's explore the status of the normative in Sellarsian philosophy, to further isolate the purported problem. First, Millikan's just quoted remark that the presence of norms turns out in the end to be "… just one more level of fact in [the] world" needs explication. Sellars is happy to acknowledge normative facts, but that is because, in general, he takes fact-talk to be material mode truth-talk, and truth, for Sellars, is simply ideal semantic assertibility. The presence of normative facts (such as that murder is wrong) comes down to the categorical ideal assertibility of certain sentences (e.g., 'Murder is wrong'), which sentences ultimately express prescriptions and proscriptions of certain kinds of action. Sellars thinks some sentences that express norms must be semantically assertible in any real language and therefore true in the language, and that means that there are normative facts. But this is not a terribly "deep" truth about the ultimate ontological status of norms, in Sellars' view, for facts are, in any case, not basic ontological items.

Linguistic rules are ought-to-be's. They are not *categorically* assertible, for they do not hold for every rational agent, but only those attempting to speak a particular language. Speaking a language entails acting *because of the rules*, even if one is not (yet) aware of behaving in accordance with such rules or of what those rules may be.

In Sellars' view, however, human languages never remain at the level of orchestrations of first-order behaviors or dispositions to behave. Human languages all contain the resources of their own meta-languages, and the possibility of reflexivity, of talk about language, is essential to mastery of one's language.

> One isn't a full-fledged member of the linguistic community until one not only conforms to linguistic ought-to-be's (and may-be's) by exhibiting the required uniformities, but grasps these ought-to-be's and may-be's themselves (i.e., knows the rules of the language.) (LTC, 101)

Such a reflexive grasp of one's language is *necessary* for a peculiar reason.

> Rule obeying behavior contains, in some sense, both a game and a metagame, the latter being the game in which belong the rules obeyed in playing the former game as a piece of rule obeying behavior. (SRLG in SPR, 327; in ISR, 34)

According to Sellars, rules of criticism (ought-to-be's) have no way of being realized in the world (other than merely accidentally) except insofar as there are *agents* who infer rules of action from them and thereupon undertake the requisite actions.[2] Unless there are agents cognizant of and acting on some rules of action, talk of *any* rules, including the apparently less committal rules of criticism, turns out to be empty.[3]

This has consequences. This is why the application of ought- or rule-talk to the activities of undomesticated animals is, in Sellars' view, always analogical.[4] Sellars also has difficulty explaining how the first 'ought's came to be, for there could be no rule-obeyers, it seems, unless they were trained to recognize and obey already existent rules, yet there could be no such rules until there are rule-obeyers, agents, ready to recognize and obey them. Most likely Sellars thinks that this problem can be resolved by a move parallel to his resolution of the threat of circularity in his epistemology. Well-entrenched patterns of behavior in a community are *retrospectively* endowed with a normative status they did not originally have, because the community comes to *endorse* that pattern of behavior as community members come to acquire the explicit conceptions of an 'ought' and a rule. Behavioral patterns selected by environmental or community pressures that could be metaphorically or analogously described as rule-governed can become recognized and endorsed by the community and thereby become literally rule-governed. The relevant notion of a rule is transformed from a summary of a historically grounded pattern to a prescriptive with normative force.

In his late article "Mental Events," Sellars discusses human language as a species of a broader genus, animal representational systems (RSs). Sellars argues there that what distinguishes distinctively human RSs from other animal RSs is not the presence of subject–predicate structure, for this is just a particular way of embodying two characteristics that are common to *every* representational event: "… one by virtue of which it represents an

---

[2] There is more work for Sellars to do hereabouts. We do not, for instance, currently have explicit knowledge of all the ought-to-be's of English. But we do have a conception of English as a rule-governed system, and I, for one, want my children to command it well, and took steps when they were growing up to ensure they did.

[3] A word of warning: this does not mean that behavior fulfilling ought-to-be's ends up as paradigmatic rule-following behavior. The basic behavior patterns of language speakers are not only "… acquired as pattern governed activity, they remain pattern governed activity. The linguistic activities which are perceptual takings, inferences and volitions never become obeyings of ought-to-do rules" (MFC, 424; in ISR, 88).

[4] Domesticated animals are often trained and therefore behave in certain ways *because of rules* that their trainers have recognized.

object in its environment (or itself); another by virtue of which it represents that object as being of a certain character" (MEV, ¶72(b), 338; in ISR, 294). What distinguishes human RSs, Sellars holds, is the presence of *explicit logical structure*: "The crucial distinction is between *logic-using* RSs and RSs which do not *use* logic, though their operations are described by *mentioning* logical operations" (MEV, ¶79, 340; in ISR, 296). The latter systems, which he calls "Humean RSs," associate one representational state, such as 'Smoke here,' with others, such as 'Fire nearby,' and possess propensities to move from the one state to the other directly. Logic-*using*, or "Aristotelian," systems can get from a 'Smoke here' to a 'Fire nearby' representation either by this direct Humean route or by means of an intermediary, standing, quantified representation "Wherever smoke, fire nearby" together with (Humean) propensities to infer in accordance with formally valid inference forms. "Thus we can say, following Leibnitz, that a Humean RS which moves directly from 'Smoke here' to 'fire nearby' apes an Aristotelian RS which syllogizes.... As Leibniz put it, '[animals have] a sort of *consecutiveness* which imitates reason'" (MEV, ¶90, 342; in ISR, 298).[5]

Logical particles in an RS make a significant difference in the expressive power of the system, particularly if one includes the modal operators as logical particles. But it is odd that Sellars does not mention reflexivity, the possibility of talk about talk and thought about thought, as at least *another* important distinguishing trait of human representational systems. We have seen already that he is committed to such a thesis, and it is important to getting a full grasp of the autonomy of reason. The relation between reflexivity and the presence of logical structure for him remains unclear.

Brandom offers a hint, treating the logical particles as enabling us to make explicit relations between representations that would otherwise remain implicit in the inferential practices involving those representations. This is not yet a new level of meta-representations, but it might be construed as the first step toward evolving practices that enable us to treat representations as representations in their own right. What Brandom does bring out strongly is that normative statuses are derived from normative attitudes, so that rulishness is tied, not directly to our *de facto* patterns of behavior, but to how we represent those patterns. Normativity or rulishness is *instituted* by such attitudes – that is how Brandom expresses the point Sellars makes by basing the reality of ought-to-be's on their being transformed into and acted upon as ought-to-do's.

Scientific realism and the causal order

The metaphor offers itself that the rule- and ought-focused perspective on language and thought is a "top-down" view that needs to be complemented by a corresponding "bottom-up" view of the relevant behaviors. Sellars' deep commitment to naturalism, and more specifically, scientific realism, requires that high-level events such as thinkings or sayings get somehow grounded in the low-level events described and explained in the natural sciences. "Somehow grounded" is, of course, vague and challenging: can Sellars

---

[5] Sellars refers here to §26 of Leibniz' *Monadology*.

articulate a sufficiently clear and powerful "grounding" relation to serve his naturalism? Still, the "top-down vs. bottom-up" picture is at least slightly misleading, since, in Sellars' view, these perspectives employ incommensurable vocabularies, guaranteeing that there will be no smooth and unproblematic melding of bottom-up and top-down views.

Even a cursory reading of Sellars shows that he rejects the idea that rules, norms, intentional states, etc. can be simply *defined* in natural scientific terms. The concepts of rules, norms, and oughts and the dependent concepts of intentional states, of linguistic utterances and performances, remain, in the end, at least partially independent of the descriptive conceptions used in the natural sciences, precisely to the extent that normative, action-motivating concepts cannot be fully analyzed in descriptive, non-action-motivating language. A rule cannot be *analyzed away* in the purely descriptive language of natural science.

Nonetheless, Sellars is clearly a scientific realist. One construal of scientific realism is that it claims that the facts expressible in the language of the empirical sciences are *all* the facts. Sellars cannot subscribe to this version of scientific realism, for, as we have seen, he is happy countenancing moral and other normative facts. Sellars' version of scientific realism is object-centered in the sense that he is committed to the thesis that the objects identified by the empirical sciences (at their ideal completion) exhaust the *basic* (kinds of) objects of the world. Other (kinds of) objects will prove to be in some appropriate sense reducible to or derivative from the basic objects identified in the empirical sciences.[6] There is, in principle, according to Sellars, a stratum in any empirically usable language (or conceptual scheme) in which the objects that are *basic* within the language are represented in configuration in a way in which no normative or prescriptive term occurs (see CDCM, 282–283).[7]

As a scientific realist, Sellars is also a normative anti-realist. This does not mean that he rejects the idea that there are normative facts. Rather, it means that there are no essentially normative *basic objects*, that normative properties are never basic, intrinsic, or non-relational properties of objects. Every normative fact is dependent on some (usually incredibly complex) descriptive *relational* facts about the object(s) involved. This also means that *persons*, objects that are essentially characterized in normative terms, cannot be naturalistically *basic* objects. So much for the manifest image.

Millikan correctly points out that according to Sellars, "[f]rom the scientific realist's standpoint, you can understand the nature of the normative practices of a community without participating in them" (Millikan 2005c, 62). That is, in explaining the practices of a community, one can adopt a standpoint outside the particular practices of the community studied, linguistic practices included. One "… can describe what patterns of

---

[6] According to Sellars, the principle here is that "If an object is *in a strict sense* a system of objects, then every [intrinsic or non-relational] property of the object must consist in the fact that its constituents have such and such qualities and stand in such and such relations or, roughly, every [intrinsic or non-relational] property of a system of objects consists of properties of, and relations between, its constituents" (PSIM in SPR, 27; in ISR, 395).

[7] In CDCM Sellars projects an ideal of a "pure description" of the world that would contain nothing either modal or normative. That idea, it turns out, won't do as it stands, and Sellars developed the idea of picturing in order to resolve the difficulties that arose. I make a first attempt to spell out the dialectic that led to his development of the notion of picturing in my "Naturalism, the Autonomy of Reason, and Pictures" (deVries 2010).

response in a language community, along with the origins of these responses in a history of language training, and training of the language trainers, and so forth, constitutes that 'rot' means red in that community" (Millikan 2005c, 62). Indeed, Sellars sums the basic principle here succinctly: "Espousal of principles is reflected in uniformities of performance" (TC, in SPR, 216).[8]

Sellars is aware of the fact that explanations of the uniformities of performance in human practices will generally account for their fixation and proliferation by citing their utility and ultimately their survival value. The analogy with the forms of evolutionary explanation is unavoidable.

> The phenomena of learning present interesting analogies to the evolution of species. (Indeed, it might be interesting to use evolutionary theory as a *model*, by regarding a single organism as a series of organisms of shorter temporal span, each inheriting disposition [*sic*] to behave from its predecessor, with new behavioural tendencies playing the role of mutations, and the 'law of effect' the role of natural selection.)' (SRLG ¶16, in SPR, 327; in ISR, 34).[9]

Millikan sees this evolutionary line of approach as a "… competing theme in Sellars' discussion of linguistic rules" (Millikan 2005c, 64). After all, if we can explain the uniformities of linguistic performance, which is just the kind of thing science is good at, won't we have thereby also explained or made simply otiose any further explanation of the "principles" involved? Since there will inevitably be slack in interpreting exactly which normative principles are "reflected in" a determinate set of uniformities of performance – or do we believe that a complete enough specification of the uniformities of performance in a community will determine a *unique* account of its 'principles' (linguistic, aesthetic, or moral)? – having a good theory of the behavioral uniformities seems likely to displace reliance on the principles altogether. This is the side of the crack in Sellars' edifice that Millikan has cultivated to great effect.

Picturing

Millikan and Sellars agree that a full understanding of intentionality requires us to believe that our mental states embody a *picture* or *map* of our environment that enables us appropriately to maneuver within and modulate our behavior in response to that environment.

---

[8] Note the vagueness of Sellars' "reflected in." This is intentional on his part; his point is intended to be minimal: "I am merely saying that the espousal of a principle or standard, *whatever else it involves*, is characterized by a uniformity of performance. And let it be emphasized that this uniformity, though not the principles of which it is the manifestation, is describable in matter-of-factual terms" (TC, 216). Millikan inveighs at some length against the idea that conventions involve *regularities* (e.g., Millikan 1998a, sections IV and V, 170–175), and this could be taken as a rejection of Sellars' principle. But the regularities that Millikan rejects seem to be *stringent*, and the uniformities that Sellars refers to are not stringent. They can be, for example, historically or culturally conditioned.

[9] A bibliographic warning: the version of SRLG reprinted in ISR is the original, unrevised, unexpanded journal article. The versions coincide pretty well through ¶30, but there is a large addition in the SPR version consisting of ¶¶31–46 as well as smaller changes in later paragraphs.

For Sellars, picturing is a non-semantic isomorphism in the natural order between the structure of the occurrences of a subset of the tokens of the singular propositional representations of a representational system and the structures of spatio-temporal objects they represent. This isomorphism is a *de facto* relationship between objects in the environment and representational events, and it serves a natural purpose in modulating the organism's interactions with its environment. Sellars' conception of picturing is often thought to be fairly obscure and has not received much attention; Millikan is one of the few who have picked up the idea and worked with it. But Millikan's conception of mapping does not seem entirely at one with Sellars'.

Millikan reports that, according to Sellars, "... the manner in which the names occur in the picture is a projection, in accordance with a fantastically complex system of rules of projection, of the manner in which the objects occur in the world" (TC, 215; NAO, V ¶93, 118–119).[10] She then claims that "These fantastic complexities are introduced mainly by the inference rules, formal and, more importantly, material, that govern 'statement–statement' (hence judgment–judgment) transitions" (Millikan 2005c, 60). The fantastic complexities in the rule of projection will certainly be reflected in the (also complex) inference rules of the language, but it is misleading to say that they are *introduced* by the inference rules, as if the inference rules somehow came first. The notions of *picturing* and *method of projection* are, remember, relations "*in rerum natura*" (TC, 222; NAO, V ¶116, 125). Inference rules are patterns in the conceptual order; methods of projection are patterns in the *ordo essendi*.

> ... [T]o say of a projection that it is *correct*, is, indeed, to use normative language[;] the principle which ... I am taking as axiomatic [that espousal of principles is reflected in uniformities of performance] assures us that corresponding to every espoused principle of correctness there is a matter-of-factual uniformity in performance. And it is such uniformities, which link natural-linguistic objects with one another and with the objects of which they are the linguistic projections, that constitute picturing as a relation of matter of fact between objects in the natural order. (TC, 222; NAO, V ¶116, 125)

The "projection relation" is rarely, if ever, a matter of a simple, straightforward law of nature. Rather, any particular item pictures only because of its position in a vastly complex *system* characterizable in terms of complex "laws operative in situ," as Millikan calls them. Wittgenstein's model was the relation between the shape of the groove in a phonograph record and the sound of a musical performance – a relation in which the similarity of the picture and the pictured is still relatively intelligible (once one has learned to think of sound itself as possessing a wave form). But the projection relation between sounds and the sequence of bits in a digital sound file is difficult to construe as a relation of resemblance and makes sense only within the context of an orchestration of complex electro-mechanical systems, none of them essentially inferential.

---

[10] Another bibliographic warning. There are two printings (both from Ridgeview) of Sellars' *Naturalism and Ontology* (NAO; Sellars 1979), the original printing of 1979 and a "corrected" printing from 1996. Though they are not marked as different editions of the work, they do not agree in pagination. I lost my copy of the original printing (the one Millikan refers to) a long time ago. My citations of NAO include the edition-neutral chapter and paragraph numbers as well as the page in the 1996 printing.

In contrast to Millikan, however, Sellars explicitly stipulates that in picturing,

> ... the correspondence for which we have been looking is limited to elementary statements, or, more accurately, to the elementary thoughts which are expressed by elementary statements and which we conceive of by analogy with elementary statements. (TC, 223)

According to this stipulation, the picturing relation, *strictu sensu*, holds only between objects (in configuration) and elementary thoughts. Logically compound or complex thoughts, thoughts that contain metalinguistic terms (perhaps in material disguise), do not *picture* anything.

Millikan, however, seems to construe the notion of a map much more expansively. She asserts, "I adopt Sellars' suggestion that adequate intentional representing is a kind of picturing or mapping" (Millikan 2005c, 67). For Sellars, she says, "The map of the world produced by a language is not found sentence by sentence but only in the whole of the living language *cum* thought running isomorphic to the whole world in sketch" (Millikan 2005c, 60). She also cites in support of this interpretation Sellars' assertion that "... the representational features of an empirical language require the presence in the language of a [whole] schematic world story" (NAO, V ¶59, 109).

But I think Millikan is in danger of misconstruing Sellars' use of the map metaphor. Sellars would not have agreed that "... adequate intentional representing is a kind of picturing or mapping," if that is supposed to mean that *all* intentional states are pictures or maps of something. Sellars wants to maintain a *distinction* between picturing and intentionality. An animal representational system that pictures its environment is not *thereby* possessed of full-fledged intentionality. What he believes is that any system capable of intentional representation must contain some intentional representations that *also* picture objects in the environment. But there is no inference from "*S* is an intentional representation" (or "*S* is an adequate intentional representation") to "*S* pictures something in the environment." In Sellars' view, picturing is a relation, but talk of intentionality is actually classificatory talk (viz., functional classification). To ascribe intentionality to something classifies it in a way that *rests* on complex relations, one of which is (often) the picturing relation, but it does not attribute any particular relation directly. Some, but not all, of the representations in an intentional system must picture objects in the world.

Second, I think it is a mistake to assimilate Sellars' notion of a world *story* to the notion of a *map* of the world, though in accordance with the previous point, any such story will *imply* a map of the world. Sellars himself invites a comparison between maps and world stories: "If one is going to compare a world story with a map, one must ponder the distinction between 'real' maps and 'fictional' maps. One doesn't try to go places with a map of Hobbit-land" (NAO, V ¶60, 109). But Sellars immediately goes on to admit that the distinction between real and fictional maps "doesn't cut deeply enough" (NAO, V ¶61, 109). What it seems to lack, in Sellars' view, is a recognition that languages permit the formulation of alternative world-stories and discussions of how to choose among them. He believes that any language at any point in time must make a commitment to *one* (sketchy) world-story, but that commitment is *provisional* and the world-story is under constant revision. The relevant passages in chapter V, section V of Sellars' *Naturalism and Ontology* (NAO) are fairly obscure. Let me quote them and then try to interpret what Sellars has in mind.

> 63. That languagings are *evoked* (in contexts) by happenings of certain kinds is a *causal* fact which is nevertheless essential to their conceptual character. This causal aspect of perceptual takings, introspective awarenesses, inferences, and volitions accounts for the selecting of *one* world story *rather than another* and connects the 'is' of this selecting with the rule-governed or 'ought to be' character of the language. The 'presence' of this unique story at each stage in the development of the language makes possible the referential framework of names, descriptions and demonstratives and, by so doing, makes possible the exploratory activity which lead [*sic*] to the story's enrichment and revision.
>
> 64. Thus, the fact that the uniformities (positive and negative) involved in language-entry, intralinguistic and language-departure transitions of a language are governed by specific ought-to-be statements in its meta-linguistic stratum, and these in turn by ought-to-bes and ought-to-dos concerning explanatory coherence, constitutes the Janus-faced character of languagings as belonging to both the causal order and the order of reasons. This way of looking at conceptual activity transposes into more manageable terms traditional problems concerning the place of intentionality in nature. (NAO, V ¶¶63–64, 110)

These paragraphs are not pellucid. What I take from them, so far as our current discussion is concerned, is that the notion of a world-*story*, with its explicit reference to language, is preferable to the notion of a *map* of the world in that the distinctively meta-representational capacities of language, which are not available in maps, are crucial to understanding the "Janus-faced character" not just of languagings, but of rational activity as such. It is only in language and thought that we worry about what we ought to say and think and whether we are speaking and thinking correctly. There are no maps of the rules governing maps or maps for correcting maps, but there are linguistic expressions of the rules governing language and rational activity.[11] If the causal account of the selection of the world-story we accept must also underpin an account that worries about whether we have a *right* to accept that story, whether it is a *justified*, even a *truthful* story, some of the causes of our acceptance must show up within the story as *reasons* embedded in a system of norms. Such a system, Sellars believes, requires the reflexive or "meta-" capacities of a language and thus goes beyond anything the map metaphor can give an account of.

The picture or map metaphor, which Millikan treats as going a long way toward helping us understand the nature of language, is treated by Sellars as useful to help us understand a transcendental condition of the meaningfulness of empirical language. That under some conditions elementary sentences picture is, he believes, a crucial element in any empirically meaningful language, but it cannot be the whole story of the language. He remains committed to the need to refer to rules and norms.

---

[11] Notice that the *legend* that gives the semantics of a map does so by pairing map-symbol types with *linguistic* counterparts. The semantics of a map, like the semantics of a language, is expressed explicitly in language.

Games, Conventions, Rules, Norms, and Essential Perspectives

The supposed fault line in Sellarsian philosophy which divides Millikan and Brandom concerns the proper treatment of the fundamental uniformities of behavior that underlie all language use and ultimately, all rational action. Brandom, about whose position I have said little, says that there are norms "all the way down," that is, that there is no point at which to place a transition from non-normatively described dispositions or uniformities to normatively characterized dispositions or uniformities. Millikan, in contrast, thinks that all we ever need are historically grounded but always naturalistically describable dispositions (where that means described in non-prescriptive-rulish terms) or uniformities. For Millikan, it is causes all the way up. But Sellars insists on the "Janus-faced character" of our dispositions: both the causal–natural and the normative–rational vocabularies are indispensable.

Brandom, like Sellars, motivates his view that rules and normative norms are indispensable in understanding language, in part, by the constant analogy between languages and games. Millikan, however, thinks her view competes, that is, *conflicts*, with the idea that linguistic rules are like rules in a game. For one thing, the dispositions or uniformities in question *matter*, for what is at stake is *survival*. Millikan disparages games as opportunities for "displaying certain social graces" (Millikan 2005c, 64), whereas "... coming to follow the patterns prescribed by the rules of one's language community is not just a game but has some broader utility for the child or for its community" (Millikan 2005c, 64). But this criticism seems off the mark; there is no conflict between something's being a *game* and its having significant utility. Some of the most highly remunerated members of our society are sports stars, and playing a decent game of golf has long had significant utility in the business community beyond providing an opportunity for "displaying certain social graces." Games are deeply woven into human social structures and often matter a great deal. In any case, it is not the lack of seriousness that is the feature of most games that the metaphor seeks to exploit, but the facts that certain kinds of acts are possible only in the context of a particular game, and that, though games are perfectly objective realities in the world, they are instituted and constituted by the acceptance of certain rules governing the participants' behavior.

A more apt target for Millikan would be the voluntarism of games. The playing of a game is almost always optional. When it becomes non-optional (or optional only at a discouragingly great cost) we declare that it is "no longer a game." This is not the same as saying that games don't matter. Many things that matter deeply are optional, such as betting the farm on drawing to an inside straight. Second, although some "games" (like peek-a-boo) seem to develop naturally, most of us learn to play various games by being taught the rules, so the moves made in playing games are paradigmatic rule-obeying behaviors, although in accomplished practitioners explicit consideration of the rules often recedes from consciousness except in highly problematic cases.

Learning a language does not seem optional for humans, and most of the rules governing language never rise to explicit consciousness at all. Of course, none of these disanalogies between language and games is news to Millikan. But she does not claim that the game metaphor is *wrong*, only that it can *mislead* us at times. The fundamental point is that what Sellars calls a causal/anthropological approach and Millikan prefers to call simply a biological approach is, in Millikan's view, neither a metaphor nor misleading; it can be elaborated into a full-blown theory, and the right one at that. Importantly, it is an approach in which rules need never be mentioned as such.

Millikan does not shy away from the consequence that, seen from this angle, the vaunted normativity of linguistic rules simply evaporates.

> The norms for language are uses that have had "survival value," as Sellars put it. As such these norms are indeed disposition transcendent, but they are not fraught with ought. They are not prescriptive or evaluative norms. Their status has nothing to do with anyone's assessments. A norm is merely a measure from which actual facts can depart; it need not be an evaluative measure. (Millikan 2005c, 64–65)

Millikan wants to understand language, not as a *rule*-governed system, but as "a sprawling mass of crisscrossing, overlapping conventions, some known to some people, others to others" (Millikan 2003a, 216), where a convention is a pattern of activity or behavior that is *reproduced* (in Millikan's technical sense of that term) "due in part to the weight of tradition, rather than due, for example, to its intrinsically superior capacity to perform some function, or due to ignorance of any superior way to perform it" (Millikan 2003a, 219).

In her view, "Language conventions are best thought of merely as lineages of behavioral patterns involving a speaker's utterance and a hearer's response. They do not correspond to rules, and certainly not to prescriptive rules" (Millikan 2005c, 67). If language is a distinctively human activity, it is not, in Millikan's view, distinctive *in principle*, but only in fact. There is a certain mass of conventions present in human societies that enable a rather striking facility for communication among the members of that society, but there is nothing in principle "deeper" going on. Millikan is willing to talk about the "natural purpose" (Millikan 2005c, 65) of these conventions and the social forms that embody them, but in her mouth the notion of a natural purpose is a historical–causal notion with no normative or evaluative content to it.

Millikan seems eager to deny natural language the kind of normativity that would justify calling it a "rule-governed system." She is not as clear about this as she ought to be. As we've seen, Millikan believes that the *real* story underlying our linguistic behavior is to be told in terms of conventions; non-prescriptive, historically transmitted, *de facto* patterns of behavior. Yet she too often uses the term 'rule' in a way that, in this context, has to be called loose, for 'rule' often contrasts with 'convention.'

Indeed, there is a use of 'rule' in which no normativity is implied, for instance "As a rule, rush hour traffic is over by 9:30." "These rules describe conventional patterns; they do not prescribe them. The bare existence of a convention neither mandates, nor gives permission for anything" (Millikan 1998a, 173). But Millikan also talks somewhat confusingly about "conventional rules" (e.g., Millikan 1998a, 166, 173), and though she explicitly denies that these have any prescriptive force, her language muddies important waters. To my ear, there is a difference between "In these circumstances, there is a convention that one A's" and "In these circumstances, there is a conventional rule that one A's."

The latter implies some normativity, evidenced by the fact that these sentences have different implications concerning whether it is appropriate to describe not A-ing in those circumstances as a *mistake*. Thus, if, as a rule, rush hour traffic is over by 9:30, then, when 10:00 rolls around and traffic is undiminished, one could say, "Something is wrong, traffic is as bad as it was earlier." But it would be very odd to say "There's been a mistake, traffic is as bad as it was earlier." However, the notion of mistake is regularly applied to linguistic performances.

The point here is not to engage in an extended critique of Millikan's conception of language as conventional. Much of her vision is very powerful, and I have no complaints about her notion of conventions. My concern is that her conception of normativity is relatively thin and without much articulation. Too many of Millikan's examples of rules with prescriptive force, unfortunately, are cases of actual legislation, cases where the coercive power of the state is placed behind the sanctions connected to the rule. If, in thinking about language, we're forced to choose between convention vs. legislation, we cannot put language on the legislated side.

But the idea that we are forced to choose between convention and explicitly legislated and enforced patterns of behavior is surely a false dichotomy. It was to avoid such a false dilemma that Sellars carefully distinguished different forms of rule-governedness. The "weight of tradition" can be brought to bear upon the reproduction of a certain pattern of behavior in humans in several different ways, depending on how conscious of the tradition the subjects are and how strongly they endorse and support the tradition. But for Millikan, apparently, one causal ancestry of pattern-governed behavior is pretty much equivalent to any other – as long as the patterns are all explained by reference to their causal ancestry and selective forces, normativity as such seems absent.

In Sellars' view, not all causal ancestries are equivalent: he reserves a special place for two kinds of behaviors:

1 Behaviors that include among their typical causal ancestry events or dispositions in the subject that are representations of rules, which representations, in conjunction with others (both cognitive and conative representations), tend to bring about behavior that accords with the rule.
2 Pattern-governed behaviors that include among their causal ancestry events or dispositions in others (the trainers or community members) that are representations of rules, which representations bring about behaviors in those others that lead to the acquisition in the trainee of pattern-governed behavior that accords with the rule.[12]

How can Sellars justify giving these kinds of behavior special treatment? In Millikan's view we need employ only the notion of a convention to understand the basics of linguistic behavior. She singles out conventions by means of a peculiarity in their causal ancestry; conventions differ from other reproduced behavioral patterns in having not only a proper function, but also being reproduced to an important degree simply because of

---

[12] This is still rough. Cases where the representation of a rule accidentally causes the behaviors in question would have to be ruled out.

tradition. (I noted above that there may not be just a single way in which traditions can be involved.) Structurally, Sellars' move is similar. He singles out a subset of pattern-governed behaviors by means of a different peculiarity in their causal ancestry, namely, having in an appropriate position in their causal ancestry a representation of a rule with which the pattern-governed behavior is supposed to accord.

Given Sellars' view of representations and representational systems, it will always be the case that some of the representations in a system will connect perception and action pretty directly. In simple RSs, most the representations will have this character. Millikan agrees and calls these "pushmi-pullyu" representations. But Millikan clouds the issues a bit when she then says:

> Already at this simple level a stringent criterion of correctness for rule following is in effect. The perceptual systems must manage systematically to deliver representations of the world that accord with a rule of correspondence to which the action systems are also adjusted. (Millikan 2005c, 68)

In simple animals, the convergence of perceptual and action systems is a matter not so much of *rules*, properly so-called, as of "laws operative in situ." In animals, the perceptual and the action systems have been designed by evolution to mesh together well enough to keep the organism functioning and reproducing.

But in more complex organisms, Millikan rightly notes, the attunement between the perceptual and the action systems is less direct and more complex.

> Beyond perception for action, humans, at least, make cognitive maps that are not dedicated in advance to the guidance of particular behaviors. We collect great quantities of information with no immediate uses in view, storing it away perhaps for later contingencies. Having separated the descriptive from the directive aspects of representation, these have to be joined together again through practical inference. But representations of fact that are not immediately tested in action and that are then used to form more representations and then still more through inference, need to be screened for accuracy and consistency in some way. Rules or patterns of belief formation need to be strictly regimented as they are developed, well in advance of practical uses for the resulting beliefs. (Millikan 2005c, 68)

But notice here the unexpanded notion of inference. I pointed out earlier Sellars' distinction between "Humean" and "Aristotelian" inferences and RSs. The Humean RS can adjust its inferential patterns only by changing old or acquiring new habits of association; it cannot consider or address its inferences *as such*. This is a slow and painful way to change one's representations. The Aristotelian RS can change its inferential patterns by discarding or acquiring new generalizations, leaving the basic inference patterns untouched. In the Aristotelian RS the *form* and the *content* of an inference can start to be teased apart, unlike the Humean RS.

It is also too simple to say that the descriptive aspects of representation have been, in humans, simply separated from the directive aspects. Humans can have representations with only indirect implications for action, but every representation must have some directive aspect. In cases of representations that are very indirectly tied to action – representations of pi-mesons, unicorns, or chiliagonality – the representations must have directive aspects in another key. Call these directive meta-aspects; they concern how we proceed to act on other representations, rather than how we act directly on the objects of the world. We isolate the descriptive aspect of a representation, not by separating it from all its directive force, but by kicking the directive force up into the meta-level, where it concerns the inferential powers of the representation, rather than ground-level activity.

This invites us to revisit the map metaphor. Clearly, we sell the metaphor short if we restrict the maps we use as models for cognitive representations to the moribund paper maps one gets from AAA. Maps can be not only 3- but 4-dimensional, that is, portraying not just where things are at a time, but also how they move and change over time. Computer simulations of an auto accident or of a weather system are essentially dynamic maps of those events. An animal representational system is a world-simulator, though in simple animals the simulation is very thin and partial. Such dynamic maps have to represent the dynamic principles operative in nature, but they need do so only in a quick and dirty fashion adequate to keep the animal functioning and reproducing. Animal systems, therefore, have approximate procedural representations of some of the dynamic principles operative in the world. Humans, however, can, in language, also represent those principles declaratively, and thereby make them an object of explicit concern. They can then seek to refine their understanding of the dynamics of nature, seeking not merely quick-and-dirty, partial, satisficing representations, but optimally correct and complete representations. Human representational systems aim at an ideal unfathomable by other animals.

While animals adjust their dynamic maps unconsciously in the school of very hard knocks, the linguistic facility of humans enables them to reprogram their maps in a methodical and conscious fashion. We can't say that natural human languages are the programming languages for our animal RSs; the relationship is not so direct. But Sellars takes seriously the idea that research in the empirical sciences will ultimately restructure the representational system we employ in very significant ways, replacing the original image[13] acquired through the blind processes of evolution and conditioning with a consciously elaborated, carefully justified, more thorough and complete scientific image of the world. This process is one that Sellars thinks can be thoroughly intelligible only to someone who has the normative concept of rule.

Some object that the normativity involved isn't ontologically real – it is only a "believed-in" normativity, for the fundamental being of a rule, on this view, depends on our ability to *represent* rules. Millikan could not make this objection, with her even harsher view of normativity, but there are still normative realists among us. I cannot address this objection fully, but Sellars thinks that within his system we can preserve everything we need to say about norms and values.[14] He simply doesn't believe that norms have any reality independent of minds that can represent them, for the whole point of a norm is prescriptive, to motivate behavior in a cognitive being. However, he also thinks we cannot fully understand the behavior of sophisticated cognitive beings without mentioning the norms they believe in and regulate themselves by.

---

[13] See PSIM, in SPR, 7, 10; in ISR, 375, 378–379.
[14] Principally in the final chapter of SM and in ORAV.

This brings us back to an earlier theme. Millikan points out that "[f]rom the scientific realist's standpoint, you can understand the nature of the normative practices of a community without participating in them" (Millikan 2005c, 62). This is true. But it is equally important to appreciate the fact that one cannot stand outside *all* normative practices and understand them only from the outside. Understanding something *is* participating in a normative practice. As cognitive beings, we can see any part of the world in non-normative terms, save that part that includes our selves as beings responsive to norms, possessed of the *right* to believe some things and not others and engaged in rational activity. To the extent that we can see the part of the world we inhabit in thoroughly non-normative terms, there would be no *we* left within it, nor any *seeing* or *talking* being done. We *could*, of course, cease to see ourselves as responsive to norms, because we could be *caused* to change our ways of interacting with each other and treating ourselves, but we couldn't validly *reason* our way to the idea that reason is an illusion.

An external point of view on any normative practice is always possible, but we cannot infer that points of view from within practices are thoroughly dispensable. To put oneself outside all normative practices is to abandon both practical discourse and practical reason as well as theoretical discourse and theoretical reason. We would no longer recognize the norms to which we feel ourselves to be responsive in reasoning of any kind. Were we thus to cease to represent rules, such representations would not play significant roles in our behavior, so descriptions and explanations of our behavior would no longer advert to oughts or rules. The picturing relation between our internal states and the external world might endure (since picturing is not "fraught with 'ought'" (TC, 212)). But any subsequent change toward more or less adequate picturing would have to be the result of unconscious, impersonal selection pressures, for, though "truth (adequacy of representation) abides as the *would be* of linguistic representation" (NAO, 111), there could be no recognition that truth or adequacy of representation might serve as the *should be* of linguistic representation.

A theoretical position that undercuts the possibility of possessing a theoretical position is not attractive. Brandom doesn't pay enough attention to the causal underpinnings of linguistic and conceptual activity, but Millikan doesn't pay enough attention to the indispensability of prescriptive, normative concepts and attitudes in any coherent conceptual framework. So I encourage Brandom and Millikan to cultivate their plots on either side of the supposed Sellarsian fault line; there is much to be learned from their efforts. But for the larger picture, I'll stick with daddy: the supposed crack is not a fault line, but a joint in the complex and articulated reality in which humans live.

Reply to deVries

A Daughter's Respectful Disagreement with Sellars

DeVries' understanding of Sellars' project is very much deeper than mine. I *especially* recommend DeVries (2010) that he cites to those interested in Sellars on picturing. (When Sellars taught at Yale, we learned from him mainly what he thought were the questions, not his answers.) DeVries shows us that Sellars had a thorough understanding of the supposed crack in his corpus to which I referred, and that his way of avoiding it was at the very heart of his enterprise. Given this more focused understanding of Sellars, let me then explain – the barest sort of sketch, given very limited space – where I depart from Sellars' views (and, I gather, from deVries').

Contemporary child language studies indicate clearly that rather than being trained into language, children learn language effortlessly without any hint of instruction – in the case of average US school children, an intricate grammar and 45,000 words by age 16. Many modern linguists believe this can be accounted for only if essential aspects of language are universal and innate (I am skeptical). But those 45,000 words? I do not see how they can be accounted for on an inferentialist theory. I do think they can be accounted for, however, if we accept two points that I have argued for in previous writings.

First and most radical perhaps, I have argued that understanding descriptive language is a form of perception of the world, as direct as is seeing what's happening via ambient light, or as hearing what's happening via other kinds of ambient sound (OCCI, ch. 6, VOM, ch. 9, LBM ch. 10, ms). Children learn what things look like from various angles, in various lights, when partially occluded in various ways, what they may feel like when being touched in various ways using various parts of the body, what they may smell and taste like, what they may sound like from various distances through various media or when probed in various ways, *and what they sound like manifested through the medium of the local language*. Obviously, living in a language community is essential for this last, but not because a community's norms of response and inference must be absorbed. Understanding a language, like understanding other perceptual inputs, is in the first instance a purely practical matter, a learned perceptual skill. Speaking a language is a skill in affecting the perceptions and activities of others. It is true, as deVries remarks, that "... the notion of a mistake is regularly applied to linguistic performances" and also, of course, to linguistic understanding. But mistakes are not misdemeanors. Making a mistake is accidentally doing something it was not one's purpose to do in place of something it was one's purpose to do. The measure of a mistake is one's own purposes, not the community's norms. If one tries to speak or understand in the way other English speakers generally do or have done but one fails in this, that will indeed be a mistake. Teachers, and others, may have certain ideas about what constitutes "correct English," but linguists have been telling us for a long time now that only social snobbery lies behind this.

Second – and underlying – I have argued for a different view of the nature of intentionality and, correlatively, of empirical concepts (for a summary on concepts, see my reply to Braddon-Mitchell, this volume). My view implies a rejection, too, of semantic holism along with the best-known aspect of Sellars' view of "the semantic" (prodigal daughter!). But there are two sides to Sellars on semantics. DeVries (2010) explains how Sellars' "picturing" was needed to explain the relation between his "realm of semantics," AKA "realm of reasons," and the "causal" or "real" order, while avoiding idealism. This seems right and important. Another way to view this move is as the very earliest, though generally unacknowledged, move into what is nowadays called an "externalism" in the semantics of language and thought. "The criterion of the correctness of the performance of asserting a basic matter-of-factual proposition is the correctness of the proposition *qua* picture, i.e. the fact that it coincides with the picture the world-*cum*-language would generate in accordance with the uniformities controlled by the semantical rules of the language. Thus the correctness of the picture is not defined in terms of the correctness of a performance but vice versa" (Sellars 1967, 136, quoted in deVries 2010, 22). This "criterion" is obviously not one wielded internally by the thinker/speaker but an external one. But in speaking of "externalism in semantics" here we are taking "semantics" to concern truth-conditions, and we are considering truth as a kind of causal-order correspondence between language, or the language of thought, and the world. "Semantics," in this sense, does not concern the realm of reasons but instead what Sellars called "representing."

It is this clean externalist move that I admire, much cleaner than in other recent philosophers. And both Sellars and I think that it is only world structure, not quality, that is pictured both by language and by thought – further, I have added, by perception. (Taking perception to picture also, and also to involve only structural mappings, I can hold that believing what one is told is akin to perceptual judgment.) Having an empirical concept and understanding what one is thinking of involves only the ability to recognize "that again; oh, that once more; hey, the same thing again," nothing fuller.

But my picturing or "mapping" is very different from Sellars'. His mapping is "a projection, in accordance with a fantastically complex system of rules of projection, of the manner in which the objects occur in the world" onto another maze consisting of the actual utterances and thoughts resulting from actual use, in accordance with its input, output, and inference rules, of some particular *whole* language. My mapping is quite Tarskian, simple and not holistic, and that is all there is to my semantics.

Learning a language is not learning input, output, and inference rules or dispositions. Empirical concepts are not nodes in an inference net that is acquired with one's language. They are capacities to reidentify objective selfsames, ideally in multiply diverse ways, via perception and/or inference. These capacities are used, in the first instance, for acquiring and implementing practical skills, in the second instance, for making empirical judgments, which may be used in inference. Vastly many empirical concepts are indeed acquired in acquiring a language, but this is because it is so easy to reidentify through linguistic media – one of the many transforming powers of language. The basic rules of a language are its mapping rules only, and these apply word by word, idiom by idiom, and grammatical construction by grammatical construction, not holistically. Just as Sellars claimed, these rules are not *internal* rules conformed to in understanding and speaking a language but *external* rules. Understanding and speaking according to the conventional patterns of a particular language may be accomplished in hugely various ways, depending on the exact means of reidentifying that support each empirical concept for that particular language user (always remember Helen Keller). And standard use is acquired, not through sensitivity to Mother and Daddy's frowns and smiles (as Sellars used to put it), but by the test of agreement with others in judgments reached (Millikan 2010) and in the final instance, in the acquisition of knowledge that proves useful in practice (see my reply to Godfrey-Smith, this volume).

DeVries says that "Sellars wants to maintain a distinction between picturing and intentionality." I do too, for picturing, in the sense merely of a definable isomorphism, obtains between absolutely everything and absolutely anything else. Recall here Putnam's model-theoretic argument (e.g., Putnam 1981, 217–218; 1983; also see Shea, this volume). The vastness and complexity of Sellars' postulated mapping could not alone determine its uniqueness. I don't know what solution to this problem Sellars intended – perhaps that only one projection from language to world would be causally necessitated regardless of accidental initial conditions, given the disciplined dispositions of the language community? The solution I intend (which receives some fine tuning in the discussion with Shea, this volume) is my version of teleosemantics. Succinctly, the projection rules are relations between signs and signifieds that the sign producers have learned to produce in the process of cooperating with their customary representation interpreters/consumers in the fulfillment of certain kinds of tasks, these sign-signified relations having consistently borne a *causal* responsibility in cases of success.

But deVries has in mind more than that. Sellars thought there was no intentionality without rationality – without reason*ing* – and that reasoning involved following rational principles, which in turn involved sensitivity to "oughts," whose essence lay, ultimately, in directing doings, not in picturing.

Now, first, it seems to me that in its basic puzzling sense, "intentionality" has to do with every kind of designed *ofness* or *aboutness*, and that there is lots of designed *ofness* and *aboutness* around prior to any rationality – for example, in animal signaling, in animal cognition, and in human perception-for-action (see also my reply to Godfrey-Smith). Second, I think that Sellars introduced a mistaken kind of asymmetry between descriptives and directives. On my view, descriptives and directives both map in accordance with historically determined projection rules when producer and consumer serve their cooperative proper functions Normally. The difference lies only in whether instantiation of the mapping relation helps to cause fulfillment of this function or whether it is, rather, the result of fulfillment. Thinking is, of course, a form of doing. If one needs to follow directives to accomplish good thinking, this doing is as fully immersed in the causal order as any other doing. This does not involve any threat to freedom. As Sellars often observed, causal laws are not causes pushing other things around. *You* are the thing that causes you to follow directives of reason, but like other doings, thinkings/inferrings are not mere happenings, of course. They are purposive, purposiveness being the most central explanandum of the biosemantic theory. Finally, correct doing, including correct thinking, shows itself in its outcome, not in whether certain prior rules were followed. Correctness of rules is evidenced by consistently correct outcomes. (In LTOBC, ch. 18 and in WQ, ch. 14, I argued that even the law of noncontradiction is, at root, just a well-confirmed hypothesis about the ubiquity of certain skeletal structures in nature that have proved extraordinarily useful to rely on; see my reply to Nussbaum, this volume.)

I suppose that the logical particles might be considered to generate pushmi-pullyu representations incorporating directives about how to think. One might suppose them, then, to be in a sense metalinguistic, meta the-language-of-thought. It remains then for me to address Sellars' thoughts on the realm of the "semantic," as he used the term, more directly. Is going metalinguistic entering another realm, or is it not? Is it just talking about another natural item in the world, or is it a different kind of talk?

My analysis of public linguistic meaning (stated most clearly, perhaps, in LBM, ch. 3) takes the *functions* of linguistic devices to be basic, the mapping of language to be secondary. In the first instance, the meaning of a linguistic form is its "stabilizing function," a function which, when performed, tends to circle around so as to encourage both its continuing use by speakers and continuing consistent reaction in hearers. Suitably syntactically combined into sentences, the combined function of some linguistic forms is to produce intentional attitudes in hearers, inner representations with satisfaction conditions, paradigmatically, beliefs and explicit intentions. These linguistic combinations have the same satisfaction conditions as the intentional attitudes they produce. But there are also linguistic combinations that don't have this effect, for example, sentences asserting existence or asserting identity (LTOBC, ch. 12, LBM, ch. 3), some of the modals, probably sentences asserting counterfactual conditionals, and so forth. Sellars' "translation rubric," the form "'X' means Y," the analysis of which lies at the center of his views of the "semantic," is another important example. According to Sellars, someone who asserts "'Und' means *and*":

> ... mentions the German vocable '*und*' and *uses* the English vocable 'and'. He uses the latter, however, in a peculiar way, a way which is characteristic of *semantic* discourse. He presents us with an instance of the word itself, not a name of it, and, making use of the fact that we belong to the same language community, indicates to us that we have only to rehearse our use of 'and' to appreciate the role of 'und' on the other side of the Rhine. (Sellars 1963b, 314–315)

In parallel fashion, I have claimed (in LTOBC, ch. 10, 2010, *inter alia*) that the stabilizing function of the "'X' means Y" rubric is not to produce, in the hearer, an intentional attitude of some kind, either about words or about any abstract entities. Its function is to produce a correct disposition – one that will serve the hearer's ends in trying to use language as others do – to use and understand the word "X" in the way he already knows to use and understand the word "Y."

Thus "believing" the form "'X' means Y" does not require forming any beliefs, nor does it require possessing concepts of words. Although the truth-conditions of "'X' means Y" involve that "X" and "Y" have similar stabilizing functions (for Sellars this would be similar "roles") it is not a statement about words. I take all this to be very close to Sellars' own position. Yet it has no tendency to take the "semantic" one step away from the causal order. Moreover, we can explain exactly how it happens that a linguistic form whose job is not to produce an intentional attitude can nonetheless have a Tarskian kind of satisfaction condition. The form "'X' means Y" cannot serve its stabilizing function (to produce a *correct* use disposition) unless it projects, as in past successful uses, onto an actual relation of similarity between the (socially/historically determined) stabilizing functions of the words "X" and "Y." In LTOBC I said that this kind of correspondence relation involved "proto reference." A proto referent is the referent of an intentional device ("intentional icon") that serves its function without that referent being identified in thought, that is, without invoking a concept of that referent. Insofar as logical particles might be supposed to be, in a sense, meta the-language-of-thought, they have thoughts as proto referents only.

Summarizing this intuitively, I agree with Sellars that sentences using "means" and perhaps also the logical particles do not state facts in the ordinary way. Their job is not to cause mental representations of anything but to spur other kinds of mental doings. But having the job of spurring these doings – being "fraught with ought" – does not remove them in any way from perfectly ordinary and full immersion in the causal order. The "ought" involved is the ought of natural teleology. It is a reference to nothing more than a (very revealing, of course) fact concerning their natural history.

AFTERWORD

RUTH MILLIKAN

Those were hard to do. And instructive! I owe many thanks to the authors of these essays for raising such helpful and challenging questions ... even for their occasional misinterpretations, which were a needed reminder to speak clearly.

I am glad that so many of the essays helped me to emphasize the centrality of my thoughts on "empirical concepts" in connecting together other parts of my work, for I have not been sufficiently clear on that before. Dan Dennett said in his foreword that some things called "unicepts" were soon to arrive on the scene. So let me tell you a story. In the aisle after a talk given at a meeting we both attended, Dan suggested that the way the speaker had used the idea of a "concept" must have bothered me and I replied, without thinking, that we should just trash that word and start over. Then recalling the woman who, when admonished to think before she spoke, replied "How do I know what I think 'til I hear what I say?," I thought to myself, perhaps I should take myself seriously. What I have been talking about all these years, since LTOBC, are not really empirical concepts at all, but *unicepts* – "uni" for one, of course, and "cept" for Latin *capere*, to take or to hold. A unicept takes in information gathered from many proximal stimulations and holds them as one distal object, property, kind, and so forth. Unicepts *should not be confused with concepts* and I propose to do so no longer. Unicepts are something nobody much has noticed before, but that deserve careful consideration. The celerity with which this has cleared cobwebs from my head is amazing!

I am very grateful to the editors of this volume, who persuaded such good people to write for me, carefully reviewed and made suggestions on the manuscripts, then sorted it all out into a coherent consecutive volume with a single bibliography and uniform style. I am not merely grateful, I am deeply touched by their willingness to do this work. Thank you. I am especially indebted to Dan Ryder, who nursed along my replies, gently pressing me to address issues he thought were most underlying, suggesting where summaries were needed, holding my hand when I had replied to drafts that I thought were final but they were not final so that the replies had to be redone, quietly persuading me to try to address, for example, the status of the modalities.... Thank you very much, Dan. If the replies come across as genuine replies, a lot of it will have been your doing.

References

Abbott, K., and Dukas, R. (2009) "Honeybees Consider Flower Danger in their Waggle Dance," *Animal Behavior*, 78: 633–635.
Antony, L. (ms) "In Praise of Loose Talk: Three Ways of Following a Rule."
Armstrong, D. (1973) *Belief, Truth and Knowledge*. Cambridge: Cambridge University Press.
Armstrong, D. (1997) *A World of States of Affairs*. Cambridge: Cambridge University Press.
Arnqvist, G., and Rowe, L. (2005) *Sexual Conflict*. Princeton: Princeton University Press.
Axelrod, R. (1984) *The Evolution of Cooperation*. New York: Basic Books.
Bazylinski, D., and Frankel, R. (2004) "Magnetosome Formation in Prokaryotes," *Nature Reviews Microbiology*, 2: 217–230.
Bergstrom, C., and Rosvall, M. (2011) "The Transmission Sense of Information," *Biology and Philosophy*, 26, 159–176.
Bermúdez, J. L. (2003) *Thinking Without Words*. Oxford: Oxford University Press.
Biesmeijer, J., and Seeley, T. (2005) "The Use of Waggle Dance Information by Honey Bees Throughout Their Foraging Careers," *Behavioral Ecology and Sociobiology*, 59: 133–142.
Billock, V. A, Gleason, G. A., and Tsou, B. H. (2001) "Perception of Forbidden Colors in Retinally Stabilized Equiluminant Images: An Indication of Softwired Cortical Color Opponency?" *Journal of the Optical Society of America*, 18 (10): 2398–2403.
BonJour, L. (1998) *In Defense of Pure Reason*. Cambridge: Cambridge University Press.
Boyd, R. (1991) "Realism, Anti-Foundationalism and the Enthusiasm for Natural Kinds," *Philosophical Studies*, 61: 127–148.
Boyd, R. (1999) "Kinds, Complexity and Multiple Realization: Comments on Millikan's 'Historical Kinds and the Special Sciences'," *Philosophical Studies*, 95: 67–98.
Braddon-Mitchell, D. (2004) "Masters of Our Meanings," *Philosophical Studies*, 118: 133–152.
Braddon-Mitchell, D., and Jackson, F. (1996) *Philosophy of Mind and Cognition*. Oxford: Blackwell.
Brandom, R. (1994) *Making It Explicit: Reasoning, Representing, and Discursive Commitment*. Cambridge, MA: Harvard University Press.
Brandom, R. (2000) *Articulating Reasons: An Introduction to Inferentialism*. Cambridge, MA: Harvard University Press.
Burge, T. (2010) *Origins of Objectivity*. Oxford: Oxford University Press.
Camilli, A., and Bassler, B. L. (2006) "Bacterial Small-Molecule Signaling Pathways," *Science*, 311: 1113–1116.

Carruthers, P. (2002) "The Cognitive Functions of Language," *Behavioral and Brain Sciences*, 25: 657–674.
Carruthers, P. (2004) "On Being Simple Minded," *American Philosophical Quarterly*, 41: 205–220.
Chalmers, D. (2006) "The Foundations of Two-Dimensional Semantics," in M. Garcia-Carpintero and J. Macia (eds.) *Two-Dimensional Semantics: Foundations and Applications*. Oxford: Oxford University Press.
Changizi, M. A., Zhang, Q., and Shimojo, S. (2006) "Bare Skin, Blood and the Evolution of Primate Colour Vision," *Biology Letters*, 2: 217–221.
Cheng, K. (2006) "Arthropod Navigation: Ants, Bees, Crabs, Spiders Finding Their Way," in E. Wasserman and T. Zentall (eds.) *Comparative Cognition: Experimental Explorations of Animal Intelligence*. Oxford: Oxford University Press.
Churchland, P. M. (1979) *Scientific Realism and the Plasticity of Mind*. Cambridge: Cambridge University Press.
Collett, T. S., and Collett, M. (2002) "Memory Use in Insect Visual Navigation," *Nature Reviews Neuroscience*, 3: 542–552.
Cover, T. M., and Thomas, J. A. (2006) *Elements of Information Theory* (second edition). Hoboken, NJ: John Wiley & Sons.
De Marco, R. J. (2006) "How Bees Tune Their Dancing According to Their Colony's Nectar Influx: Re-examining the Role of the Food-receivers' 'Eagerness'," *Journal of Experimental Biology*, 209: 421–432.
De Marco, R. J., and Menzel, R. (2008) "Learning and Memory in Communication and Navigation in Insects," in R. Menzel and J. Byrne (eds.) *Learning and Memory: A Comprehensive Reference*. New York: Elsevier.
Dennett, D. (1987) *The Intentional Stance*. Cambridge, MA: MIT Press.
Dennett, D. (1991) *Consciousness Explained*. Boston: Little, Brown.
Dennett, D. (1995) *Darwin's Dangerous Idea: Evolution and the Meanings of Life*. New York: Simon and Schuster.
Dennett, D. (1996) *Kinds of Minds: Toward an Understanding of Consciousness*. New York: Basic Books.
Dennett, D. (2009) "Darwin's 'Strange Inversion of Reasoning'," *Proceedings of the National Academy of Sciences*, 196 (1): 110061–110065.
Devitt, M. (2008) "Resurrecting Biological Essentialism," *Philosophy of Science*, 75: 344–382.
Devitt, M., and Sterelny, K. (1999) *Language and Reality*. Cambridge, MA: MIT Press.
deVries, W. (2010) "Naturalism, the Autonomy of Reason, and Pictures," *International Journal of Philosophical Studies*, 18 (3): 395–413.
Dretske. F. (1981) *Knowledge and the Flow of Information*. Cambridge, MA: MIT Press.
Dretske, F. (1986) "Misrepresentation," in R. Bogdan (ed.) *Belief: Form, Content and Function*. Oxford: Oxford University Press.
Dretske, F. (1988) *Explaining Behavior: Reasons in a World of Causes*. Cambridge, MA: MIT Press.
Dummett, M. (2007) "Reply to John McDowell,' in R. E. Auxier and L. E. Kahn (eds.) *The Philosophy of Michael Dummett*. Chicago: Open Court.
Dyer, F. (2002) "The Biology of the Dance Language," *Annual Review of Entomology*, 47: 917–949.
Eco, U. (1976) *A Theory of Semiotics*. Bloomington: Indiana University Press.
Elder, C. (1998) "What vs. How in Naturally Selected Representations," *Mind*, 107: 349–363.
Elder, C. (2004) *Real Natures and Familiar Objects*. Cambridge, MA: MIT Press.
Elder, C. (2006) "Conventionalism and Realism-Imitating Counterfactuals," *Philosophical Quarterly*, 56: 1–15.
Elder, C. (2007a) "Conventionalism and the World as Bare Sense-Data," *Australasian Journal of Philosophy*, 85: 261–275.
Elder, C. (2007b) "Realism and the Problem of *Infimae Species*," *American Philosophical Quarterly*, 44: 111–127.

Ereshefsky, M. (1992) "Eliminative Pluralism," *Philosophy of Science*, 59: 671–690.
Ereshefsky, M., and Matthen, M. (2005) "Taxonomy, Polymorphism, and History: An Introduction to Population Structure Theory," *Philosophy of Science*, 72: 1–21.
Fales, E. (1990) *Causation and Universals*. New York: Routledge.
Field, H. (1986) "Mental Representation," *Erkenntnis*, 13: 9–61.
Field, H. (2001) *Truth and the Absence of Fact*. Oxford: Clarendon Press.
Findlay, J. N. (1963) *Meinong's Theory of Objects and Values*. Oxford: Oxford University Press.
Fodor, J. (1981) *Representations*. Cambridge MA: MIT Press.
Fodor, J. (1984) "Semantics, Wisconsin Style," *Synthese*, 59: 231–250.
Fodor, J. (1987) *Psychosemantics*. Cambridge, MA: MIT Press.
Fodor, J. (1990) *A Theory of Content and Other Essays*. Cambridge, MA: MIT Press.
Fodor, J. (1991) "Replies," in B. Loewer and G. Rey (eds.) *Meaning in Mind: Fodor and His Critics*. Oxford: Blackwell, 255–319.
Fodor, J. (2010) *LOT 2: The Language of Thought Revisited*. Cambridge, MA: MIT Press.
Fodor, J., and Lepore, E. (1996) "The Red Herring and the Pet Fish: Why Concepts Still Can't Be Prototypes," *Cognition*, 58 (2): 253–270.
Fumerton, R. (1989) "Russelling Causal Theories of Reference," in *Rereading Russell*. Minneapolis: University of Minnesota Press, 108–118.
Gallistel, C. R. (1990) *The Organization of Learning*. Cambridge, MA: MIT Press.
Gallistel, C. R. (1998) "Symbolic Processes in the Brain: The Case of Insect Navigation," in D. Scarborough and S. Sternberg (eds.) *An Invitation to Cognitive Science, Vol. 4: Methods, Models and Conceptual Issues* (2nd ed.). Cambridge, MA: MIT Press.
Gallistel, C. R. (2006) "The Nature of Learning and the Functional Architecture of the Brain," in Q. Jing et al. (eds.) *Psychological Science Around the World, Vol. 1: Proceedings of the 28th International Congress of Psychology*. Brighton: Psychology Press, 63–71.
Gallistel, C. R. (2008) "Insect Navigation: Brains as Symbol-Processing Organs," in D. Osherson, D. Scarborough, and S. Sternberg (eds.) *Invitation to Cognitive Science*, *Vol.* 4. Cambridge, MA: MIT Press.
Gallistel, C. R. (2009) "The Foundational Abstractions," in M. Piattelli-Palmarini, J. Uriagereka, and P. Salaburu (eds.) *Of Minds and Language*. Oxford: Oxford University Press.
Gallistel, C. R., and King, A. (2009) *Memory and the Computational Brain*. Oxford: Wiley-Blackwell.
Ghiselin, M. (1974) "A Radical Solution to the Species Problem," *Systematic Zoology*, 23: 536–544.
Giurfa, M., and Capaldi, E. (1999) "Vectors, Routes, and Maps: New Discoveries about Navigation in Insects," *Trends in Neuroscience*, 22: 237–242.
Godfrey-Smith, P. (1989) "Misinformation," *Canadian Journal of Philosophy*, 19: 533–550.
Godfrey-Smith, P. (1992) "Information and Adaptation," *Synthese*, 92: 283–312.
Godfrey-Smith, P. (1996) *Complexity and the Function of Mind in Nature*. Cambridge: Cambridge University Press.
Godfrey-Smith, P. (2006a) "Mental Representation, Naturalism and Teleosemantics," in G. Macdonald and D. Papineau (eds.) *Teleosemantics: New Philosophical Essays*. Oxford: Oxford University Press.
Godfrey-Smith, P. (2006b) "The Strategy of Model-Based Science," *Biology and Philosophy*, 21: 725–740.
Gould, J. L., and Gould, C. G. (1982) "The Insect Mind: Physics or Metaphysics?" in D. Griffin (ed.) *Animal Mind–Human Mind*. Berlin: Springer.
Grüter, C., Sol Balbuena, M., and Farina, W. (2008) "Informational Conflicts Created by the Waggle Dance," *Proceedings of the Royal Society*, 275: 1321–1327.
Hacking, I. (1991) "A Tradition of Natural Kinds," *Philosophical Studies*, 61: 109–126.
Hacking, I. (2007) "Natural Kinds: Rosy Dawn, Scholastic Twilight," *Royal Institute of Philosophy Supplement*, 82. Cambridge: Cambridge University Press, 203–239.

Haig, D. (1996) "Placental Hormones, Genomic Imprinting, and Maternal–Fetal Communication," *Journal of Evolutionary Biology*, 9: 357–380.
Haig, D. (2008) "Conflicting Messages: Genomic Imprinting and Internal Communication," in P. D'Ettorre and D. P. Hughes (eds.) *Sociobiology of Communication*. Oxford: Oxford University Press, 209–223.
Haig, D. (2010) "Transfers and Transitions: Parent–Offspring Conflict, Genomic Imprinting, and the Evolution of Human Life History," *Proceedings of the National Academy of Sciences*, 107: 1731–1735.
Hamilton, W. D. (1975) "Innate Social Aptitudes of Man: An Approach from Evolutionary Genetics," in R. Fox (ed.) *Biosocial Anthropology*. London: Malaby Press, 133–155.
Hamilton, W. D., and Zuk, M. (1982) "Heritable True Fitness and Bright Birds: A Role for Parasites?" *Science*, 218: 384–387.
Harms, W. (2004) "Primitive Content, Translation, and the Emergence of Meaning in Animal Communication," in D. Oller and U. Griebel (eds.) *Evolution of Communication Systems: A Comparative Approach*. Cambridge, MA: MIT Press, 31–48.
Hawley, K. (2001) *How Things Persist*. Oxford: Oxford University Press.
Hegel, G. W. F. (1807/1977) *Phenomenology of Spirit*, trans. A. V. Miller. New York: Oxford University Press.
Hegel, G. W. F. (1811–1816/1976) *Science of Logic*, trans. A. V. Miller. New York: Humanities Press.
Hegel, G. W. F. (1830/1991) *Encyclopedia Logic*, trans. T. F. Geraets, W. A. Suchting and H. S. Harris. Indianapolis: Hackett.
Heim, I., and Kratzer, A. (1998) *Semantics in Generative Grammar*. Oxford: Blackwell.
Hohwy, J., Roepstorff, A., and Friston, K. (2008) "Predictive Coding Explains Binocular Rivalry: An Epistemological Review," *Cognition*, 108 (3): 687–701.
Holland, B., and Rice, W. R. (1998) "Chase Away Sexual Selection: Antagonistic Seduction vs. Resistance," *Evolution*, 52: 1–7.
Holldobler, B., and Wilson, E. O. (2008) *The Superorganism: The Beauty, Elegance and Strangeness of Insect Societies*. New York: W. W. Norton.
Hull, D. (1978) "A Matter of Individuality," *Philosophy of Science*, 45: 335–360.
Jackson, F. (1998) *From Metaphysics to Ethics: A Defence of Conceptual Analysis*. Oxford: Oxford University Press.
Jackson, F. (2006) "The Epistemological Objection to Opaque Teleological Theories of Content," in G. Macdonald and D. Papineau (eds.) *Teleosemantics: New Philosophical Essays*. Oxford: Oxford University Press.
Jackson, F., and Braddon-Mitchell, D. (1996) *Philosophy of Mind and Cognition*. Oxford: Blackwell.
Johnson, W. E. (1921/1964) *Logic, Part I*. New York: Dover.
Joyce, R. (2001) *The Myth of Morality*. Cambridge: Cambridge University Press.
Jubien, M. (1993) *Ontology, Modality, and the Fallacy of Reference*. Cambridge: Cambridge University Press.
Kant, I. (1787/1998) *Critique of Pure Reason* (2nd ed.), trans. P. Guyer and A. Wood. Cambridge: Cambridge University Press.
Keeley, B. L. (2002) "Making Sense of the Senses: Individuating Modalities in Humans and Other Animals," *Journal of Philosophy*, 99: 5–28.
Knill, D. (2007) "Robust Cue Integration: A Bayesian Model and Evidence from Cue Conflict Studies with Stereoscopic and Figure Cues to Slant," *Journal of Vision*, 7: 1–24.
Knill, D., and Richards, W. (1996) *Perception as Bayesian Inference*. Cambridge: Cambridge University Press.
Kripke, S. (1982) *Wittgenstein on Rules and Private Language: An Elementary Exposition*. Cambridge, MA: Harvard University Press.
Lachmann, M., Szamado, S., and Bergstrom, C. T. (2001) "Cost and Conflict in Animal Signals and Human Language," *Proceedings of the National Academy of Sciences*, 98: 13189–13194.

Lewis, D. (1969) *Convention*. Cambridge, MA: Harvard University Press.
Lewis, D. (1986) *On the Plurality of Worlds*. Oxford: Blackwell.
Lewis, D. (1994) "Reduction of Mind," in S. Guttenplan (ed.) *A Companion to the Philosophy of Mind*. Oxford: Blackwell, 412–431.
Loewer, B., and Rey, G., eds. (1991) *Meaning in Mind: Fodor and His Critics*. Oxford: Blackwell.
Lycan, W. G. (1988) "Review of Lewis' *On the Plurality of Worlds*," *Journal of Philosophy*, 85 (1): 42–47.
McDowell, J. (1977) "On the Sense and Reference of a Proper Name," *Mind*, 342: 159–185. Reprinted in McDowell (1998), 171–198.
McDowell, J. (1984) "*De Re* Senses," *Philosophical Quarterly*, 34: 283–294. Reprinted in McDowell (1998), 214–217.
McDowell, J. (1986) "Singular Thought and the Extent of Inner Space," in P. Pettit and J. McDowell (eds.) *Subject, Thought, and Context*. Oxford: Clarendon Press, 137–168. Reprinted in McDowell (1998), 228–259.
McDowell, J. (1998) *Meaning, Knowledge and Reality*. Cambridge, MA: Harvard University Press.
McDowell, J. (1999) "Responses," in M. Willaschek (ed.) *John McDowell: Reason and Nature*. Münster: LIT, 91–114.
McDowell, J. (2007) "Dummett on Truth Conditions and Meaning," in R. E. Auxier and L. E. Kahn (eds.) *The Philosophy of Michael Dummett*. Chicago: Open Court, 351–363.
McDowell, J. (2009) "Naturalism in the Philosophy of Mind," in *The Engaged Intellect*. Cambridge, MA: Harvard University Press, 257–275.
McKinnon, J. S. et al. (2004) "Evidence for Ecology's Role in Speciation," *Nature*, 429: 294–298.
Matthen, M. (1988) "Biological Functions and Perceptual Content," *Journal of Philosophy*, 85 (1): 5–27.
Matthen, M. (1998) "Biological Universals and the Nature of Fear," *Journal of Philosophy*, 95: 105–132.
Matthen, M. (1999) "Evolution, Wisconsin Style: Selection and the Explanation of Individual Traits," *British Journal for the Philosophy of Science*, 50: 143–150.
Matthen, M. (2000) "What is a Hand? What is a Mind?" *Revue Internationale de Philosophie*, 214: 123–142.
Matthen, M. (2005) *Seeing, Doing, and Knowing: A Philosophical Theory of Sense Perception*. Oxford: Clarendon Press.
Matthen, M. (2006) "Teleosemantics and the Consumer," in G. Macdonald and D. Papineau (eds.) *Teleosemantics: New Philosophical Essays*. Oxford: Clarendon Press, 146–166.
Matthen, M. (2009) "Chickens, Eggs, and Speciation," *Noûs*, 43: 94–115.
Maynard Smith, J. (2000) "The Concept of Information in Biology," *Philosophy of Science*, 67: 177–194.
Maynard Smith, J., and Harper, D. (2003) *Animal Signals*. Oxford: Oxford University Press.
Medin, D. L., and Atran, S. (1999) *Folkbiology*. Cambridge, MA: MIT Press.
Menzel, R. (2008) "Insect Minds for Human Minds," in M. Guadagnoli, A. Benjamin, J. S. de Belle, B. Etnyre, and T. A. Polk (eds.) *Human Learning*. San Diego: Elsevier.
Menzel, R., Geifer, K., Müller, U., Joerges, J., and Chittka, L. (1998) "Bees Travel Novel Homeward Routes by Integrating Separately Acquired Vector Memories," *Animal Behavior*, 55: 139–152.
Menzel, R., and Giurfa, M. (2006) "Dimensions of Cognition in an Insect, the Honeybee," *Behavioral and Cognitive Neuroscience Reviews*, 5: 24–40.
Millikan, R. G. (1969) *Empirical Identity*. Dissertation in philosophy, Yale University (Sterling Library).
Millikan, R. G. (1984) *Language, Thought, and Other Biological Categories: New Foundations for Realism*. Cambridge, MA: Bradford/MIT Press. ("LTOBC")
Millikan, R. G. (1986) "Thoughts Without Laws: Cognitive Science With Content," *Philosophical Review*, 95: 47–80.

Millikan, R. G. (1989a) "Biosemantics," *Journal of Philosophy*, 86: 281–297. Reprinted in B. McLaughlin (ed.) (2000) *The Oxford Handbook of the Philosophy of Mind*. Oxford: Oxford University Press.
Millikan, R. G. (1989b) "In Defense of Proper Functions," *Philosophy of Science*, 56 (2): 288–302.
Millikan, R. G. (1990) "Truth Rules, Hoverflies, and the Kripke–Wittgenstein Paradox," *Philosophical Review*, 99 (3): 323–353. Reprinted in Millikan (1993a), 211–239.
Millikan, R. G. (1991) "Speaking up for Darwin," in B. Loewer and G. Rey (eds.) *Meaning in Mind: Fodor and His Critics*. Oxford: Blackwell, 151–164.
Millikan, R. G. (1993a) *White Queen Psychology and Other Essays for Alice*. Cambridge, MA: Bradford/MIT Press. ("WQ")
Millikan, R. G. (1993b) "White Queen Psychology," in R. G. Millikan (1993a), 279–363.
Millikan, R. G. (1993c) "Knowing What I'm Thinking Of," *Proceedings of the Aristotelian Society, supplementary volume* 67, 109–124.
Millikan, R. G. (1995) "Reply: A Bet with Peacocke," in C. Macdonald and G. Macdonald (eds.) *Philosophy of Psychology: Debates on Psychological Explanation*. Oxford: Blackwell, 285–292.
Millikan, R. G. (1996a) "Pushmi-Pullyu Representations," in J. Tomberlin (ed.) *Philosophical Perspectives, Vol. 9*. Atascadero, CA: Ridgeview Publishing, 185–200. Also in L. May and M. Friedman (eds.) (1996) *Mind and Morals*. Cambridge, MA: MIT Press, 145–161. Also in S. Nuccetelli and G. Seay (eds.) (2008) *Philosophy of Language: The Central Topics*. Lanham, MD: Rowman and Littlefield, 363–374.
Millikan, R. G. (1996b) "Swampkinds," *Mind and Language*, 11: 103–117.
Millikan, R. G. (1998a) "Language Conventions Made Simple," *Journal of Philosophy*, 95 (4): 161–180.
Millikan, R. G. (1998b) "Cognitive Luck: Externalism in an Evolutionary Frame," in P. Machamer and M. Carrier (eds.) *Philosophy and the Sciences of Mind*. Pittsburgh: Pittsburgh University Press and Konstanz: Universitätsverlag Konstanz, 207–219.
Millikan, R. G. (1999a) "Historical Kinds and the Special Sciences," *Philosophical Studies*, 95: 45–65.
Millikan, R. G. (1999b) "Reply to Boyd," *Philosophical Studies*, 95: 99–102.
Millikan, R. G. (2000a) *On Clear and Confused Ideas: An Essay about Substance Concepts*. Cambridge: Cambridge University Press. ("OCCI")
Millikan, R. G. (2000b) "Naturalizing Intentionality," in B. Elevitch (ed.) *Philosophy of Mind: Proceedings of the Twentieth World Congress of Philosophy, Vol. 9*. Philosophy Documentation Center.
Millikan, R. G. (2001) "What has Natural Information to do with Intentional Representation?" in D. Walsh (ed.) *Evolution, Naturalism and Mind*. Cambridge: Cambridge University Press, 105–126. Reprinted as Appendix B of Millikan 2000a.
Millikan, R. G. (2003a) "In Defense of Public Language," in L. M. Antony and N. Hornstein (eds.) *Chomsky and His Critics*. Oxford: Blackwell.
Millikan, R. G. (2003b) "Teleological Theories of Mental Content," in L. Nadel (ed.) *The Encyclopedia of Cognitive Science*. New York: Macmillan.
Millikan, R. G. (2004a) *Varieties of Meaning*. Cambridge, MA: Bradford/MIT Press. ("VOM")
Millikan, R. G. (2004b) "On Reading Signs: Some Differences Between Us and the Others," in D. K. Oller and U. Griebel (eds.) *Evolution of Communication Systems: A Comparative Approach*. Cambridge, MA: MIT Press.
Millikan, R. G. (2005a) *Language: A Biological Model*. Oxford: Clarendon Press. ("LBM")
Millikan, R. G. (2005b) "Meaning, Meaning, and Meaning," in Millikan (2005a), 53–76.
Millikan, R. G. (2005c) "The Son and the Daughter: On Sellars, Brandom, and Millikan," *Pragmatics and Cognition*, 13: 59–72. Reprinted in Millikan (2005a), 77–91.

Millikan, R. G. (2006) "Reply to Elisabetta Lalumera," *SWIF Philosophy of Mind Review*, 5 (2), http://www.swif.uniba.it/lei/mind/swifpmr.htm.
Millikan, R. G. (2007) "An Input Condition for Teleosemantics? Reply to Shea (and Godfrey-Smith)," *Philosophy and Phenomenological Research*, 75 (2): 436–455.
Millikan, R. G. (2009a) "Embedded Rationality," in M. Aydede and P. Robbins (eds.) *The Cambridge Handbook of Situated Cognition*. Cambridge: Cambridge University Press, 171–181.
Millikan, R. G. (2009b) "Language Without a Theory of Mind," unpublished manuscript delivered to the Society for Philosophy and Psychology Conference, University of Indiana, June 12, 2009.
Millikan, R. G. (2010) "On Knowing the Meaning; With a Coda on Swampman," *Mind*, 119 (473): 43–81.
Millikan, R. G. (2011) "Loosing the Word–Concept Tie," *Proceedings of the Aristotelian Society, Supplementary Volume* 85: 125–143.
Millikan, R. G. (2012) "Natural Signs," in S. B. Cooper, A. Dawar, and B. Löwe (eds.) "How the World Computes: Seventh Conference on Computability in Europe (CiE 2012)," *Lecture Notes in Computer Science*. Heidelberg: Springer.
Millikan, R. G. (forthcoming a) "Natural Information, Intentional Signs and Animal Communication," in U. Stegmann (ed.) *Animal Communication Theory: Information and Influence*. Cambridge: Cambridge University Press.
Millikan, R. G. (forthcoming b) "An Epistemology for Phenomenology?," in R. Brown (ed.) *Consciousness Inside and Out: Phenomenology, Neuroscience, and the Nature of Experience*. Berlin: Springer.
Millikan, R. G. (ms) "Learning Language."
Neander, K. (1991) "Functions as Selected Effects," *Philosophy of Science*, 58 (2): 168–184.
Neander, K. (1995) "Misrepresenting and Malfunctioning," *Philosophical Studies*, 79 (2): 109–141.
Neander, K. (1996) "Swampman Meets Swampcow," *Mind and Language*, 11: 118–129.
Neander, K. (2006) "Content for Cognitive Science," in G. Macdonald and D. Papineau (eds.) *Teleosemantics: New Philosophical Essays*. Oxford: Oxford University Press, 167–194.
O'Keefe, J., and Nadel, L. (1978) *The Hippocampus as a Cognitive Map*. Oxford: Clarendon Press.
Papineau, D. (1979) *Theory and Meaning*. Oxford: Oxford University Press.
Papineau, D. (1984) "Representation and Explanation," *Philosophy of Science*, 51 (4): 550–572.
Papineau, D. (1987) *Reality and Representation*. Oxford: Blackwell.
Papineau, D. (1998) "Teleosemantics and Indeterminacy," *Australasian Journal of Philosophy*, 77 (1): 1–14.
Pietroski, P. (1992) "Intentionality and Teleological Error," *Pacific Philosophical Quarterly*, 73: 267–281.
Price, C. (1998) "Determinate Functions," *Noûs*, 32 (1): 54–75.
Priest, G., Beall, J. C., and Armour-Garb, B. (2004) *The Law of Non-Contradiction: New Philosophical Essays*. Oxford: Clarendon Press.
Putnam, H. (1975) "The Meaning of 'Meaning'," in K. Gunderson (ed.) *Minnesota Studies in the Philosophy of Science, Vol. 7: Language, Mind and Knowledge*. Minneapolis: University of Minnesota Press, 131–193.
Putnam, H. (1981) *Reason, Truth and History*. Cambridge: Cambridge University Press.
Putnam, H. (1983) "Why There Isn't a Ready-Made World," in *Realism and Reason*. Cambridge: Cambridge University Press, 205–228.
Quine, W. V. (1960) *Word and Object*. Oxford: John Wiley.
Quine, W. V. (1969) "Natural Kinds," in *Ontological Relativity and Other Essays*. New York: Columbia University Press, 114–138.
Ramsey, F. P. (1929/1965) "General Propositions and Causality," in *The Foundations of Mathematics and other Logical Essays*, ed. R. B. Braithwaite. London: Routledge and Kegan Paul, 237–255.

Rea, M. (2000) "Naturalism and Material Objects," in J. P. Moreland and W. L. Craig (eds.) *Naturalism: A Critical Analysis*. London: Routledge, 110–132.
Rea, M. (2002) *World without Design*. Oxford: Oxford University Press.
Redding, P. (2007) *Analytic Philosophy and the Return of Hegelian Thought*. Cambridge: Cambridge University Press.
Reid, A., and Staddon, J. (1997) "A Reader for the Cognitive Map," *Information Sciences*, 100: 217–228.
Reid, A., and Staddon, J. (1998) "A Dynamic Route-Finder for the Cognitive Map," *Psychological Review*, 105: 585–601.
Rendall, D., Owren, M., and Ryan, M. (2009) "What Do Animal Signals Mean?" *Animal Behaviour*, 78, 233–240.
Rescorla, M. (2009) "Cognitive Maps and the Language of Thought," *British Journal for the Philosophy of Science*, 60: 377–407.
Rescorla, M. (forthcoming) "Bayesian Perceptual Psychology," in M. Matthen (ed.) *The Oxford Handbook of the Philosophy of Perception*. Oxford: Oxford University Press.
Riley, J., Greggers, U., Smith, A. D., Reyonds, D. R., and Menzel, R. (2005) "The Flight Paths of Honeybees Recruited by the Waggle Dance," *Nature*, 435: 205–207.
Rosenberg, A., and McShea, D. (2008) *Philosophy of Biology: A Contemporary Introduction*. London: Routledge.
Rowe, L., and Houle, D. (1996) "The Lek Paradox and the Capture of Genetic Variance by Condition-Dependent Traits," *Proceedings of the Royal Society*, 263: 1415–1421.
Rundle, H. D. et al. (2000) "Natural Selection and Parallel Speciation in Sympatric Sticklebacks," *Science*, 287: 306–308.
Russell, B. (1948) *Human Knowledge: Its Scope and Limits*. London: George Allen and Unwin.
Ryle, G. (1949) *The Concept of Mind*. New York: Barnes and Noble.
Sainsbury, M. (1997) "Fregean Sense," in T. Childers, P. Kolář, and V. Svoboda (eds.) *Logica '96*. Prague: Filosofia, 261–276. Reprinted in Sainsbury (2002), 125–136.
Sainsbury, M. (2002) *Departing from Frege*. London: Routledge.
Sainsbury, M. (2008) "Fly Swatting: Davidsonian Truth Theories and Context," in M. C. Amoretti and N. Vassallo (eds.) *Knowledge, Language and Interpretation: On the Philosophy of Donald Davidson*. Frankfurt: Ontos.
Schluter, D., and Nagel. L. M. (1995) "Parallel Speciation by Natural Selection," *American Naturalist*, 146: 292–301.
Searle, J. (1992) *The Rediscovery of the Mind*. Cambridge, MA: MIT Press.
Seefeldt, S., and De Marco, R. J. (2008) "The Response of the Honeybee Dance to Uncertain Rewards," *Journal of Experimental Biology*, 211: 3392–3400.
Sellars, W. S. (1954) "Some Reflections on Language Games," *Philosophical Studies*, 21: 204–228. Reprinted with extensive additions in Sellars (1963b). ("SRLG")
Sellars, W. S. (1957) "Counterfactuals, Dispositions, and the Causal Modalities," in H. Feigl, M. Scriven, and G. Maxwell (eds.) *Minnesota Studies in the Philosophy of Science, Vol. 2*. Minneapolis: University of Minnesota Press. ("CDCM")
Sellars, W. S. (1960) "Being and Being Known," *Proceedings of the American Catholic Philosophical Association*: 28–49. Reprinted in Sellars (1963b). ("BBK")
Sellars, W. S. (1962a) "Philosophy and the Scientific Image of Man," in R. Colodny (ed.) *Frontiers of Science and Philosophy*. Pittsburgh: University of Pittsburgh Press, 35–78. Reprinted in Sellars (1963b). ("PSIM")
Sellars, W. S. (1962b) "Truth and Correspondence," *Journal of Philosophy*, 59: 29–56. Reprinted in Sellars (1963b). ("TC")
Sellars, W. S. (1963a) "Abstract Entities," *Review of Metaphysics*, 16: 627–671. Reprinted in Sellars (1967a). ("AE")

Sellars, W. S. (1963b) *Science, Perception and Reality*. New York: Humanities Press. ("SPR")
Sellars, W. S. (1963c) "Is There a Synthetic Apriori?" in Sellars (1963b): 298–320.
Sellars, W. S. (1963d) "Empiricism and the Philosophy of Mind," in Sellars (1963b): 127–196.
Sellars, W. S. (1967a) *Philosophical Perspectives*. Springfield, IL: Charles C. Thomas. ("PP")
Sellars, W. S. (1967b) *Science and Metaphysics: Variations on Kantian Themes*. London: Routledge and Kegan Paul.
Sellars, W. S. (1969) "Language as Thought and as Communication," *Philosophy and Phenomenological Research*, 29: 506–527. ("LTC")
Sellars, W. S. (1974) "Meaning as Functional Classification (A Perspective on the Relation of Syntax to Semantics)," *Synthese*, 27: 417–437. ("MFC")
Sellars, W. S. (1979) *Naturalism and Ontology*. Atascadero, CA: Ridgeview. ("NAO")
Sellars, W. S. (1980) "On Reasoning About Values," *American Philosophical Quarterly*, 17: 81–101. ("ORV")
Sellars, W. S. (1981) "Mental Events," *Philosophical Studies*, 39: 325–345. ("MEV")
Sellars, W. S. (2007) *In the Space of Reasons: Selected Essays of Wilfrid Sellars*, ed. K. Scharp and R. B. Brandom. Cambridge, MA: Harvard University Press. ("ISR")
Seyfarth, R., Cheney, D., and Marler, P. (1980) "Monkey Responses to Three Different Alarm Calls: Evidence for Predator Classification and Semantic Communication," *Science*, 210: 801–803.
Shannon, C. (1948) "A Mathematical Theory of Communication," *Bell System Technical Journal*, 27: 379–423. Reprinted in C. Shannon and W. Weaver (eds.) *The Mathematical Theory of Communication*. Urbana: University of Illinois Press.
Shea, N. (2007a) "Representation in the Genome and in Other Inheritance Systems," *Biology and Philosophy*, 22: 313–331.
Shea, N. (2007b) "Consumers Need Information: Supplementing Teleosemantics with an Input Condition," *Philosophy and Phenomenological Research*, 75 (2): 404–435.
Sidelle, A. (1989) *Necessity, Essence, and Individuation*. Ithaca: Cornell University Press.
Sidelle, A. (1998) "A Sweater Unraveled: Following One Thread of Thought for Avoiding Coincident Entities," *Noûs*, 32: 423–448.
Sider, T. (1996) "All the World's a Stage," *Australasian Journal of Philosophy*, 74: 433–453.
Sider, T. (2001) *Four-Dimensionalism*. Oxford: Oxford University Press.
Skyrms, B. (1995) *Evolution of the Social Contract*. Cambridge: Cambridge University Press.
Skyrms, B. (2009) *Signals: Evolution, Learning, and Information*. Cambridge: Cambridge University Press.
Sober, E. (1984) *The Nature of Selection*. Cambridge, MA: MIT Press.
Srinivasan, M. (2010) "Honey Bees as a Model for Vision, Perception, and Cognition," *Annual Review of Entomology*, 55: 267–284.
Srinivasan, M. V., Zhang, S. W., and Bidwell, N. J. (1997) "Visually Mediated Odometry in Honeybees," *Journal of Experimental Biology*, 200: 2513–2522.
Stalnaker, R. (2004) "Assertion Revisited: On the Interpretation of Two-Dimensional Modal Semantics," *Philosophical Studies*, 118: 299–322.
Stampe, D. (1979) "Toward a Causal Theory of Representation," in P. French, T. Uehling, Jr., and H. Wettstein (eds.) *Contemporary Perspectives in the Philosophy of Language*. Minneapolis: University of Minnesota Press, 81–102.
Stegmann, U. (2009) "A Consumer-Based Teleosemantics for Animal Signals," *Philosophy of Science*, 76: 864–875.
Stich, S. (1983) *From Folk Psychology to Cognitive Science*. Cambridge, MA: MIT Press.
Strawson, P. F. (1974) *Subject and Predicate in Logic and Grammar*. London: Methuen.
Tetzlaff, M., and Rey, G. (2009) "Systematicity and Intentional Realism in Honeybee Navigation," in R. Lurz (ed.) *The Philosophy of Animal Minds*. Cambridge: Cambridge University Press.
Thomasson, A. (2007a) *Ordinary Objects*. New York: Oxford University Press.

Thomasson, A. (2007b) "Artifacts and Human Concepts," in E. Margolis and S. Laurence (eds.) *Creations of the Mind: Theories of Artifacts and Their Representations*. Oxford: Oxford University Press.

Tolman, E. C. (1948) "Cognitive Maps in Rats and Men," *Psychological Review*, 55: 189–208.

von Frisch, K. (1967) *The Dance Language and Orientation of the Bees*. Cambridge, MA: Harvard University Press.

von Hayek, F. A. (1952) *The Sensory Order: An Inquiry into the Foundations of Theoretical Psychology*. Chicago: University of Chicago Press.

Wehner, R. (1994) "The Polarization–Vision Project: Championing Organismic Biology," *Fortschritte der Zoölogie*, 39: 103–143.

Wehner, R., Boyer, M., Loertscher, F., Sommer, S., and Menzi, U. (2006) "Ant Navigation: One-Way Routes Rather than Maps," *Current Biology*, 16: 75–79.

Weisberg, M. (2007) "Who is a Modeler?" *British Journal for the Philosophy of Science*, 58 (2): 207–233.

White, J. (1979) "The Plant as a Metapopulation," *Annual Review of Ecology and Systematics*, 10: 109–145.

Williams, B. (1973) "Are Persons Bodies?" in *Problems of the Self: Collected Papers 1956–1972*. Cambridge: Cambridge University Press, 64–81.

Wilson, R. A., Barker, M. J., and Brigandt, I. (2007) "When Traditional Essentialism Fails: Biological Natural Kinds," *Philosophical Topics*, 35: 189–215.

Wittgenstein, L. (2009) *Philosophical Investigations* (4th ed.). Oxford: Wiley-Blackwell.

Wray, M., Klein, B., Mattila, H., and Seeley, T. (2008) "Honeybees Do Not Reject Dances for 'Implausible' Locations: Reconsidering the Evidence for Cognitive Maps in Insects," *Animal Behavior*, 76: 261–269.

Wright, L. (1976) *Teleological Explanations*. Berkeley: University of California Press.

Yegnashankaran, K. (2010) *Reasoning as Action*. Harvard University PhD thesis.

Index

abilities, 9, 10, 52, 55, 89, 107ff., 119–122, 124–127, 212, 224, 237, 256, 274, 277
analytic (vs. synthetic), 15, 17, 191, 217, 241, 242, 252, 253
antirealism, 13, 178
a priori, 17, 139, 157, 161, 181, 183, 184, 241, 242, 251, 257
a posteriori, 210
Aristotle, 140, 141, 166, 178, 188, 190, 191
Armstrong, David, 53n, 178n, 188
assertability, 186

belief, x, 5, 7, 15, 21, 40, 53, 54, 57, 58, 60–62, 71, 76, 85, 99, 100, 102, 104–106, 112, 113, 116, 117, 121, 159–161, 173, 186, 187, 211–213, 244, 252, 279, 280
biosemantic, *see* teleosemantics
Bradley's Paradox, 85
Brandom, Robert, 2, 18, 19, 181n, 185, 186, 189–194, 201, 259–262, 264, 270, 275
Brentano, Franz, 81
Burge, Tyler, 9, 92, 94, 95, 97, 104, 105, 106, 200

Castañeda, Hector-Neri, xi
causal theory of representation (content), 29, 39, 128
causation, 26, 27, 36, 128, 201, 210
Chalmers, David, 247
Churchland, Paul, 186

competence, xi, xii, 128, 187
concepts, ix, xii, 2, 9–18, 32, 33, 36, 54, 61, 62, 76, 107ff., 119ff., 123, 124, 126–130, 131ff., 137, 172, 173, 175, 182, 193, 194, 198ff., 218ff., 224, 237, 240ff., 254ff., 278, 281
conceptions, ix, xii, 14–17, 123, 209, 210, 223, 224, 228, 237, 238, 240, 254, 255–257
conceptual analysis, 15, 258
conceptualist (account of natural kinds), 156, 157, 163
content
  aboutness of, 21, 22
  animal, 249
  attributions/ascriptions of, 6, 31, 48, 53, 63, 235, 249, 253
  causal theory of, 29
  conceptual, 11, 17, 108, 111, 112, 199, 213, 241, 243, 244, 248, 252
  descriptive, 99, 134, 202
  determination of, 63, 71, 74, 201, 214–215
  directive, 60, 99
  distal-proximal, 23, 33–35
  externalism, *see* externalism
  false content, 49, 51
  identity/difference of, 116, 120, 225, 226
  imperative, 50, 71
  indeterminacy of, 33
  indicative, 30, 32, 33, 50, 71
  informational, 59, 61, 114n

informational accounts of, 6
intentional, 59, 60, 92, 99, 111, 222, 225
internalism, *see* internalism
intuitions about, 51, 52, 59
knowledge of, 221, 226, 238
linguistic, 180, 199
mental, 2, 16, 18, 21, 22, 135, 198–204, 222
non-conceptual, 180
normativity of, 21, 22, 271
of assertions, 235, 236, 238
of attitudes, 221, 238
of belief, 212
of desire, 5
of icons, 50–52, 57, 61
of inferences, 273
of messages, 48, 58
of quantified statements, 126
of signals and signs, 6, 48–52, 58, 59
of thought, 17, 41, 42, 52, 53, 58, 113, 200, 225, 226, 228, 229, 243, 244
perceptual, 135
propositional, 48, 49, 115, 180, 183, 185
representational, 4, 5, 22, 36, 53, 68, 70, 71, 73–77, 79, 80, 88, 104, 106, 120, 128, 135, 173, 212
semantic, 6, 41–43, 53, 62, 160, 165, 200, 201, 209, 210
sender–receiver approach to, 6, 7, 48
sensory, 21, 23, 30, 32, 33, 36
sentential, 224
similarity of, 250
sub-personal, 7, 58
teleosemantic theory of, 21, 32, 63
theories of, xii, 6, 22, 41, 49, 59, 60, 68, 70–74, 77, 79, 80, 81, 221, 232, 253
truth-conditional, 94, 96, 98, 106
what content, 31
where content, 31
contrariety, 13, 181–185, 188n, 193
correspondence (rules, relations, etc.), 5, 7, 8, 10, 19, 38, 39, 65, 67–70, 73, 77–79, 82–86, 186, 268, 273, 277, 280
covariation (theories of content), 72

Darwin, Charles, xi, xii, 176, 177, 186, 212
Davidson, Donald, 208, 222, 232
de re (senses, information), x, 17, 183n, 222, 226n, 228–231, 233n, 234, 235

demonstratives, 15, 125, 202, 269
Dennett, Daniel, x, xi, 6, 58n, 93, 94, 176–178, 186, 187n, 281
Descartes, René, 201, 240
descriptions, (theory of) definite, 15, 17, 18, 83, 128, 129, 156–158, 172, 202, 203, 209, 243, 244, 246, 255, 256, 269
desire, 5, 97–100, 102, 104, 106, 113, 159, 160, 173, 211–213, 252
determinable/determinate, 13, 24, 34, 132, 174, 181–184, 188
determinacy/indeterminacy (of content), 33, 35, 209, 257
determinate negation (property exclusion), 181, 182, 187, 189, 191, 192
dialetheism, 177n, 178
disjunction problem, 159
dispositions (dispositional analyses, accounts), 2, 9, 11, 17, 18, 24, 26, 74, 75, 108, 109–111, 121, 122, 133, 159, 169, 174, 186, 223, 244, 246–253, 254, 256, 257, 258, 261, 262, 266, 270–272, 278, 280
Dretske, Fred, xi, 21, 22n, 28, 29, 39, 48, 49, 53n, 78, 104, 199, 200, 214
Dummett, Michael, 186, 235n

empirical
  case studies, 87
  claims, 125
  concept(s), ix, xii, 9, 61, 62, 133, 172, 193, 194, 220, 221, 237, 238, 241, 254–257, 277, 278, 281
  constraints, 115
  criteria, 139
  discovery (matter), 194, 199, 201, 218, 221, 248, 257
  error, 116
  evidence, 11, 126, 240
  factors, 221
  hypotheses, 54
  judgment(s), 182, 210, 278
  knowledge, 1, 128
  language, 268, 269
  meaning, 17, 241, 247
  philosophy, 248
  science(s), 88, 156, 163, 265, 274
  semantics, 92, 93, 95–97, 105, 106
  speculation, 129

empirical (cont'd)
  status, 182
  terms, 17, 217, 241, 242, 245, 247, 257
  tests, 240
epistemology, 2, 10, 11, 19, 123, 131–133, 154, 178, 194, 220, 221, 233, 239, 240, 263
evolution, x, xi, 3, 5, 14, 25, 31, 37, 42, 44, 45, 47, 48, 54, 62, 69, 71, 76, 79, 80, 93, 95, 105, 110, 129, 133, 137, 141, 146, 148, 149, 150, 172, 176, 177, 181, 183, 184, 185, 186, 191, 196, 209, 220, 230, 250, 261, 266, 273, 274
extension, 15, 18, 32, 112, 113, 117, 120, 121, 126, 132, 152, 159, 162, 165, 175, 202, 207–210, 213, 214, 217, 218, 232–234, 236, 237, 241–253, 254, 257, 258
externalism (content, meaning, semantic), 2, 14, 16, 199–202, 208, 211, 216, 217, 221, 222, 227, 231, 238, 253, 277

Fales, Evan, 184
feature compatibility/incompatibility, 183, 184, 195
Fodor, Jerry, 21n, 27n, 35n, 49n, 53n, 92, 104, 120, 159n, 199, 200, 203, 211–214, 219
Frege, Gottlob (Fregean sense), 2, 10, 11, 15–17, 92, 136, 198, 201, 203, 213, 214, 222–236, 237, 238, 240
function, see proper function; teleofunction

Grice, H. P., 28, 222, 239n

Hayek, Friedrich, 183
Hegel, G. W. F., ix, 18, 19, 189–192
holism (meaning), 187, 277
holism (metaphysical), 188, 190
Hume, David, 173, 184, 188, 189n, 252, 264, 273

icons, (intentional), 43, 50–52, 54, 57, 60, 61, 69, 71, 82–85, 103–105, 182, 184–186, 191, 280
idealism, 178, 192, 277
identity
  beliefs, 116, 121
  conditions, 114, 187–190, 201, 202, 208

informative, identity claims, 17, 125, 223, 279
"is" of, 197
judgments, 9–11, 113, 114, 115, 118, 120, 121, 237
law of, 191, 194, 255
of concepts, 14
of contents, 116, 120
of individuals, 188, 189
of kinds, 13, 150
of vehicles, 116
over time, 167, 172, 173, 175, 208
property, 120, 132, 188, 189
referential, 120
relation, 96n, 114, 167, 189, 193
substance, 133, 173, 174
indexicals, 15, 125, 156
individuals, xii, 10, 12, 13, 39, 70, 115, 123, 124, 127, 129, 131–133, 136, 141, 143, 147, 150, 153, 158, 159, 161, 163, 168, 172–175, 187, 188, 190, 208, 212, 237, 254
inferentialism, 193, 201
information (channels, natural, processing, etc.), 4–6, 8, 10, 11, 15, 17, 22, 26–29, 31, 43, 46, 49, 51, 55, 59–61, 78–80, 83, 91, 95, 99, 105, 117, 173, 186, 200, 228, 230, 235, 237, 239, 240, 281
information theory, 42, 43, 48
information-theoretic semantics, 6, 21, 25, 32, 49, 52, 72, 73, 210
informational teleosemantics, 5, 6, 21ff., 37ff.
intensions (A-intensions, primary intensions), xii, 17, 18, 217, 234, 237, 238, 242–244, 246, 247, 250, 251, 253, 254–257
intentionality, 15, 19, 60, 72, 74, 81, 105, 117, 197, 260, 266, 268, 269, 277, 278
  derived, 41
  intrinsic, 41
  original, 94, 106
internalism (content, meaning, semantic), 2, 201, 202, 211, 225, 227, 228
isomorphism, 7–9, 51, 63ff., 81ff., 96, 97, 104, 105, 180, 267, 278

Jackson, Frank, 53n, 221n, 247

Kant, Immanuel, ix, 191, 193, 239, 260
kinds, kind terms
  biological, 12, 135–137, 140–142, 145, 150, 151
  eternal, 131, 154
  functional, 12, 248, 249, 254, 255
  historical, 10, 12, 13, 131, 135ff., 152–154, 174, 175, 206, 212, 219
  natural, 1, 2, 13, 58, 128, 129, 136–140, 142–144, 151, 155, 156–161, 163–167, 170, 171, 173, 188, 204, 205, 212, 214, 216, 243, 247, 248–250, 256
Kripke, Saul, 16, 128, 186, 199, 200, 204, 216, 248
Kripke–Wittgenstein Paradox, 186

Language of Thought (LOT), 9–11, 55, 107, 113–115, 120, 121, 277, 279, 280
laws of nature (natural laws), 13, 28, 189n
law of noncontradiction, 176ff., 193ff.
learning, 19, 37, 45, 60, 100, 104, 105, 121, 127, 133, 172, 185, 194, 196, 220, 243, 244, 255, 260, 261, 266, 270, 278
Leibniz, G. W. (Leibniz's Law), 132, 188, 189n, 264
Lewis, David, 44, 45, 167, 170n, 177n, 178–180, 181n, 188, 248
linguistic meaning, 14–16, 18, 197, 203, 204, 221, 223, 226n, 237, 279
linguistics, 55, 92
Locke, John, 204

manifest image, 265
mapping (rule, etc.), 4, 5, 7, 8, 14, 15, 18, 50, 55, 63, 64, 66, 70–76, 82–84, 93, 104, 183, 186, 197, 203, 212, 223, 224, 267, 268, 277–279
McDowell, John, 17, 201, 222, 226–229, 231, 235n, 238
meaning rationalism, xii, 16, 198, 200, 201, 226, 240
Meinong, Alexius, 162
metaphysics, 1, 157, 187, 199
Millianism, 198, 200, 201, 203, 210
modality, 14, 176–180, 193, 197, 281
mode(s) of presentation, xii, 10, 113, 114n, 120, 200, 201, 203, 213, 225, 226, 228

model-theoretic argument (Putnam's), 278
Morris, Charles, 190
Myth of the Given, 178, 190, 198, 200

natural selection, xi, xii, 24–26, 28, 29, 35, 37, 50, 54, 59, 97, 104, 121, 122, 143, 144, 154, 157–160, 182, 194, 207, 266
normal
  conditions, 4, 5, 9, 31, 32n, 50, 69, 77, 82, 83, 106, 108, 110, 111, 121, 134, 183
  effects, 120
  explanations, 3–6, 23n, 31, 39, 40, 50, 51, 60–62, 71–73, 76, 77, 79, 81–84, 93
  functional isomorphism, 9, 93, 185, 279
  functions, ix, 26, 31, 33, 105, 212, 227
  system, 31
  use, 120
normativity, 22, 117, 262, 264, 271, 272, 274
necessity
  causal, 189
  logical, 178, 179
  natural, 13, 174, 178, 181, 187–189, 197

ontology, 123, 125, 131, 132, 168, 169, 172, 177–179, 185, 187, 190, 192, 197, 207, 211, 213, 214, 221, 262, 274

Peirce, Charles Sanders, 190
perception, 14, 17, 99, 100, 105, 120, 195, 196, 231, 238, 239, 273, 277, 279
picturing, 53, 72, 73, 260, 265n, 266–268, 275, 276–278
Plato, 53
Platonism, 178, 179
possibility (logical, natural, etc.), 14, 176–178
possible worlds, x, xi, 126, 157, 159, 177–182, 187–189, 199, 218, 241–243, 250, 257, 258
proper function, ix, xi, 1–6, 9, 14, 15, 19, 25, 33, 37, 39, 40, 61, 62, 69–71, 75, 76, 80, 82, 93, 97, 101, 104, 105, 160, 184, 191, 211, 212, 272, 279
properties, 11, 12, 13, 33, 72–75, 77, 78, 123ff., 131ff., 136, 137–145, 149, 153, 154, 157, 158, 162–171, 173–175, 181–185, 187–189, 194–197, 206, 210, 213, 218, 219, 228, 234, 235, 237, 239, 240, 252, 255–257

Putnam, Hilary, 16, 156, 162n, 186, 199, 200, 201, 204, 213, 218, 221, 244, 248, 254, 278

quality space, 220
Quine, W. V. O., 126, 172, 178, 182, 183n, 184, 189, 194, 204, 220

realism
  about individuals, 10, 12, 13, 155ff.
  about kinds, 199, 201, 211, 213, 214
  internal, 201
  modal, 13, 178, 179
  moral, 206
  scientific, 259, 264, 265
reason (autonomy of), 19, 193, 259, 260, 264
reference, 16, 17, 120, 125, 128, 129, 180, 197, 200–211, 213, 214, 220, 222–225, 229, 230, 233, 242, 243, 254, 257, 280
relations, 3, 7–9, 12, 14, 65, 66, 68ff., 82–84, 120, 140ff., 152–154, 156, 164, 167, 168, 201, 230, 265, 267, 275, 278
representation(s), x, 2, 3, 41, 63, 71, 72, 81, 92, 95, 97, 103–106, 107, 115, 172, 173, 212, 273
  analog, 180
  biosemantic theory of, 172
  causal theory of, 128
  cognitive, 274
  consumer, 4, 6, 8, 32, 38, 43, 54, 76, 79, 185
  content of, 74–76, 79, 88, 104, 106, 120, 128, 134, 212
  descriptive, 82, 99
  directive, 99
  discursive, 113
  empty, 197
  feature, 212, 213, 268
  functions of, 4, 6, 21, 22, 73, 135
  imperative, 4
  indicative, 4, 5, 135
  intentional, 37, 62n, 73, 84, 268
  internal (inner), 14, 42, 53, 59, 61, 99, 106, 160, 161, 279
  linguistic, 179, 180, 194, 275
  mapping rules of, 5, 55, 71, 73
  medium of, 113
  mental, 7, 21, 37, 73, 87, 88, 96, 114, 119, 134, 197, 280
  meta-representations, 264, 269
  misrepresentation, 8, 9, 21, 22, 36, 49, 63, 69, 81, 116
  of rules, 272, 273
  perceptual, 195
  pictorial, 113
  producer, 4–7, 21–23, 31, 33, 37–40, 43, 54, 61, 77–78, 184, 185
  productivity, 68, 70
  propositional, 267
  pushmi-pullyu, 5, 98–101, 273, 279
  sensory, 21, 23, 29, 30, 32, 33, 36
  sentential, 186
  system of (representational system), 68–70, 75–77, 79, 82, 83, 85, 86, 120, 121, 173, 184, 195, 263, 264, 267, 268, 273, 274
  visual, 104, 105, 183n
representational capacities, 87
representational theory of mind, 53, 113, 239, 240
representational vehicle, 119–121
representationalism, 53, 54, 239
rigid designators, 128, 243, 247, 248
Russell, Bertrand (Russellian), 125, 126, 136, 137, 139, 142, 144, 148, 152, 230

Schelling, F. W. J., 190
scientific image, 259n, 274
Searle, John, 41
Sellars, Wilfrid, 2, 16, 19, 186, 190–192, 193, 197, 204, 259ff., 276–280
sense modalities, 17, 230, 231
sensory
  bias, 146
  channel, 231
  cues, 158, 159, 161
  experience, 124
  functions, 183
  impression, 209, 220
  input, 55, 88, 99, 219
  pathways, 35
  representations, 4, 6, 21, 30, 32, 33, 36
  responses, 186
  spaces, 184
  stimuli (stimulations), 56, 88, 90, 91, 100, 194

surfaces, 133, 255
systems, 6, 23–29, 32–34, 36, 182–184, 186, 219
Shannon, Claude, 42, 43, 78
signals, 4, 6, 41ff., 60, 61, 101, 104, 239, 278
signs (natural, intentional), 6, 7, 22, 23n, 27n, 28, 39, 40, 41, 43, 47, 48, 51–53, 57, 58, 59–61, 63, 81, 105, 172, 219, 239, 278
Skyrms, Brian, 6, 42, 45–50, 52, 59, 60
somatosensory map, 57
space of reasons, 16, 186, 192
speaker meaning, 211, 237
species, 12, 136, 138n, 140, 141, 143–151, 152–154, 174, 209, 266
Strawson, P. F., 11, 113, 114, 120, 132, 182, 185, 200, 210
substance
 and property ranges, 184, 187, 192
 "clots," 219
 concepts, 2, 10, 11, 32, 107, 108, 112, 116, 117, 119, 120, 123–128, 130, 131–133, 175, 184, 187, 208, 219, 245
 historical, 190n
 identity of, 10, 18, 173, 174, 189, 194, 233, 235, 242–244, 246, 248, 249, 257
 primary, 188n
 real, 18, 173
 recognition of, 11
 reidentification of, 119
 relativity of, 188, 195
 representation of, 128

"rough," 174, 218
secondary, 170, 190
template, 212, 213, 256
terms, 18, 216, 218, 219, 245, 248
symbols, 43, 95, 114–116

teleofunction, xii, 52, 107
teleology, 3, 22, 63, 72, 74, 81, 82, 201, 210, 280
teleosemantics, 2, 3, 6, 16, 19, 21–25, 28, 31, 35, 36, 38, 40, 61, 64, 71, 72, 81, 93–96, 120, 172, 195, 197, 222, 246, 278, 279
tracking, 117, 212, 213, 220, 252
truth, 8, 9, 14, 15, 38, 69–71, 76, 85, 134, 136, 186, 189, 197, 222, 223, 225, 232, 235, 237, 242, 262, 275, 277, 280
truth-conditional content, 92ff., 103ff.
truth-makers, 164, 167
Twin Earth, 17, 18, 152, 153, 199, 242, 247, 258
two-dimensional semantics, 15, 17, 18, 242, 243, 245–251, 253, 258

unicept, ix, 14, 281

verificationism, 32, 173, 186, 187, 189

Weiss, Paul, 190
Wittgenstein, Ludwig, 85, 132, 138, 186, 190, 260, 267